Problem Books in Mathematics

Series Editor

Peter Winkler, Department of Mathematics, Dartmouth College, Hanover, NH, USA

Books in this series are devoted exclusively to problems - challenging, difficult, but accessible problems. They are intended to help at all levels - in college, in graduate school, and in the profession. Arthur Engels "Problem-Solving Strategies" is good for elementary students and Richard Guys "Unsolved Problems in Number Theory" is the classical advanced prototype. The series also features a number of successful titles that prepare students for problem-solving competitions.

Edmundo Capelas de Oliveira • José Emílio Maiorino

Analytical Methods in Applied Mathematics

 Springer

Edmundo Capelas de Oliveira (iD)
Applied Mathematics
Universidade Estadual de Campinas
Campinas, São Paulo, Brazil

José Emílio Maiorino (iD)
Universidade Estadual de Campinas
Campinas, São Paulo, Brazil

ISSN 0941-3502   ISSN 2197-8506   (electronic)
Problem Books in Mathematics
ISBN 978-3-031-74796-0   ISBN 978-3-031-74794-6   (eBook)
https://doi.org/10.1007/978-3-031-74794-6

To our parents: Conceição and Manoel
Anna Lydia and José Maurício

Preface

The present book has been thought as a bibliographical complement for the disciplines of Methods of Applied Mathematics taught in one or two semesters in courses of Mathematics, Applied Mathematics, Physics and Engineering. Besides, we believe that it may well serve as a didactic support for several other disciplines.

For the sake of simplicity and clarity we have chosen a philosophy of **learning by doing**: it is our belief that if the student does not practice while attempting a regular course, he/she will hardly be able to apply his/her knowledge when necessary.

For this reason, chapters have been written in increasing levels of difficulty, as well as the solved exercises presented in each chapter. Proposed exercises, however, are not ordered the same way. All proposed exercises are accompanied by their solutions or by some hint for solving them, so that the student will be able to measure his/her own progress. With the exception of Chap. 11, all chapters are written according to the following structure: after a brief introduction in which we present a **résumé** of the theory, we propose and solve some examples and then solved exercises are discussed step-by-step; each chapter ends with about 50 proposed exercises, with answer and/or suggestion, to be solved by the student.

In Chap. 1 we discuss a few simple linear, first-order and second-order ordinary differential equations. Some nonlinear ordinary differential equations are also discussed, namely the Riccati and Bernoulli equations. In Chap. 2, after a brief review of power series, we discuss the general method for solving linear ordinary differential equations with variable coefficients known as Frobenius method. In Chap. 3 we study some properties of functions of a complex variable in order to introduce, in Chap. 4, the so-called special functions, which are constructed by means of the hypergeometric function. Having studied such functions, we present in Chap. 5 the Fourier, Fourier-Bessel, and Fourier-Legendre series.

Chapter 6 is entirely devoted to Fourier and Laplace integral transforms, with emphasis on the evaluation of inverse Laplace and Fourier transforms. In Chap. 7 we study Sturm-Liouville systems and introduce the concept of Green's function. In Chap. 8 we present linear, second-order partial differential equations with emphasis on their classification before studying, in Chap. 9, the method of separation of variables and the way to find solutions of such equations satisfying given boundary and initial value conditions. In Chap. 10 we present a brief introduction to fractional calculus, discussing two types of fractional derivatives and the Mittag-Leffler functions. Finally, in Chap. 11 we present some classical applications of the theory previously developed, solving completely three or more examples and leaving a list of exercises to be solved by the student.

It would be useless to try to name all the people who, in one way or another, helped this work. First of all, we wish to express our gratitude to our wives, Ivana and Maria. We also would like to mention especially Professors Waldyr Alves Rodrigues Júnior, *in memoriam*, Erasmo Recami, *in memoriam*, Jayme Vaz Júnior, Márcio José Menon, Quintino Augusto Gomes de Souza, and Hamilton Germano Pavão, whose support and incentive were fundamental for the accomplishment of this work. Also, we would like to express our gratitude to the many students whose suggestions and/or corrections contributed a lot to improve this work. Finally, we would like to express our sincere thanks to the anonymous *ad hoc* advisors and the editor Robinson dos Santos for several useful suggestions.

Campinas, São Paulo, Brazil   Edmundo Capelas de Oliveira
June 2024   José Emílio Maiorino

Contents

Chapter 1
Ordinary Differential Equations

Mathematics is the queen of sciences.

1777 – Johann Carl Friedrich Gauss – 1855

In this chapter, we present some basic material on linear ordinary differential equations of the first and second order. Some special first-order nonlinear ordinary differential equations are also mentioned, namely the Bernoulli [1654 – Jacob Bernoulli – 1705] and Riccati [1676 – Jacopo Francesco Riccati – 1754] equations.

1.1 Preliminaries and General Concepts

In this section, we introduce the concepts of ordinary differential equation, order of a differential equation, linearity, solution of an ordinary differential equation, and the Cauchy [1789 – Augustin Louis Cauchy – 1857] problem.

1.1.1 Ordinary Differential Equation

Let $x \in \mathbb{R}$ denote an independent variable and $y \in \mathbb{R}$ the dependent variable, which we call the unknown function. A differential equation is an equation involving the independent variable x, the unknown function y, and its derivatives y', y'', y''', ..., $y^{(n)}$; it can be written in the general form

$$F\left(x, y', y'', y''', \ldots, y^{(n)}\right) = 0.$$

As the unknown function depends on just one independent variable, x, the differential equation is called an ordinary differential equation. If the unknown

© The Author(s), under exclusive license to Springer Nature Switzerland AG 2025
E. Capelas de Oliveira, J. E. Maiorino, *Analytical Methods in Applied Mathematics*,
Problem Books in Mathematics, https://doi.org/10.1007/978-3-031-74794-6_1

function depends on more than one independent variable, we will have a partial differential equation, a concept presented in Chap. 8.

Example 1.1 These are some examples of ordinary differential equations:

$$\text{(a)} \ \ \frac{d}{dx}y(x) \equiv \frac{dy}{dx} \equiv y' = 0, \quad \text{(b)} \ \ \frac{d^2}{dx^2}y(x) + y(x) = 2022,$$

$$\text{(c)} \ \ y\,y'' - 4y' + 4y = x^2, \qquad \text{(d)} \ \ (x^2 + 1)dy + x^3\,dx = 0,$$

$$\text{(e)} \ \ y'' + 8y' + y^4 = 0, \qquad \text{(f)} \ \ y\,\frac{d^5 y}{dx^5} + x\,\frac{d^2 y}{dx^2} = 2021.$$

Here, x is the independent variable and $y = y(x)$ is the dependent variable, that is, the unknown function. $\qquad\square$

1.1.2   Order

The order of an ordinary differential equation is the order of the derivative of highest order that appears in the equation.

Example 1.2 The equations in Example 1.1 have orders (a) First, (b) Second, (c) Second, (d) First, (e) Second, and (f) Fifth, respectively. $\qquad\square$

Another importat criterion for classifying differential equations is their linearity or nonlinearity.

1.1.3   Linearity

Let $x, y \in \mathbb{R}$ and $y = y(x)$. The differential equation $F(x, y, y', \ldots, y^{(n)}) = 0$ is called linear if F is a linear function of variables $y, y', \ldots, y^{(n)}$; otherwise, the differential equation is called nonlinear.

Example 1.3 In what concerns their linearity, the equations in Example 1.1 are, respectively, (a) Linear, (b) Linear, (c) Nonlinear, (d) Linear, (e) Nonlinear, and (f) Nonlinear. $\qquad\square$

1.1.4   Solutions

The solution of an ordinary differential equation is a function $y = f(x)$, defined on an open interval $a < x < b$, such that when we substitute $y = f(x)$ along with its derivatives into the expression of the differential equation, it becomes an identity on the interval $a < x < b$. In what follows, we distinguish, a general solution, containing arbitrary constants, and a particular solution.

Example 1.4 Considering again Example 1.1, we have that:

Equation (a) has the general solution $y = f(x) = C$, where C is an arbitrary constant, and a particular solution $y = f(x) = 2022$

Equation (b) has the general solution $y = f(x) = C_1 \cos x + C_2 \sin x + 2022$, with C_1 and C_2 arbitrary constants. A particular solution is $y = f(x) = 2022$. $\square$

1.1.5 Initial Value Problem

The general form (implicit form) of a first-order ordinary differential equation is $F(x, y, y') = 0$. Sometimes, it is possible to isolate the derivative y', writing the equation in the equivalent form (explicit form)

$$y' = f(x, y). \tag{1.1}$$

We are thus writing the first derivative y' in terms of the dependent variable—the unknown function—and of the independent variable.

In real problems, we usually need to find more than just a function satisfying Eq. (1.1). It is also necessary that the function assumes a certain value y_0—the *initial value*—when the independent variable x is equal to a given value x_0. We call this condition, a condition imposed on $y(x)$ at point x_0, the *initial condition*.

The problem of finding a solution of Eq. (1.1) satisfying the initial condition (condition imposed at a point) $y_{x=x_0} = y_0$ is known as Cauchy problem. Geometrically this means that we are searching for a curve, named integral curve, that passes through the point $P(x_0, y_0)$ of the xOy plane.[1]

The general solution of Eq. (1.1) is a function $y = f(x, C)$, depending on an arbitrary constant C, such that (i) y satisfies Eq. (1.1) for all values of constant C and (ii) whatever the initial condition $y(x_0) = y_0$, we can always find a value C_0 for the constant C such that the function $y = f(x, C_0)$ satisfies the initial condition.

Given a general solution, a particular solution of Eq. (1.1) is a function obtained from the general solution by assigning a specific value to the arbitrary constant C.

Example 1.5 Let $x \in \mathbb{R}$. Show that the solution of the initial value problem

$$\begin{cases} y' - 2y = -4x, \quad y = y(x), \\ y(0) = 1 \end{cases}$$

is given by $y(x) = 2x + 1$.

---

[1] Theorems on the existence and uniqueness of solutions to the Cauchy problem can be found in references [1–3].

First, we verify that $y(x) = 2x + 1$ satisfies the linear, first-order ordinary differential equation aforementioned. Evaluating its derivative we find that $y'(x) = 2$; substituting the result into the differential equation, we get

$$2 - 2(2x + 1) = 2 - 4x - 2 = -4x.$$

As for the initial condition, we see that

$$y(0) = 2 \cdot 0 + 1 = 1.$$

Since $y(x) = 2x + 1$ satisfies the differential equation and the initial condition, it is the solution of the initial value problem. $\square$

1.2 First-Order Ordinary Differential Equations

We now turn our attention to the linear first-order ordinary differential equation, for which we discuss a few elementary integration methods. We shall also mention some specific nonlinear first-order differential equations, the Bernoulli and Riccati equations.

1.2.1 The Linear Equation

Let $x \in \mathbb{R}$. The first-order ordinary differential equation

$$\frac{dy}{dx} + p(x)\, y = q(x), \tag{1.2}$$

where $p(x)$ and $q(x)$ are known functions defined on the open interval (a, b), is called a linear first-order ordinary differential equation on the interval (a, b). In order to solve this ordinary differential equation, we assume that functions $p(x)$ and $q(x)$ are continuous on the interval (a, b). If $p(x) = 0$, the ordinary differential equation can be solved by direct integration, as shown as follows.

1.2.2 Direct Integration

Let $x \in \mathbb{R}$. Consider the linear first-order ordinary differential equation

$$\frac{d}{dx} y(x) = f(x) \tag{1.3}$$

where $f(x)$ is a nonhomogeneous—i.e. nonnull—term, independent of the dependent variable y. Then, integrating both sides of Eq. (1.3) we obtain

$$y(x) = \int^x f(\xi)\, \mathrm{d}\xi + C \equiv F(x) + C,$$

where C is an arbitrary constant. This expression is the general solution of the linear ordinary differential equation, as it contains an arbitrary constant.

Notice that an entirely analogous reasoning can be used, for example, for a linear second-order ordinary differential equation of the type

$$\frac{\mathrm{d}^2 y}{\mathrm{d}x^2} = g(x).$$

In this case, the general solution contains two arbitrary contants because we have two sucessive integrations. The generalization for an arbitrary order $n \in \mathbb{N}$ is imediate.

Example 1.6 Solve the initial value problem (uniform motion)

$$\begin{cases} \dfrac{\mathrm{d}}{\mathrm{d}t} s(t) = v, \ \ v = \text{constant,} \\[2mm] s(0) = s_0 = \text{constant.} \end{cases}$$

Integrating both sides of the linear first-order ordinary differential equation we have

$$s(t) = vt + C,$$

with C an arbitrary constant. Imposing the initial condition, i.e., substituting $t = 0$ into the last equation, $s(0) = v \cdot 0 + C = s_0$, we get

$$s(t) = s_0 + vt,$$

which is the well-known horary equation of uniform motion, characterized by having a constant speed. $\qquad\qquad\square$

1.2.3   Integrating Factor

If $p(x) \neq 0$ in Eq. (1.2), we introduce the so-called integrating factor, denoted by $\mu(x)$, so as to turn the ordinary differential equation in a form in which we can use direct integration. Indeed, multiplying both sides of Eq. (1.2) by

$$\mu(x) = \exp\left(\int^x p(\xi)\,\mathrm{d}\xi\right),$$

we can write, after simplification,

$$\frac{\mathrm{d}}{\mathrm{d}x}\left[y(x)\,\mu(x)\right] = q(x)\,\mu(x).$$

Integrating both sides with respect to variable x and solving for $y(x)$, we obtain the general solution (it contains an arbitrary constant) of the linear first-order ordinary differential equation:

$$y(x) = \frac{1}{\mu(x)}\int^x q(\xi)\,\mu(\xi)\,\mathrm{d}\xi + \frac{C}{\mu(x)},$$

where C is an arbitrary constant. $\mu(x)$, which transforms the left side of Eq. (1.2) into an exact derivative, is known as the *integrating factor* to this differential equation.

It is worth mentioning that the uniqueness, the existence, and the stability of the solution for the linear initial value problem, i.e., the linear first-order ordinary differential equation with a given initial condition, can be proved as a theorem [4].

Example 1.7 Let $x \in \mathbb{R}$. Solve the linear initial value problem

$$\begin{cases} \dfrac{\mathrm{d}}{\mathrm{d}x}y(x) - y(x) = x, \\ y(0) = 0. \end{cases}$$

Comparing this problem with the general form of the integrating factor, we can see that, in this case, the integrating factor is

$$\mu(x) = \exp\left(\int^x (-1)\cdot\mathrm{d}\xi\right).$$

Calculating the integral, we have $\mu(x) = \mathrm{e}^{-x}$. Multiplying both sides of the ordinary differential equation by $\mu(x)$ and simplifying, we get

$$\frac{\mathrm{d}}{\mathrm{d}x}\left[y(x)\,\mathrm{e}^{-x}\right] = x\,\mathrm{e}^{-x}.$$

Integrating with respect to variable x we get the equation $y(x)\,\mathrm{e}^{-x} = \int^x \xi\,\mathrm{e}^{-\xi}\,\mathrm{d}\xi$, whose integration by parts furnishes $y(x)\,\mathrm{e}^{-x} = -x\,\mathrm{e}^{-x} - \mathrm{e}^{-x} + C$ with C a constant. It then follows that

$$y(x) = -x - 1 + C\,\mathrm{e}^{x}.$$

Imposing the initial condition, $y(0) = -1 + C = 0$, we find $C = 1$ and we finally get

$$y(x) = -x - 1 + e^x,$$

which is the solution of the linear initial value problem. $\square$

1.2.4   Separable Equations

A first-order ordinary differential equation is said to be separable if it is possible to express it in one of the forms

$$\frac{dy}{dx} = \frac{f(x)}{g(y)} \qquad \text{or} \qquad \frac{dx}{dy} = \frac{h(x)}{i(y)},$$

with $g(y) \neq 0$ $[i(y) \neq 0]$.

In order to solve an ordinary differential equation of the first form, we rewrite it as

$$g(y)dy = f(x)dx$$

whose integration furnishes

$$G(y) = \int^y g(\eta)d\eta = \int^x f(\xi)d\xi = F(x) + C,$$

where C is an arbitrary constant. $F(x)$ and $G(y)$ are the so-called primitives of $f(x)$ and $g(y)$, respectively. For the second form, see Example 1.8.

Example 1.8 Let $x \in \mathbb{R}^*$. Solve the ordinary differential equation, with $y(x) \neq 0$,

$$\frac{dy}{dx} = \frac{y}{x}.$$

This ordinary differential equation is separable and can be written as

$$\frac{dy}{y} = \frac{dx}{x},$$

whose integration furnishes

$$\ln |y| = \ln |x| + C_1,$$

where C_1 is an arbitrary constant. This equation can be put in the form

$$|y| = C \cdot |x|,$$

with $C = e^{C_1}$, another arbitrary constant. $\square$

1.2.5 Exact Equation

Let $M = M(x, y)$ and $N = N(x, y)$. An ordinary differential equation written in the form

$$M(x, y) + N(x, y)\frac{dy}{dx} = 0 \tag{1.4}$$

is called exact if the condition

$$\frac{\partial M}{\partial y} = \frac{\partial N}{\partial x} \tag{1.5}$$

is satisfied at each point of the domain $D = \{(x, y) \in \mathbb{R}^2 / a < x < b, c < y < d\}$.

Then, there exists a function $F(x, y)$ such that

$$\frac{\partial F}{\partial x} = M(x, y) \qquad \text{and} \qquad \frac{\partial F}{\partial y} = N(x, y).$$

The solution of the ordinary differential equation will be

$$F(x, y) = C,$$

with C an arbitrary constant.

It is important to note that, when condition Eq. (1.5) is not satisfied, it is possible to obtain, in certain cases, an integrating factor, which will turn the original ordinary differential equation into an exact equation. For the general case, see reference [4].

Example 1.9 Let $x \in \mathbb{R}$ and $y \neq 0$. Solve the ordinary differential equation

$$\left(\frac{2xy + 1}{y}\right) dx + \left(\frac{y - x}{y^2}\right) dy = 0.$$

Comparing this expression with Eq. (1.4), we find

$$M(x, y) = \frac{2xy + 1}{y} \qquad \text{and} \qquad N(x, y) = \frac{y - x}{y^2}.$$

Evaluating the partial derivatives, Eq. (1.5), we get

$$\frac{\partial M}{\partial y} = -\frac{1}{y^2} \qquad \text{and} \qquad \frac{\partial N}{\partial x} = -\frac{1}{y^2}.$$

Thus, the condition is satisfied and the ordinary differential equation is exact.

In order to find $F(x, y)$, we integrate the first relation, $M(x, y)$, with respect to variable x, obtaining

$$F(x, y) = x^2 + \frac{x}{y} + g(y),$$

where $g(y)$ is an arbitrary function depending on y only.

Differentiating $F(x, y)$ with respect to variable y and using the expression for $N(x, y)$ we find

$$-\frac{x}{y^2} + g'(y) = N(x, y) = \frac{y - x}{y^2},$$

that is,

$$g'(y) = \frac{1}{y}$$

and so

$$g(y) = \ln|y| + D,$$

where D is an arbitrary constant. Going back to the expression for $F(x, y)$, we finally find the general solution

$$x^2 + \frac{x}{y} + \ln|y| = C,$$

with C another arbitrary constant. $\qquad\qquad\qquad\qquad\qquad\qquad\qquad\qquad\qquad\square$

In many cases, it is possible to introduce a change of variable (substitution) that would reduce the original ordinary differential equation to an equation that can be more easily solved. We present here some examples of this method. One of them, in particular, is used to convert a nonlinear differential equation, the so-called Bernoulli equation, into a linear ordinary differential equation.

I. Separable equation Let $a, b, c \in \mathbb{R}$. Consider the (linear or nonlinear) first-order ordinary differential equation

$$y' = f(ax + by + c)$$

with $b \neq 0$. If we introduce the change of variable $v = ax + by + c$, it becomes the ordinary differential equation

$$\frac{dv}{dx} = b\,f(v) + a,$$

which is separable and can be written in the form

$$\frac{dv}{b\,f(v) + a} = dx.$$

II. Homogeneous type Ordinary differential equations of the type

$$\frac{dy}{dx} = f\left(\frac{y}{x}\right),$$

are sometimes called ordinary differential equations of the homogeneous type. We introduce the substitution $v = y/x$ to get

$$\frac{dv}{f(v) - v} = \frac{dx}{x},$$

which is also separable.

III. Bernoulli equation Let $P(x)$ and $Q(x)$ be continuous functions. Consider the ordinary differential equation

$$\frac{dy}{dx} + P(x)y = y^{\alpha} Q(x),$$

with $y = y(x)$ and α a real constant. This differential equation is known as Bernoulli equation. If $\alpha = 0$, this is a linear nonhomogeneous equation, and if $\alpha = 1$ it is linear and homogeneous. For all other values of α it will be nonlinear. Assume that $\alpha \neq 0$ and $\alpha \neq 1$ and introduce the substitution

$$v = y^{1-\alpha}.$$

This converts the nonlinear ordinary differential equation into a linear ordinary differential equation:

$$\frac{dv}{dx} + (1 - \alpha)P(x)v = (1 - \alpha)Q(x).$$

Example 1.10 Let $y = y(x)$. Solve the following initial value problem:

$$\begin{cases} (2x - y)dy + (x - 2y + 3)dx = 0, \\ y(0) = 0. \end{cases}$$

Note that this ordinary differential equation is not an exact equation. So, we introduce a linear transformation to convert it into an equation of the homogeneous type. To this end, consider the changes

$$x = z + \alpha \qquad \text{and} \qquad y = t + \beta,$$

where α and β are constants to be determined. In the new variables z and t, the ordinary differential equation has the form

$$[2z - t + (2\alpha - \beta)]dt + [z - 2t + (\alpha - 2\beta + 3)]dz = 0.$$

To reduce this equation to an equation of the homogeneous type, we must find the values of α and β which satisfy the system

$$\begin{cases} 2\alpha - \beta = 0, \\ \alpha - 2\beta = -3. \end{cases}$$

The solution is $\alpha = 1$ and $\beta = 2$. Thus, the ordinary differential equation can be written as

$$\frac{dt}{dz} = \frac{2t - z}{2z - t},$$

which can be recognized as a differential equation of the homogeneous type. Introducing the substitution $v = t/z$ we get

$$\frac{v - 2}{1 - v^2}dv = \frac{dz}{z}.$$

Using partial fractions, we obtain

$$\int \frac{-1/2}{1 - v}dv + \int \frac{-3/2}{1 + v}dv = \int \frac{dz}{z},$$

whose integration furnishes

$$\frac{1}{2}\ln|1 - v| - \frac{3}{2}\ln|1 + v| = \ln|z| + A,$$

where A is an arbitrary constant. A few simple manipulations allow us to write for the solution

$$(1 - v) = C\, z^2 (1 + v)^3,$$

where C is another arbitrary constant. Going back to the old variables x and y, we obtain

$$x - y + 3 = C\,(x + y - 1)^3.$$

Imposing the initial condition $y(0) = 0$, we find that $C = -3$ and we finally get

$$x - y + 3 = -3 (x + y - 1)^3,$$

which is the solution of the initial value problem. $\square$

1.3   Second-Order Ordinary Differential Equations

Let $A(x)$, $B(x)$ and $C(x)$ be coefficients, which will be considered real, and $y = y(x)$, the dependent variable. A linear, second-order ordinary differential equation for the function $y(x)$ is given, in its most general form, by

$$A(x)y''(x) + B(x)y'(x) + C(x)y(x) = F(x),$$

with $A(x) \neq 0$ and where the primes denote differentiation with respect to the independent variable x. Assuming that $A(x)$, $B(x)$, $C(x)$ and $F(x)$ are continuous functions on an interval I, the ordinary differential equation aforementioned may be transformed into

$$y''(x) + p(x)y'(x) + q(x)y(x) = f(x). \tag{1.6}$$

If $f(x) = 0$, it is called a *homogeneous* differential equation. For this reason, the equation obtained from Eq. (1.6) by suppressing the independent term $f(x)$ (also called the *nonhomogeneous term*) is called the *homogeneous equation associated* with Eq. (1.6).

The general solution of the ordinary differential equation Eq. (1.6) is

$$y(x) = y_H(x) + y_P(x),$$

where $y_H(x)$ is the general solution of the homogeneous differential equation and $y_P(x)$ is a particular solution of the nonhomogeneous equation. For a linear and homogeneous second-order ordinary differential equation, its general solution will be given by the sum of two *linearly independent* functions $y_1(x)$ and $y_2(x)$, each of which multiplied by an arbitrary constant, i.e., $y_H(x)$ will have the form

$$y_H(x) = Ay_1(x) + By_2(x),$$

where A and B are arbitrary constants. Each function $y_i(x)$, $i = 1, 2$, is a solution of the homogeneous ordinary differential equation, and any possible solution may be written as a linear combination of them.

In order to solve the nonhomogeneous differential equation, it is enough to know *one* solution of the associated homogeneous equation. From this solution, by reduction of order (PE. 1.16), we may obtain the second, linearly independent solution for the same homogeneous differential equation. Once both linearly independent solutions of the homogeneous differential equation are known, we may find a particular solution of the corresponding nonhomogeneous equation by using the *method of variation of parameters*, also called *Lagrange* [1736 – Joseph Louis Lagrange – 1813] *method*.

The complete solution of an ordinary differential equation involves also the imposition of constraints on the general solution, by imposing that the function obtained should satisfy certain *boundary conditions* or *initial conditions* appearing in the real problem that has given rise to the equation.

Example 1.11 Let $y = y(x)$. Classify the differential equation

$$(1 - x^2)\frac{\mathrm{d}^2}{\mathrm{d}x^2}y(x) - 2x\frac{\mathrm{d}}{\mathrm{d}x}y(x) + \ell(\ell + 1)y(x) = 2022$$

with $\ell = 0, 1, 2, \ldots$

This is an ordinary (only one independent variable) differential equation, second-order (highest order of the derivative), linear (all coefficients are function of the independent variable and $y(x)$ and its derivatives are linear), nonhomogeneous (the second member is not zero), with nonconstant (i.e. variable) coefficients. The corresponding homogeneous differential equation is known as Legendre equation; its general solution is given by Legendre polynomials and Legendre functions, both depending on the parameter ℓ, as we will see in Chap. 4.   □

As they are very common and useful, we will present the methods for solving differential equations with constant coefficients and differential equations of the Euler [1707 – Leonhard Euler – 1783] type.

1.3.1 Equations with Constant Coefficients

We want to solve a linear homogeneous second order ordinary differential equation with constant coefficients,

$$y''(x) + ay'(x) + by(x) = 0,$$

where $x \in \mathbb{R}$ and a and b are constants. As the derivative of an exponential function is equal to the same exponential function multiplied by a constant factor, we suppose a solution of the form

$$y(x) = \mathrm{e}^{mx},$$

with m a constant parameter.

Substituting $y(x)$ and its derivatives into the equation given, we obtain an algebraic equation of the second degree in variable m, called *auxiliary equation*, also known as *characteristic equation*, given by

$$m^2 + am + b = 0\,,$$

and whose roots are m_1 and m_2. If $m_1 \neq m_2$, we immediately have two linearly independent solutions, $y_1(x) = e^{m_1 x}$ and $y_2(x) = e^{m_2 x}$. When the algebraic equation found has only one root, i.e., a double root m, we suppose a solution with the form $y = e^{mx} v(x)$ and look for another linearly independent solution by reduction of order.

Example 1.12 Let $x \in \mathbb{R}$. Solve the initial value problem

$$\begin{cases} y'' - 6y' + 5y = 0, \\ y(0) = 1 \quad \text{and} \quad y'(0) = 5. \end{cases}$$

Let $y(x) = e^{\lambda x}$, with λ a constant parameter. The characteristic equation is

$$\lambda^2 - 6\lambda + 5 = 0$$

and its roots are $\lambda_1 = 1$ and $\lambda_2 = 5$. The roots are different; hence, the solution of the ordinary differential equation can be written as

$$y(x) = A\,e^x + B\,e^{5x},$$

with A and B arbitrary constants. Using the initial conditions, we get

$$y(0) = A + B = 2 \qquad \text{and} \qquad y'(0) = A + 5B = 5$$

whose solution is $A = 0$ and $B = 1$. So,

$$y(x) = e^{5x}$$

is the solution of the initial value problem. $\square$

1.3.2   Equations of the Euler Type

Let $x \in \mathbb{R}$. We call *ordinary differential equation of the Euler type* all equations of the form

$$x^2 y''(x) + axy'(x) + by(x) = 0,$$

where a and b are constants. The solution of such ordinary differential equation is obtained by supposing that it has the form

$$y(x) = x^m,$$

where m is a constant parameter. This leads, as in the case of the ordinary differential equation with constant coefficients, to an algebraic equation for m:

$$m(m-1) + am + b = 0.$$

If this algebraic equation has two distinct roots m_1 and m_2, then $y_1 = x^{m_1}$ and $y_2 = x^{m_2}$ are the two linearly independent solutions sought. If $m_1 = m_2$, we just need to find a second linearly independent solution by reduction of order.

Example 1.13 Let $x \in \mathbb{R}^*$. Solve the ordinary differential equation

$$x^2 \frac{d^2}{dx^2} y(x) + 4x \frac{d}{dx} y(x) + 2y(x) = 0.$$

This ordinary differential equation can be identified with an ordinary differential equation of the Euler type. So, we search a solution of the form $y(x) = x^m$, where m is a parameter to be determined. Introducing this solution into the ordinary differential equation, we obtain an algebraic equation given by

$$m^2 + 3n + 2 = 0,$$

whose roots are $m = -1$ and $m = -2$.

As the roots of the algebraic equation are different, we have obtained the general solution (containing two arbitrary constants) of the ordinary differential equation,

$$y(x) = A x^{-1} + B x^{-2},$$

where A and B are two arbitrary constants. $\qquad\square$

1.4 Solved Exercises

SE 1.1 (Integranting Factor) As we have seen, an integrating factor, denoted $\mu = \mu(x, y)$, if it exists, converts an ordinary differential equation into an exact ordinary differential equation. It is possible to demonstrate that a function $\mu(x, y)$ defined on an interval I, with continuous first order partial derivatives, is an integrating factor if

$$\mu \left(\frac{\partial M}{\partial y} - \frac{\partial N}{\partial x} \right) = N \frac{\partial \mu}{\partial x} - M \frac{\partial \mu}{\partial y}$$

is valid on I, where $M = M(x, y)$ and $N = N(x, y)$ are given in Eq. (1.4). In the particular case in which $\mu = \mu(y)$ (a similar argument is valid for $\mu = \mu(x)$) is a function depending on y only, we obtain, using the preceding equation,

$$\mu \left(\frac{\partial M}{\partial y} - \frac{\partial N}{\partial x} \right) = -M \frac{\partial \mu}{\partial y}$$

which can be written in the following form

$$-\frac{\partial \mu}{\partial y} = \frac{1}{M} \left(\frac{\partial M}{\partial y} - \frac{\partial N}{\partial x} \right) \mu, \tag{1.7}$$

which is a separable ordinary differential equation. As a particular case, discuss the equation

$$\frac{\mathrm{d}}{\mathrm{d}x} y(x) = -\frac{3y^2 + 2}{3xy^2}.$$

Solution To discuss this particular case, involving the nonlinear first order ordinary differential equation

$$\frac{\mathrm{d}}{\mathrm{d}x} y(x) = -\frac{3y^3 + 2}{3xy^2},$$

we first write it in a more adequate form,

$$(3y^3 + 2)\, \mathrm{d}x + 3xy^2 \, \mathrm{d}y = 0.$$

Comparing this last equation with Eq. (1.4), we see that $M(x, y) = 3y^3 + 2$ and $N(x, y) = 3xy^2$. Evaluating the first order partial derivatives, we obtain

$$\frac{\partial M}{\partial y} - \frac{\partial N}{\partial x} = 9y^2 - 3y^2 = 6y^2$$

and we conclude that the ordinary differential equation is not exact, and its integrating factor depends on variable y only.

Using Eq. (1.7), we write this separable ordinary differential equation

$$-\frac{\mathrm{d}\mu}{\mathrm{d}y} = \frac{6y^2}{3y^3 + 2} \mu.$$

Solving this ordinary differential equation, we find the integrating factor

$$\mu(y) = C\,(3y^3 + 2)^{-\frac{2}{3}},$$

where C is an arbitrary constant. Multiplying the ordinary differential equation by the integrating factor, we can write

$$(3y^3 + 2)^{\frac{1}{3}}\,\mathrm{d}x + (3y^3 + 2)^{-\frac{2}{3}}3xy^2\,\mathrm{d}y = 0.$$

Identifying this last equation with Eq. (1.4), we can write new functions M and N, $\bar{M}(x, y) = (3y^3 + 2)^{\frac{1}{3}}$ and $\bar{N}(x, y) = (3y^3 + 2)^{-\frac{2}{3}}3xy^2$. Using these values for $\bar{M}$ and $\bar{N}$, we have

$$\frac{\partial \bar{M}}{\partial y} - \frac{\partial \bar{N}}{\partial x} = \frac{1}{3}(3y^3 + 2)^{-\frac{2}{3}} \cdot 3 \cdot 3y^2 - 3 \cdot y^2 \cdot (3y^3 + 2)^{-\frac{2}{3}} = 0.$$

Thus, multiplying the ordinary differential equation by the integrating factor, it was converted into an exact ordinary differential equation, whose integration is shown in Example 1.9.

SE 1.2 (Bernoulli Equation) Let $x \in \mathbb{R}^*$ and $y(x) = y$. Solve the nonlinear first-order ordinary differential equation

$$\frac{\mathrm{d}y}{\mathrm{d}x} + \frac{y}{x} = y^4 x.$$

Solution As the unknown, dependent variable is raised to the fourth power in the right-hand side of this equation, the equation is obviously nonlinear. Identifying it with a Bernoulli equation, we find $\alpha = 4$, $p(x) = 1/x$ and $q(x) = x$. In order to transform the Bernoulli equation, a nonlinear ordinary differential equation, into a linear first-order differential equation, we introduce the following change of dependent variable:

$$v = \frac{1}{y^3}.$$

Evaluating the derivative, substituting it into the Bernoulli equation and simplifying, we get

$$\frac{\mathrm{d}v}{\mathrm{d}x} - 3\frac{v}{x} = -3x,$$

which is a linear first-order nonhomogeneous ordinary differential equation. In order to solve this differential equation, we proceed as when we got the general form of the solution of a linear first order ordinary differential equation.

The methodology to be employed will be clearer when we study linear second-order ordinary differential equations, as proposed in PE. 1.16. We show that a linear second-order ordinary differential equation, under certain conditions, can be transformed into a linear first-order ordinary differential equation, which we are supposed to be able to solve.

We start with the corresponding homogeneous ordinary differential equation,

$$\frac{dv}{dx} - 3\frac{v}{x} = 0 .$$

This is a separable equation whose general solution is given by

$$v(x) = c\,x^3 ,$$

where c is an arbitrary constant. Now, we look for a solution of the respective nonhomogeneous ordinary differential equation, which we denote $v(x)$, such that we have $v(x) = c(x)\,x^3$, that is, we now consider the arbitrary constant as a function of the independent variable. Evaluating the derivatives with the help of the chain rule, substituting into the nonhomogeneous ordinary differential equation and simplifying, we obtain an ordinary differential equation for $c(x)$,

$$c'(x) = -\frac{3}{x^2} ,$$

whose solution is given by $c(x) = \frac{3}{x} + c_1$ where c_1 is another arbitrary constant.

Using this expression in $v(x)$, we have $v(x) = 3x^2 + c_1 x^3$. Finally, with $v(x)$ known, we obtain, formally, the general solution of the Bernoulli differential equation

$$y(x) = \frac{1}{3x^2 + c_1 x^3} ,$$

with c_1 an arbitrary constant, which can be determined by imposing a specific condition on the solution.

SE 1.3 (Method of Variation of Parameters) Find a solution for the nonhomogeneous linear second-order ordinary differential equation

$$y'' + y = \sin x ,$$

satisfying the conditions $y(0) = 1$ and $y'(0) = 1/2$.

Solution The solution of the respective homogeneous ordinary differential equation (an equation with constant coefficients)

$$y'' + y = 0$$

is given by $y_H(x) = A \sin x + B \cos x$, where A and B are two arbitrary constants.

We will use the *method of variation of parameters* to obtain a particular solution for the nonhomogeneous ordinary differential equation, even though in this case it would be easier to start with a linear combination of sines and cosines and substitute it into the nonhomogeneous differential equation in order to find the coefficients which would make that linear combination a solution of that equation (this is the *method of undetermined coefficients*). See an explicit example of the application of this method in Chap. 11, SE 11.2.

Let us then suppose that constants A and B are functions of x, i.e., suppose that $A = u_1(x)$ and $B = u_2(x)$; a particular solution of the nonhomogeneous ordinary differential equation will have the form

$$y_P(x) = u_1(x)\sin x + u_2(x)\cos x, \tag{1.8}$$

where $u_1(x)$ and $u_2(x)$ will be determined from two conditions. The first condition to be imposed is, of course, that $y_P(x)$ be a solution of the nonhomogeneous ordinary differential equation. However, this condition is not sufficient to univocally determine $u_1(x)$ and $u_2(x)$, and this leaves us free to impose another arbitrary condition upon these functions, which will allow us to simplify the solution of this problem.

Differentiating the expression for $y_P(x)$ we have

$$y_P'(x) = u_1'(x)\sin x + u_1(x)\cos x + u_2'(x)\cos x - u_2(x)\sin x.$$

As we are free to impose a second condition, we choose that

$$u_1'(x)\sin x + u_2'(x)\cos x = 0,$$

whence it follows that

$$y_P'(x) = u_1(x)\cos x - u_2(x)\sin x. \tag{1.9}$$

The second derivative of $y_P(x)$ then becomes

$$y_P''(x) = u_1'(x)\cos x - u_1(x)\sin x - u_2'(x)\sin x - u_2(x)\cos x. \tag{1.10}$$

Now, using the first condition, we replace Eqs. (1.9) and (1.10) into the nonhomogeneous ordinary differential equation and we get

$$u_1'(x)\cos x - u_2'(x)\sin x = \sin x.$$

Thus, in order to determine $u_1'(x)$ and $u_2'(x)$ we must solve the following linear system:

$$\begin{cases} u_1'(x)\sin x + u_2'(x)\cos x = 0, \\ u_1'(x)\cos x - u_2'(x)\sin x = \sin x. \end{cases}$$

Applying Cramer's [1704 – Gabriel Cramer – 1752] rule as if this system were a system of linear algebraic equations with constant coefficients, we find that

$$u_1'(x) = \sin x \cos x \qquad \text{and} \qquad u_2'(x) = -\sin^2 x,$$

and integrating these equations we obtain

$$u_1(x) = -\frac{1}{4}\cos 2x; \tag{1.11}$$

$$u_2(x) = -\frac{x}{2} + \frac{1}{4}\sin 2x. \tag{1.12}$$

Note that it is not necessary to add the integration constants, as we are searching for *one* particular solution of the nonhomogeneous ordinary differential equation.

The solution of the nonhomogeneous ordinary differential equation is obtained substituting Eqs. (1.11) and (1.12) into Eq. (1.8), and adding the result to the general solution of the homogeneous ordinary differential equation. The result is

$$y(x) = C_1 \sin x + C_2 \cos x - \frac{x}{2}\cos x,$$

where C_1 and C_2 are arbitrary constants. Now, using the conditions given, we have

$$y(0) = C_2 = 1 \Rightarrow C_2 = 1 \qquad \text{and} \qquad y'(0) = C_1 - \frac{1}{2} = \frac{1}{2} \Rightarrow C_1 = 1.$$

From this result, it follows that the function that satisfies the ordinary differential equation and the initial conditions is given by

$$y(x) = \sin x + \cos x - \frac{x}{2}\cos x.$$

SE 1.4 (Mass-Spring Problem with Forced Vibrations) Consider the undamped motion of a mass m coupled to the extremity of a spring with elastic constant k. Suppose that an external periodical force given by $f_0 \sin \mu t$, with f_0 and μ real constants, is applied to the mass. Using Newton's [1642 – Isaac Newton – 1727] second law, the ordinary differential equation (PE. 1.25) describing the motion of the mass is given by

$$\frac{d^2}{dt^2}x(t) + \omega^2 x(t) = \frac{f_0}{m}\sin \mu t, \tag{1.13}$$

where $\omega^2 = k/m$ is the frequency of the motion (harmonic oscillator).

(a) Show that the position $x(t)$ of mass m is given by

$$x(t) = c_1 \cos \omega t + c_2 \sin \omega t + c_3 \sin \mu t,$$

where c_1 and c_2 are arbitrary constants and $c_3 = f_0/m(\omega^2 - \mu^2)$, with $\mu \neq \omega$, and (b) discuss the case in which $\omega = \mu$.

Solution (a) First, let us consider the corresponding homogeneous linear second order ordinary differential equation

$$\frac{\mathrm{d}^2}{\mathrm{d}t^2}x(t) + \omega^2 x(t) = 0,$$

whose general solution is given by

$$x(t) = c_1 \cos \omega t + c_2 \sin \omega t,$$

where c_1 and c_2 are arbitrary constants.

In order to obtain a particular solution of the nonhomogeneous differential equation, let us consider the combination

$$x_p(t) = A \cos \mu t + B \sin \mu t,$$

as the second member of Eq. (1.13) is equal to the product of a constant by $\sin \mu t$.

Then, differentiating this expression two times and replacing the result into the nonhomogeneous ordinary differential equation, we may write

$$-A\mu^2 \cos \mu t - B\mu^2 \sin \mu t + A\omega^2 \cos \mu t + B\omega^2 \sin \mu t = \frac{f_0}{m} \sin \mu t,$$

whence we obtain the following algebraic linear system:

$$\begin{cases} -A\mu^2 + A\omega^2 = 0 \\ -B\mu^2 + B\omega^2 = \dfrac{f_0}{m} \end{cases}$$

whose solution is $A = 0$ and $B = f_0/m(\omega^2 - \mu^2)$. Thus, a particular solution of the nonhomogeneous ordinary differential equation is

$$x_p(t) = \frac{f_0}{m(\omega^2 - \mu^2)} \sin \mu t,$$

with $\omega \neq \mu$. Combining this solution with the corresponding homogeneous ordinary differential equation, we find for the displacement

$$x(t) = c_1 \cos \omega t + c_2 \sin \omega t + c_3 \sin \mu t,$$

where $c_3 = f_0/m(\omega^2 - \mu^2)$.

(b) On the other hand, when $\omega = \mu$, the nonhomogeneous term is also a solution of the respective homogeneous ordinary differential equation. Here, we will use directly the method of variation of parameters in order to obtain a particular solution for the nonhomegeneous ordinary differential equation.

From the solution of the homogeneous ordinary differential equation, with $\omega = \mu$, we write the following equation

$$x_p(t) = u(t)\cos\mu t + v(t)\sin\mu t,$$

where $u(t)$ and $v(t)$ are to be determined.

Differentiating the previous expression with respect to t, we get

$$x_p'(t) = u'\cos\mu t - \mu u\,\sin\mu t + v'\sin\mu t + \mu v\,\cos\mu t,$$

where we have omitted the explicit dependence on the independent variable. By imposing the (free) condition

$$u'\cos\mu t + v'\sin\mu t = 0,$$

we get

$$x_p'(t) = -\mu u\,\sin\mu t + \mu v\,\cos\mu t.$$

Differentiating again with respect to t and inserting the result into the nonhomogeneous ordinary differential equation, we obtain

$$u'\sin\mu t - v'\cos\mu t = -\frac{f_0}{\mu m}\sin\mu t$$

or rather, the following algebraic linear system for u' and v':

$$\begin{cases} u'\cos\mu t + v'\sin\mu t = 0, \\ u'\sin\mu t - v'\cos\mu t = -\dfrac{f_0}{\mu m}\sin\mu t, \end{cases}$$

whose solution is given by

$$u' = -\frac{f_0}{\mu m}\sin^2\mu t \quad\text{and}\quad v' = \frac{f_0}{\mu m}\sin\mu t\,\cos\mu t.$$

Integrating these expressions, we may write

$$u = -\frac{f_0}{2\mu m}t + \frac{f_0}{4\mu^2 m}\sin 2\mu t \quad\text{and}\quad v = -\frac{f_0}{4\mu^2 m}\cos 2\mu t.$$

Then, going back to solution $x_p(t)$, we find a particular solution we are searching for, given by

$$x_p(t) = -\frac{f_0}{2\mu m} t \cos \mu t,$$

from which we can write the general solution of the nonhomogeneous linear second-order ordinary differential equation,

$$x(t) = c_1 \cos \mu t + c_2 \sin \mu t - \frac{f_0}{2\mu m} t \cos \mu t,$$

where c_1 and c_2 are arbitrary constants.

SE 1.5 (The Radial Equation)   When we use the method of separation of variables (Chap. 9) for solving the Laplace [1739 – Pierre Simon Laplace – 1827] equation (a partial differential equation satisfied, e.g. by electric and gravitational potentials) in spherical coordinates, we obtain the so-called radial equation (a linear second order ordinary differential equation that does not depend on the angle variables), given by

$$\frac{d^2}{dr^2} R(r) + \frac{2}{r} \frac{d}{dr} R(r) - \frac{\ell(\ell+1)}{r^2} R(r) = 0,$$

where the separation constant ℓ is a nonnegative integer.

(a) Solve the radial equation. (b) What shall we have to consider if we require (a physical condition of the problem) that the (radial) solution be regular (Chap. 3) at the origin?

Solution: (a) This ordinary differential equation is easily identified as an equation of the Euler type, for which we must search for a solution with the form

$$R(r) = r^\alpha,$$

where α is a parameter that has to be determined.

Differentiating with respect to r and substituting the result into the ordinary differential equation, we get an algebraic equation

$$[\alpha(\alpha - 1) + 2\alpha - \ell(\ell + 1)]\, r^\alpha = 0,$$

whence it follows that $\alpha_1 = \ell$ and $\alpha_2 = -\ell - 1$.

As ℓ is a nonnegative integer, $\alpha_1 \neq \alpha_2$. Hence, the general solution is given by

$$R(r) = c_1 r^\ell + \frac{c_2}{r^{\ell+1}},$$

where c_1 and c_2 are arbitrary constants.

(b) As we are searching for solutions that are regular at the origin, $r = 0$, we must impose that $c_2 = 0$, whence it follows that

$$R(r) = c_1 r^\ell,$$

where c_1 is an arbitrary constant.

1.5   Proposed Exercises

PE 1.1 Let $y = y(x)$. Solve the linear ordinary differential equation

$$y' + 2 \sin x \cdot y = 2 \sin x.$$

PE 1.2 Solve the initial value problem

$$\begin{cases} \dfrac{d}{dx} y(x) - 2 \tan x \cdot y(x) = 1, \\ \qquad\qquad\qquad\qquad y(0) = 1/2. \end{cases}$$

PE 1.3 Let $y(x) = y > 0$. Solve the ordinary differential equation

$$\frac{dy}{dx} = \frac{3}{2} \frac{x^2 e^{x^3}}{y}.$$

PE 1.4 Let $y = y(x)$. Solve the linear ordinary differential equation

$$x^2 y \, dy + (y^2 x + x) dx = 0.$$

PE 1.5 Determine the solution for the initial value problem

$$\begin{cases} \dfrac{x}{2} dx + \left(\dfrac{3y^2 - x^2}{y} \right) dy = 0, \\ \qquad\qquad\qquad\qquad y(1) = 1, \end{cases}$$

with $y = y(x) \neq 0$.

PE 1.6 Let $y = y(x)$. Solve the ordinary differential equation

$$\frac{dx}{dy} = \frac{\sqrt{x+y} - \sqrt{x-y}}{\sqrt{x+y} + \sqrt{x-y}}, \qquad x > 0, \quad x \geq |y|.$$

PE 1.7 Let $y = y(x)$. Obtain the solution for the ordinary differential equation

$$\frac{dy}{dx} = \cos^2(x - y).$$

PE 1.8 Let $y = y(x)$ and $x \in \mathbb{R}^*$. Let $f(y)$ be a continuous and integrable function. Solve the ordinary differential equation

$$y^2 dx - [2xy + f(y)] dy = 0.$$

PE 1.9 Let $x = x(y) \neq 0$. Obtain the solution for the initial value problem

$$\begin{cases} \dfrac{dx}{dy} + \dfrac{x}{y} = 2, \\ \quad\; x(1) = 2. \end{cases}$$

PE 1.10 Let $y = y(x)$. Solve the first-order nonlinear ordinary differential equation

$$(x - y)dy - (2x + y - 3)dx = 0.$$

PE 1.11 Find a general solution for each of the following ordinary differential equations with constant coefficients ($y = y(x)$):

(a) $y'' + \omega^2 y = 0$ with $\omega^2 = $ a positive constant.
(b) $y'' + 2y' + y = 0$.
(c) $y'' + 5y' + 4y = 0$.
(d) $y'' + 2y' + 2y = 0$.

PE 1.12 Discuss, according to whether the parameter λ is positive, null or negative, the possible solutions of the ordinary differential equation for $y = y(x)$,

$$y'' + \lambda y = 0.$$

PE 1.13 With a^2 a positive constant, solve the ordinary differential equation (assume $x \neq 0$ and $y = y(x)$)

$$x^2 y'' + 2xy' + \frac{a^2}{x^2} y = 0.$$

PE 1.14 Find the general solution for the following equation of the Euler type ($y = y(x)$):

$$x^2 y'' - 4xy' + 4y = 0.$$

PE 1.15 Solve the ordinary differential equation

$$(1 + x^3)y'' - 3x^2 y' = 0,$$

satisfying the conditions $y(0) = 0$ and $y(1) = 5$.

PE 1.16 (Reduction of Order) Assume that $y_1(x) \neq 0$ is a solution of

$$y'' + p(x)y' + q(x)y = 0.$$

The method of reduction of order consists of searching for a second, linearly independent solution of the form $y_2(x) = y_1(x)v(x)$. Replace $y_2(x)$ into the equation given and from the equation obtained for $v(x)$ show that

$$v(x) = \int^x \frac{\exp\{-\int^{x'} p(x'')\mathrm{d}x''\}}{[y_1(x')]^2}\mathrm{d}x',$$

and thus that a second solution is given by

$$y_2(x) = y_1(x) \int^x \frac{\exp\{-\int^{x'} p(x'')\mathrm{d}x''\}}{[y_1(x')]^2}\mathrm{d}x'.$$

PE 1.17 Knowing that $y_1(x) = x^3$ is a solution of

$$x^2 y'' - 5xy' + 9y = 0,$$

obtain its general solution.

PE 1.18 Knowing that $y_1(x)$ and $y_2(x)$ are two linearly independent solutions of the ordinary differential equation

$$A(x)y'' + B(x)y' + C(x)y = 0,$$

show that the Wronskian [1778 – Józef Maria Hoene Wrónski – 1853] W, defined as

$$W = \det \begin{vmatrix} y_1(x) & y_2(x) \\ y_1'(x) & y_2'(x) \end{vmatrix} = y_1(x)y_2'(x) - y_2(x)y_1'(x),$$

is equal to

$$W = K \exp\left(-\int^x \frac{B(x')}{A(x')}\mathrm{d}x'\right),$$

where K is a constant.

PE 1.19 (Bessel [1784 – Friedrich Wilhelm Bessel – 1846] Equation)
Knowing that

$$y_1(x) = x^{-1/2} \cos x$$

is a solution of the Bessel equation (Chap. 4) of order 1/2,

$$x^2 y'' + xy' + \left(x^2 - \frac{1}{4}\right) y = 0, \quad (x > 0),$$

obtain another linearly independent solution. Calculate the Wronskian.

PE 1.20 (Legendre [1752 – Adrien Marie Legendre – 1833] Equation) Knowing that $y_1(x) = x$ is a solution of the Legendre equation (Chap. 4) of order 1,

$$(1 - x^2)y'' - 2xy' + 2y = 0$$

find the other linearly independent solution. Calculate the Wronskian.

PE 1.21 Let $y = y(x)$. Solve the nonhomogeneous ordinary differential equation

$$xy'' + y' = x^2.$$

PE 1.22 Let $y = y(x)$. Find a particular solution for the equation

$$y'' - 4y = -x^2 + 2x - 3.$$

PE 1.23 Let $y = y(x)$. Solve the nonhomogeneous linear third-order ordinary differential equation

$$y''' + y' = \sec x,$$

satisfying $y(0) = y(\pi) = 0$, using the method of variation of parameters.

PE 1.24 Let $y = y(x)$. Solve the ordinary differential equation

$$y'' + \omega^2 y = \sin \omega_0 x,$$

where ω and ω_0 are positive constants, analysing the cases (a) $\omega \neq \omega_0$ and (b) $\omega = \omega_0$.

PE 1.25 (Damped Harmonic Oscillator) Solve the problem of a damped harmonic oscillator (Chap. 11), whose differential equation is

$$m\ddot{x} + \lambda\dot{x} + kx = f(t),$$

where $x = x(t)$, and the dots denote differentiation with respect to time t. The parameters m, λ and k are positive constants. Discuss the results obtained in terms of the possible values of m, λ and k.

PE 1.26 (RLC Electrical Circuit) Discuss the RLC electrical circuit shown in Fig. 1.1, where R is the value of the resistance (measured in ohm—Ω); L is the

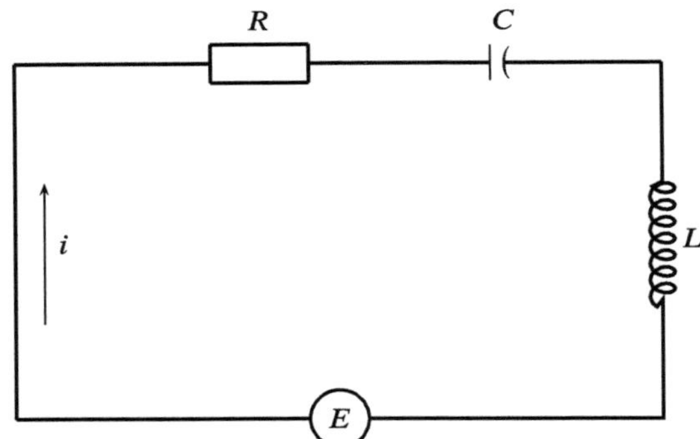

Fig. 1.1 RLC electrical circuit—PE 1.26

inductance (given in henry—H); C is the capacitance (measured in faraday—F) and E represents the electromotive force (in volt—V).

PE 1.27 Let $y = y(x)$. Find a particular solution for $y'' + y = \operatorname{tg} x$.

PE 1.28 (Whittaker [1873 – Edmund Taylor Whittaker – 1956] Equation) The linear second order ordinary differential equation

$$xy'' + (c - x)y' - ay = 0,$$

with a and c constants and $y = y(x)$, is called *confluent hypergeometric equation*. With a change of dependent variable of the form

$$y(x) = x^{\alpha} e^{\beta x} F(x),$$

with α and β constants to be determined, obtain the *Whittaker equation* (Chap. 4),

$$F'' + \left(-\frac{1}{4} + \frac{k}{x} + \frac{\frac{1}{4} - m^2}{x^2} \right) F = 0,$$

where k and m are constants.

PE 1.29 (Green's [1793 – George Green – 1841] Function) Supposing that $y_1(x)$ and $y_2(x)$ are two linearly independent solutions of the homogeneous linear ordinary differential equation associated with the nonhomogeneous linear ordinary differential equation

$$y'' + P(x)y' + Q(x)y = f(x),$$

show that the expression

$$y_p(x) = -y_1(x) \int^x \frac{y_2(x')f(x')}{W(x')}\mathrm{d}x' + y_2(x) \int^x \frac{y_1(x')f(x')}{W(x')}\mathrm{d}x',$$

which is a particular solution of the nonhomogeneous equation, can be written in the form

$$y_p(x) = \int_{x_0}^{x} G(x, t) f(t) \, dt,$$

where x_0 is a fixed point and $G(x, t)$, called *Green's function* (Chap. 7), is given by

$$G(x, t) = \frac{1}{W(t)} \begin{vmatrix} y_1(t) & y_1(x) \\ y_2(t) & y_2(x) \end{vmatrix},$$

where $W(t)$ is the Wronskian of functions $y_1(t)$ and $y_2(t)$.

PE 1.30 Using Green's function, **PE 1.29**, solve the nonhomogeneous linear second-order ordinary differential equation

$$y'' + y = f(x),$$

with $y(a) = 0$ and $y'(a) = 0$, obtaining the expression

$$y(x) = \int_{a}^{x} f(t) \sin(x - t) \, dt.$$

PE 1.31 Knowing that $y(x) = x^2$ is a solution of the ordinary differential equation

$$x(x - 2)y'' - 2(x - 1)y' + 2y = 0,$$

find the general solution of the nonhomogeneous differential equation

$$x(x - 2)y'' - 2(x - 1)y' + 2y = 2x^2 - 4x.$$

PE 1.32 Let $y = y(x)$. Solve, using a method of your choice, the nonhomogeneous linear second-order ordinary differential equation

$$x^2 y'' - x(x + 2)y' + (x + 2)y = -x^3 - x - 2,$$

showing that its general solution is

$$y(x) = c_1 x + c_2 x\, e^x + x^2 - 1,$$

where c_1 and c_2 are arbitrary constants.

PE 1.33 Let $y = y(x)$. Knowing that $y = \sin x$ is a solution of

$$\text{tg}^2 x \, y'' - 2\,\text{tg}\, x \, y' + (2 + \text{tg}^2 x)y = 0,$$

show that its general solution is

$$y(x) = c_1 \sin x + c_2 x \sin x,$$

where c_1 and c_2 are arbitrary constants.

PE 1.34 Find the general solution

$$y(x) = c_1 x\, e^x + c_2 x\, e^{-2x},$$

where c_1 and c_2 are arbitrary constants, of the ordinary differential equation

$$x^2 y'' + (x^2 - 2x)y' - (2x^2 + x - 2)y = 0,$$

beginning with the solution $y(x) = x\, e^x$.

PE 1.35 Let $y = y(x)$. Show that

$$xy'' - 2(x - 1)y' + 2(x - 1)y = 0$$

is equivalent, for a certain function $v(x)$, to the ordinary differential equation $v'' + v = 0$. Solve the differential equation for $y(x)$.

PE 1.36 Let $y = y(x)$. Solve the equation

$$x^2 \frac{d^2 y}{dx^2} = 1.$$

PE 1.37 Let $y = y(x)$. Solve the ordinary differential equation for

$$x^2 y'' + xy' - y = 0.$$

PE 1.38 Let $y = y(x)$. Using the result of the previous exercise, show that the general solution of the nonhomogeneous linear second-order ordinary differential equation

$$x^2 y'' + xy' - y = -3x^2 - 1$$

is given by

$$y = c_1 x + c_2 x^{-1} + 1 - x^2,$$

where c_1 and c_2 are arbitrary constants.

PE 1.39 Let $y = y(x)$. Find the solution to

$$(x - 1)y'' - xy' + y = 1,$$

knowing that $y(x) = e^x$ satisfies the corresponding homogeneous differential equation.

PE 1.40 Let $y = y(x)$. Solve the nonhomogeneous linear second-order ordinary differential equation

$$(1 + x^2)y'' - 2xy' + 2y = 1 + x^2,$$

using the fact that $y(x) = x$ is a solution of the ordinary differential equation

$$(1 + x^2)y'' - 2xy' + 2y = 0.$$

PE 1.41 Find the form of $Q(x)$ for which the ordinary differential equation

$$y'' + ay' + Q(x)y = 0,$$

where $y = y(x)$ and a is a constant, can be transformed, with the substitution $y = v(x)f(x)$, into a Bessel differential equation of order n,

$$v'' + \frac{1}{x}v' + \left(1 - \frac{n^2}{x^2}\right)v = 0,$$

where $v = v(x)$. Obtain the expression for $f(x)$.

PE 1.42 Apply the transformation

$$y(x) = v(x)\exp\left(-\frac{1}{2}\int^x P(x')\mathrm{d}x'\right)$$

to the ordinary differential equation

$$y'' + P(x)y' + Q(x)y = 0.$$

Obtain in this way the so-called *normal form*,

$$v'' + Iv = 0,$$

where $v = v(x)$ and $I(x) = Q(x) - \frac{1}{2}P'(x) - \frac{1}{4}P^2(x)$ is called the *invariant* of the ordinary differential equation.

PE 1.43 Find the invariant in the normal form of the *hypergeometric differential equation* (Chap. 4)

$$x(1-x)\omega'' + [c - (a+b+1)x]\omega' - ab\omega = 0,$$

where a, b and c are constants and $\omega = \omega(x)$.

PE 1.44 Find the invariant associated with the *confluent hypergeometric differential equation* (Chap. 4)

$$x\omega'' + (c-x)\omega' - a\omega = 0,$$

where a and c are constants and $\omega = \omega(x)$.

PE 1.45 The *self-adjoint form* of a linear second-order ordinary differential equation is that in which the equation for $y = y(x)$ is written as

$$\frac{\mathrm{d}}{\mathrm{d}x}[A(x)y'] + C(x)y = 0,$$

where $A(x)$ and $C(x)$ are functions of the coefficients of the original equation. Obtain the self-adjoint form of Legendre's equation

$$(1-x^2)y'' - 2xy' + n(n+1)y = 0,$$

where $y = y(x)$ and $n = 0, 1, 2, \ldots$

PE 1.46 Calculate the Wronskian of the two linearly independent solutions of the homogeneous ordinary differential equation in **PE 1.40**.

PE 1.47 We call *generalized Riccati equation* the ordinary differential equation

$$\frac{\mathrm{d}y}{\mathrm{d}x} = A_0(x) + A_1(x)y + A_2(x)y^2,$$

with $y = y(x)$. Classify this equation according to its type, order, and linearity.

PE 1.48 Introduce the change of dependent variable

$$y(x) = -\frac{\omega'(x)}{\omega(x)A_2(x)}$$

into the generalized Riccati differential equation and get for $\omega(x)$ the equation

$$A_2\omega'' - (A_2' + A_1 A_2)\omega' + A_2^2 A_0\omega = 0,$$

which is a linear second-order ordinary differential equation.

PE 1.49 Using the results shown in **PE 1.47** and **PE 1.48**, put the differential equation

$$xy'' + y' + xy = 0,$$

$y = y(x)$, in the form of a Riccati equation.

PE 1.50 Transform the Riccati equation

$$y' = 1 - x^2 + y^2,$$

with $y = y(x)$, into a linear second-order ordinary differential equation.

References

1. G. Birkhoff, G. Rota. *Ordinary Differential Equations* (Blaisdell, New York, 1969)
2. W.E. Boyce, R.C. Diprima, *Elementary Differential Equations and Boundary Value Problems* (John Wiley, New York, 2007)
3. M. Braun, *Differential Equations and their Applications* (Springer-Verlag, New York, 1983)
4. E. Capelas de Oliveira, M. Tygel, *Métodos Matemáticos para Engenharia. Mathematical Methods for Engineering*, 2nd edn. (Sociedade Brasileira de Matemática, Rio de Janeiro, 2012)

Chapter 2
Power Series and the Frobenius Method

The mathematics is not there till we put it there.

1882 – Arthur Stanley Eddington – 1944

We present in this chapter some concepts and results on power series, which are used in the study of the Frobenius [1849 – Ferdinand Georg Frobenius – 1917] method for solving homogeneous linear second-order ordinary differential equations with nonconstant coefficients. The Frobenius method plays an important role in the solution of this kind of equation, as it will always provide at least one solution, and as we already know, another linearly independent solution can in principle be obtained from the first one by reduction of order [1–3].

2.1 Preliminaries and General Concepts

The power series expansion of a solution of a homogeneous linear second-order ordinary differential equation in the neighborhood of a point $x = x_0$ is an example of a perturbative series expansion, that is, a power series of a small parameter, in this case the distance to point $x = x_0$, the center of the series.

2.1.1 Power Series

A series of the form

$$a_0 + a_1(x - x_0) + a_2(x - x_0)^2 + \cdots = \sum_{n=0}^{\infty} a_n (x - x_0)^n,$$

where $x_0, a_0, a_1, \ldots, a_n$ are constants (coefficients), is called a *power series*.

© The Author(s), under exclusive license to Springer Nature Switzerland AG 2025
E. Capelas de Oliveira, J. E. Maiorino, *Analytical Methods in Applied Mathematics*,
Problem Books in Mathematics, https://doi.org/10.1007/978-3-031-74794-6_2

A power series usually converges for $\mid x - x_0 \mid < R$ and diverges for $\mid x - x_0 \mid > R$ for some real constant R, which is called the *radius of convergence of the series*. For $\mid x - x_0 \mid = R$ the series may converge or not. The interval $\mid x - x_0 \mid < R$ is called the *interval of convergence* of the series. If $R = 0$, the series converges only for $x = x_0$, its center, and if $R = \infty$, the series converges for all values of x.

The Frobenius method is then a method for finding the solutions of a homogeneous linear second-order ordinary differential equation based on the assumption that they can be written as power series.

Example 2.1 Let $x \in \mathbb{R}$. Consider the power series

$$\frac{1}{1 + x^2} = \sum_{n=0}^{\infty} (-1)^n x^{2n} .$$

Obtain its radius of convergence.

We have here the quotient of two polynomials, $p_1(x) = 1$ and $p_2(x) = 1 + x^2$. Since the polynomial in the denominator, $p_2(x)$, is nonnull for every real x, the quotient is well defined and infinitely differentiable in $\mathbb{R}$. The power series around $x_0 = 0$, however, does not converge at all points of $\mathbb{R}$. Using the ratio test, we can show that the convergence radius is unitary. This result can also be obtained directly, using a result that says that the convergence radius of the series representing the division of two polynomials is given by the distance from the center of the series to the nearest root in the complex plane of the polynomial in the denominator. In this case, both the roots of $p_2(x) = 0$, $x = \pm i$, are at a distance to the origin equal to one; thus, the convergence radius is unitary. $\square$

2.1.2 Ordinary and Singular Points

We consider a homogeneous linear second-order ordinary differential equation written as

$$\frac{d^2}{dx^2} y(x) + p(x) \frac{d}{dx} y(x) + q(x) y(x) = 0 ,$$

where $p(x)$ and $q(x)$ are rational functions, i.e. the quotient of two polynomials.

Definition 2.1.1 (Ordinary and Singular Points)

If the coefficients $p(x)$ and $q(x)$ are rational functions and the limits

$$\lim_{x \to x_0} p(x) \quad \text{and} \quad \lim_{x \to x_0} q(x)$$

exist, then $x = x_0$ is called an ordinary point of the ordinary differential equation. In the case in which one of the limits does not exist, $x = x_0$ is called a singular point.

Theorem 2.1 *If x_0 is an ordinary point of the ordinary differential equation*

$$\frac{d^2}{dx^2} y(x) + p(x) \frac{d}{dx} y(x) + q(x) y(x) = 0,$$

then there exist two linearly independent solutions, which can be obtained by means of a Taylor [1685 – Brook Taylor – 1731] series, as we will see below. These series converge on the interval $|x - x_0| < R$ for some $R > 0$.

Example 2.2 Consider the homogeneous, linear second-order ordinary differential equation with nonconstant coefficients, called confluent hypergeometric equation,

$$x \frac{dy^2}{dx^2} + (c - x) \frac{dy}{dx} - ay = 0, \tag{2.1}$$

with $y = y(x)$ and $a, c \in \mathbb{R}$ constant parameters. Show that the point $x_0 = 0$ may be an ordinary or a singular point, depending on the values of a and c.

Let $x \neq 0$. Identifying the coefficients, we have

$$p(x) = \frac{c - x}{x} \qquad \text{and} \qquad q(x) = -\frac{a}{x}.$$

We separate in two cases. First, we assume a and c different from zero. Taking the limits

$$\lim_{x \to 0} \frac{c - x}{x} \qquad \text{and} \qquad \lim_{x \to 0} \left(-\frac{a}{x}\right)$$

we conclude that neither limit exists; thus, in this case, $x_0 = 0$ is a singular point. On the other hand, for $a = 0 = c$, the differential equation is a reducible differential equation with constant coefficients and $x_0 = 0$ is an ordinary point. $\square$

In order to address the general case, we begin with a brief discussion concerning the concept of series convergence and the definition of analytic function, which will be formally presented in the next chapter.

We also present the concepts of regular and irregular singular points and finally the Frobenius method. We present only the convergence at a point and the absolut convergence in order to discuss the interval on which power series converge or diverge.

Definition 2.1.2 (Convergence at a Point)

A power series $\sum\limits_{k=0}^{\infty} a_k (x - x_0)^k$ is convergent at a point x if the limit

$$\lim_{N \to \infty} \sum_{k=0}^{N} a_k (x - x_0)^k$$

exists. For $x = x_0$, its center, the power series converges and its limit is a_0.

Definition 2.1.3 (Absolute Convergence)

The power series $\sum_{k=0}^{\infty} a_k (x - x_0)^k$ is absolutely convergent at a point x if the series

formed by the absolute values of their terms, $\sum_{k=0}^{\infty} |a_k (x - x_0)^k|$, converges.

It is important to note that if a power series is absolutely convergent, then it converges. On the other hand, the reciprocal is not necessarily true.

Definition 2.1.4 (Interval of Convergence)

Let $R > 0$ be the convergence radius of a power series. This means that the power series is absolutely convergent if $|x - x_0| < R$, and it is divergent if $|x - x_0| > R$. If the power series converges only for $x = x_0$, the convergence radius is zero, and if it converges for all x, the convergence radius is infinite.

Definition 2.1.5 (Analytic Function)

If a function $f(x)$ admits a representation in power series with center at $x = x_0$ and a convergence radius $R > 0$, then the coefficients are unique. The series representing $f(x)$ is a Taylor series (a Maclaurin [1698 – Colin Maclaurin – 1746] series if $x_0 = 0$),

$$f(x) = \sum_{k=0}^{\infty} \frac{f^{(k)}(x_0)}{k!} (x - x_0)^k,$$

for $|x - x_0| < R$. The series is determined by the values of the function and of all its derivatives at a single point, the center of the series. In this case, we say that $f(x)$ is an analytic function at x_0.

Definition 2.1.6 (Regular Singular Point)

Consider a homogeneous, linear second-order ordinary differential equation of the form

$$A(x)\frac{d^2}{dx^2}y(x) + B(x)\frac{d}{dx}y(x) + C(x)y(x) = 0,$$

with $A(x)$, $B(x)$ and $C(x)$ polynomial functions. A point $x = x_0$ is called a regular singular point of this equation if the limits

$$\lim_{x \to x_0} (x - x_0)\frac{B(x)}{A(x)} \quad \text{and} \quad \lim_{x \to x_0} (x - x_0)^2 \frac{C(x)}{A(x)}$$

are finite; otherwise, it is called an irregular singular point.

Example 2.3 Classify the point $x_0 = 0$ relatively to the confluent hypergeometric differential equation given by Eq. (2.1).

Identifying the coefficients, we see that $A(x) = x$, $B(x) = c-x$ and $C(x) = -a$. As the limits for $x \to 0$ of the quotients $x[B(x)/A(x)]$ and $x^2[C(x)/A(x)]$ are finite, we conclude that $x_0 = 0$ is a regular singular point. $\square$

2.2   Expansion in Power Series

Suppose that a function $f(x)$ and its derivatives $f'(x), \ldots, f^{(n)}(x)$ exist and are continuous functions on the closed interval $[a, b]$ and that $f^{(n+1)}(x)$ exists on the open interval $]a, b[$. Then, $f(x)$ can be written as

$$f(x) = f(x_0) + f'(x_0)(x-x_0) + \frac{1}{2!}f''(x_0)(x-x_0)^2 + \cdots + \frac{f^{(n)}(x_0)}{n!}(x-x_0)^n + R_n,$$

where R_n is the *remainder of the series* and is given in one of the following two forms:

$$R_n = \frac{f^{(n+1)}(\xi)}{(n+1)!}(x - x_0)^{n+1} \qquad \text{(Lagrange)};$$

$$R_n = \frac{f^{(n+1)}(\xi)}{n!}(x - \xi)^n(x - x_0) \quad \text{(Cauchy)};$$

where ξ is a real number between x_0 and x. Usually, the value of ξ is different in the two forms presented earlier for the remainder.

If for every x and ξ in $[a, b]$ we have $\lim_{n \to \infty} R_n = 0$, then on this interval the following equality holds:

$$f(x) = f(x_0) + f'(x_0)(x-x_0) + \frac{1}{2!}f''(x_0)(x-x_0)^2 + \cdots + \frac{f^{(n)}(x_0)}{n!}(x-x_0)^n + \cdots \tag{2.2}$$

This is called the Taylor series for $f(x)$ or the Taylor expansion of $f(x)$. If $x_0 = 0$ we have the so-called Maclaurin series, a series centered at the origin.

Example 2.4   Consider the function $f(x) = (1 + x^2)^{-1}$ as in Example 2.1. Obtain the corresponding Maclaurin series.

In this case, $x_0 = 0$. We then have for the coefficients $f(0) = 1$, $f'(0) = 0$, $f''(0) = -2$, $f'''(0) = 0$, and so on. Substituting into Eq. (2.2) we can write

$$f(x) = 1 + \frac{1}{2!}(-2!)x^2 + \frac{1}{4!}(4!)x^4 + \ldots$$

$$= 1 - x^2 + x^4 - \ldots$$

or in the form

$$f(x) = \sum_{n=0}^{\infty} (-1)^n x^{2n}$$

which is the result shown in Example 2.1. Using the ratio test, we can show that the convergence radius is unitary, i.e., the power series converges on $-1 < x < 1$. $\square$

2.3   The Frobenius Method

The Frobenius method for solving homogeneous, linear second-order ordinary differential equations consists of searching for a solution with the form of a series of the form

$$y(x) = \sum_{n=0}^{\infty} c_n x^{n+s}, \qquad c_0 \neq 0,$$

where s is an arbitrary parameter. Here, without loss of generality, we used $x_0 = 0$, · a series centered at the origin; otherwise, a simple translation $x \rightarrow x - x_0$ will put the center at the origin.

Substituting this expression and its derivatives into the original differential equation and factorizing the resulting expression in terms of powers of x, we obtain a system of algebraic equations relating the various coefficients c_n of the series, one equation for each power of x. This leads us to a second degree algebraic equation called the *indicial* or *auxiliary equation*, whose solution furnishes the value of the parameter s. Depending on this value, the resulting power series can furnish two linearly independent solutions of the given homogeneous differential equation.

In the case in which $s = 0$ we have a Maclaurin series, that is, analytic functions nonnull at the origin, as we will see in Chap. 3. For $s = m$, with m a positive integer, the solutions are still analytic functions, with a zero of order m at the origin. In the case where $s = -m$, where m is a positive integer, the functions have a pole of order m at the origin, as we will discuss in Chap. 3; finally, for noninteger s, we have functions with certain kinds of branch points at the origin. For branch points of logarithmic type (Chap. 3), we must begin with a Frobenius series of the form

$$y(x) = \ln x \sum_{n=0}^{\infty} c_n x^{n+s} + \sum_{m=0}^{\infty} a_m x^{m+r} \ , \ \text{with } c_0 \neq 0 \text{ and } a_0 \neq 0,$$

which is called a *generalized Frobenius series*.

In short, the Frobenius method always provides one linearly independent solution. As mentioned earlier, depending on the roots of the auxiliary equation, we may also have the second linearly independent solution. As an example, we present here the case in which the roots of the auxiliary equation are equal, that is, the case in which we have a double root. We then search for the second linearly independent solution of the zero-order Bessel equation.

Example 2.5 Let $x \in \mathbb{R}$. Consider the classical zero order Bessel equation

$$x^2 \frac{d^2}{dx^2} y(x) + x \frac{d}{dx} y(x) + x^2 y(x) = 0.$$

A first linearly independent solution of this differential equation can be obtained with the Frobenius method. It is denoted $J_0(x)$, the first kind Bessel function of order zero, and is given by

$$J_0(x) = \sum_{n=0}^{\infty} \frac{(-1)^n}{(n!)^2} \left(\frac{x}{2}\right)^{2n}.$$

Using the generalized Frobenius method, obtain the general solution of the Bessel equation, i.e., a solution with two arbitrary constants.

A second solution of the Bessel equation, linearly independent from the first one, will be constructed in the form of the generalized Frobenius series,

$$y_2(x) = C J_0(x) \ln x + \sum_{n=1}^{\infty} b_n x^n, \qquad x > 0, \tag{2.3}$$

where the coefficients C and b_n are to be determined. Note that the above expression is, in fact, a generalization of the Frobenius series previously defined, since choosing $C = 0$ reduces it to the previous one, because $s = 0$ is a double root.

Differentiating twice Eq. (2.3), substituting the result in the zero-order Bessel equation and rearranging the terms, we can write

$$b_1 + 4b_2 x + \sum_{n=3}^{\infty} [n^2 b_n + b_{n-2}] x^{n-1} = -2C J_0'(x),$$

where the prime indicates the first derivative of the zero-order Bessel function. Evaluating the first derivative of the zero-order Bessel function and substituting the result into the last equation, we obtain

$$b_1 + 4b_2 x + \sum_{n=3}^{\infty} [n^2 b_n + b_{n-2}] x^{n-1} = -C \sum_{n=1}^{\infty} (-1)^n \frac{4n}{2^{2n} (n!)^2} x^{2n-1}.$$

In order to simplify the calculation of coefficients b_n, we multiply the above expression by x and separate the sum of the first member into two sums, one containing odd powers and the other containing even powers, so that we have

$$\left\{ b_1 x + \sum_{n=1}^{\infty} [(2n+1)^2 b_{2n+1} + b_{2n-1}] x^{2n+1} \right\} + \left\{ 4b_2 x^2 + \sum_{n=2}^{\infty} [(2n)^2 b_{2n} + b_{2n-2}] x^{2n} \right\}$$

$$= C x^2 + C \sum_{n=2}^{\infty} (-1)^{n+1} \frac{4n}{2^{2n}(n!)^2} x^{2n} .$$

Using the last expression, we conclude that $b_1 = b_3 = \cdots = 0$, that is, all odd coefficients are null. On the other hand, for the even coefficients we have

$$4b_2 = C$$

$$4n^2 b_{2n} + b_{2n-2} = C(-1)^{n+1} \frac{4n}{2^{2n}(n!)^2}, \quad \text{for} \quad n \geq 2,$$

whence there follow the recurrence relations for coefficients b_{2n}:

$$b_2 = \frac{C}{2^2}$$

$$b_4 = -\frac{C}{2^2 \cdot 4^2} \left(1 + \frac{1}{2} \right) = -\frac{C}{2^4 (2!)^2} \left(1 + \frac{1}{2} \right)$$

$$b_{2n} = (-1)^{n+1} \frac{C}{2^{2n}(n!)^2} \left(1 + \frac{1}{2} + \ldots + \frac{1}{n} \right) .$$

Substituting these results into Eq. (2.3) we obtain the second solution,

$$y_2(x) = C J_0(x) \ln x + C \sum_{n=1}^{\infty} (-1)^{n+1} \frac{1}{(n!)^2} \left(1 + \frac{1}{2} + \ldots + \frac{1}{n} \right) \left(\frac{x}{2} \right)^{2n} .$$

Taking $C = 1$ (normalization), this second solution is called second kind zero-order Bessel function, which is denoted $Y_0(x)$ and is given by

$$Y_0(x) = J_0(x) \ln x + \sum_{n=1}^{\infty} (-1)^{n+1} \frac{1}{(n!)^2} \left(1 + \frac{1}{2} + \ldots + \frac{1}{n} \right) \left(\frac{x}{2} \right)^{2n} .$$

Thus, the general solution of zero-order Bessel equation is given by

$$y(x) = C_1 J_0(x) + C_2 Y_0(x),$$

with C_1 and C_2 arbitrary constants. $\square$

2.4 Solved Exercises

SE 2.1 Let $x \in \mathbb{R}$ and a, b and c three real parameters. The so-called hypergeometric differential equation is given by

$$x(1-x)\frac{d^2}{dx^2}y(x) + [c-(a+b+1)x]\frac{d}{dx}y(x) - aby(x) = 0.$$

(a) Classify the hypergeometric differential equation. (b) Discuss the nature of points $x_0 = 0$ and $x_0 = 1$.

Solution (a) The hypergeometric differential equation is a homogeneous, linear, and second-order ordinary differential equations with nonconstant coefficients. (b) In order to discuss the nature of the points, we identify the coefficients

$$A(x) = x(1-x), \quad B(x) = c-(a+b+1)x, \quad \text{and} \quad C(x) = -ab.$$

First of all, these two points are singular points because $A(0) = 0 = A(1)$. Thus, we must evaluate two limits in order to know if the points are regular singular or irregular singular. For $x_0 = 0$, evaluating the limits

$$\lim_{x \to 0} x\frac{c-(a+b+1)x}{x(1-x)} \quad \text{and} \quad \lim_{x \to 0} x^2\frac{-ab}{x(1-x)},$$

we find that both are zero, that is, finite. Then, this singular point is regular. On the other hand, for $x_0 = 1$ we must evaluate the limits

$$\lim_{x \to 1} (x-1)\frac{c-(a+b+1)x}{x(1-x)} \quad \text{and} \quad \lim_{x \to 1} (x-1)^2\frac{-ab}{x(1-x)}.$$

Also in this case, the limits are finite, then we have again a regular singular point.

We conclude that the hypergeometric differential equation is a second-order ordinary differential equation and has three singular points $x_0 = 0$, $x_0 = 1$ and $x_0 = \infty$. Two of them were discussed earlier and the point $x_0 = \infty$ is left as a proposed exercise.

SE 2.2 Let $x \in \mathbb{R}$. Expand $f(x) = \sin x$ in a Maclaurin series. Obtain the interval of convergence for this expansion.

Solution We have for the derivatives:

$$f'(x) = \cos x, \ f''(x) = -\sin x, \ f'''(x) = -\cos x, \ f^{IV}(x) = \sin x \ \dots$$

$$(2.4)$$

Hence, $f'(0) = 1$, $f''(0) = 0$, $f'''(0) = -1$, $f^{IV}(0) = 0 \ldots$, and according to Eq. (2.2) we have

$$f(x) = \frac{x}{1!} - \frac{x^3}{3!} + \frac{x^5}{5!} - \cdots = \sum_{n=1}^{\infty} (-1)^{n+1} \frac{x^{2n-1}}{(2n-1)!},$$

or also

$$f(x) = \sum_{n=0}^{\infty} (-1)^n \frac{x^{2n+1}}{(2n+1)!}.$$

In order to obtain the interval of convergence, we use the *ratio test*. According to this test, we have three possibilities. Assume that

$$\lim_{n \to \infty} \left| \frac{a_{n+1}}{a_n} \right| = L.$$

Then, the series diverges if $L > 1$ and converges if $L < 1$. If $L = 1$ the test fails, that is, it does not allow us to say anything about the convergence of the given series. In the present case,

$$
\begin{aligned}
L &= \lim_{n \to \infty} \left| \frac{a_{n+1}}{a_n} \right| = \lim_{n \to \infty} \left| \frac{(-1)^{n+1} x^{2n+3}}{(2n+3)!} \frac{(2n+1)!}{(-1)^n x^{2n+1}} \right| \\
&= x^2 \lim_{n \to \infty} \left| \frac{(2n+1)!}{(2n+3)(2n+2)(2n+1)!} \right| = 0.
\end{aligned}
\tag{2.5}
$$

Hence, the series converges for every x, i.e., we have an *infinite* radius of convergence.

SE 2.3 Let $v \in \mathbb{R}$ and $y = y(x)$. Discuss the Bessel equation of order v, which we will discuss in Chap. 4,

$$x^2 y'' + xy' + (x^2 - v^2)y = 0,$$

using the Frobenius method.

Solution Consider the following series:

$$y(x) = \sum_{n=0}^{\infty} c_n x^{n+s}, \quad c_0 \neq 0,$$

where s is a free parameter. Differentiating it with respect to x we get

$$y'(x) = \sum_{n=0}^{\infty} (n+s)c_n x^{n+s-1};$$

(2.6)

$$y''(x) = \sum_{n=0}^{\infty} (n+s)(n+s-1)c_n x^{n+s-2}.$$

(2.7)

Substituting these expressions into the Bessel equation, we obtain

$$\sum_{n=0}^{\infty} (n+s)(n+s-1)c_n x^{n+s} + \sum_{n=0}^{\infty} (n+s)c_n x^{n+s} +$$

$$+ \sum_{n=0}^{\infty} c_n x^{n+s+2} - v^2 \sum_{n=0}^{\infty} c_n x^{n+s} = 0.$$

(2.8)

Now, we may change the index in the third sum, that is, we can write

$$\sum_{n=0}^{\infty} c_n x^{n+s+2} = \sum_{n=2}^{\infty} c_{n-2} x^{n+s},$$

so that we may rewrite Eq. (2.8) as

$$s(s-1)c_0 + (s+1)sc_1 x + sc_0 + (s+1)c_1 x - v^2 c_0 - v^2 c_1 x +$$

$$+ \sum_{n=2}^{\infty} \left\{ [(n+s)^2 - v^2]c_n + c_{n-2} \right\} x^{n+s} = 0.$$

(2.9)

By hypothesis, the initial constant c_0 is nonnull; then, the coefficients of power $x^0 = 1$ furnish the *indicial equation*

$$s(s-1) + s - v^2 = 0.$$

(2.10)

For powers in x we have

$$[s(s+1) + (s+1) - v^2]c_1 = 0,$$

and for the relation between c_n and c_{n-2},

$$[(n+s)^2 - v^2]c_n + c_{n-2} = 0, \quad \text{for } n \geq 2,$$

which is known as *recurrence relation*.

Solving the indicial equation Eq. (2.10), we find $s = \pm v$. Substituting these values for s into the equation involving c_1 we have

$$s = v \quad \Rightarrow \quad [(v+1)^2 - v^2]c_1 = 0 \quad \Rightarrow \quad (2v+1)c_1 = 0; \qquad (2.11)$$

$$s = -v \quad \Rightarrow \quad [(-v+1)^2 - v^2]c_1 = 0 \quad \Rightarrow \quad (-2v+1)c_1 = 0. \quad (2.12)$$

Thus, we conclude that:

$$s = v = -1/2 \Rightarrow c_1 \text{ is arbitrary}; \qquad (2.13)$$

$$s = -v = 1/2 \Rightarrow c_1 \text{ is arbitrary}; \qquad (2.14)$$

$$s \neq \pm 1/2 \Rightarrow c_1 = 0. \qquad (2.15)$$

SE 2.4 (Bessel Equation of Order 1/2) To exemplify the use of the Frobenius method, let us consider the results of **SE 2.3** for the case in which $v = -1/2$ and c_1 is arbitrary. In this case, $s = -1/2$, and we have from this result the *recurrence relation*

$$\left[\left(n - \frac{1}{2}\right)^2 - \frac{1}{4}\right] c_n + c_{n-2} = 0,$$

$$n(n-1)c_n + c_{n-2} = 0 \Rightarrow c_n = -\frac{c_{n-2}}{n(n-1)}, \quad n \geq 2.$$

Developing the first terms we find:

$$c_2 = -c_0/2;$$

$$c_3 = -c_1/3 \cdot 2 = -c_1/3!;$$

$$c_4 = -c_2/4 \cdot 3 = c_0/4!;$$

$$c_5 = -c_3/5 \cdot 4 = c_1/5!;$$

$$c_6 = -c_4/6 \cdot 5 = -c_0/6!;$$

$$c_7 = -c_5/7 \cdot 6 = -c_1/7!$$

$$\vdots$$

Therefore, we have for our series

$$y(x) = x^{-1/2}\{c_0 + c_1 x - \frac{c_0}{2!}x^2 - \frac{c_1}{3!}x^3 + \frac{c_0}{4!}x^4 + \frac{c_1}{5!}x^5 - \frac{c_0}{6!}x^6 - \frac{c_1}{7!}x^7 + \cdots\}$$

or

$$y(x) = x^{-1/2}c_0 \left\{1 - \frac{x^2}{2!} + \frac{x^4}{4!} - \frac{x^6}{6!} + \cdots \right\} + x^{-1/2}c_1 \left\{x - \frac{x^3}{3!} + \frac{x^5}{5!} - \frac{x^7}{7!} + \cdots \right\}$$

$$= x^{-1/2} c_0 \sum_{n=0}^{\infty} (-1)^n \frac{x^{2n}}{(2n)!} + x^{-1/2} c_1 \sum_{n=0}^{\infty} (-1)^n \frac{x^{2n+1}}{(2n+1)!}, \tag{2.16}$$

where c_0 and c_1 are arbitrary constants.

Comparing this result with SE 2.2, we identify these two series as $\sin x$ and $\cos x$; hence

$$y(x) = x^{-1/2}(c_0 \cos x + c_1 \sin x)$$

and as we have *two* arbitrary constants, this is the general solution for the Bessel ordinary differential equation of order $1/2$.

If we perform the calculations for the other root of the indicial equation ($s = 1/2$) we will find a linear combination of the two linearly independent solutions $x^{-1/2} \cos x$ and $x^{-1/2} \sin x$ found above. This phenomenon, in which the least root of the indicial equation generates by itself the general solution, occurs whenever the two roots are not integers but their difference is an integer. Introducing an adequate normalization for Bessel functions (see Chap. 4), we get

$$x^{-1/2} \cos x = \sqrt{\frac{\pi}{2}} \, J_{-1/2}(x); \tag{2.17}$$

$$x^{-1/2} \sin x = \sqrt{\frac{\pi}{2}} \, J_{1/2}(x). \tag{2.18}$$

Thus, the general solution of the Bessel differential equation of order $1/2$,

$$x^2 y'' + xy' + (x^2 - 1/4)y = 0,$$

is

$$y(x) = A \, J_{1/2}(x) + B \, J_{-1/2}(x),$$

where A and B are arbitrary constants. $J_{1/2}(\cdot)$ is the first kind Bessel function of order $1/2$ and $Y_{1/2}(\cdot)$ is the second kind Bessel function of order $1/2$.

SE 2.5 (Whittaker Equation) The Schrödinger [1887 − Erwin Schrödinger − 1961] equation, which will be discussed in Chap. 11, can be used to describe the motion of a quantum particle in a potential field. If we write the resulting equation in spherical coordinates and apply the method of separation of variables—which is explained in Chap. 8—we arrive at a radial equation of the form

$$\frac{d^2}{dr^2} R(r) + \frac{2}{r} \frac{d}{dr} R(r) + \left[-\beta V(r) + \beta E - \frac{\ell(\ell+1)}{r^2} \right] R(r) = 0,$$

where β and ℓ are constants, $\ell = 0, 1, 2, \ldots$, $V(r)$ is the potential and E is the energy associated with the particle.

(a) Suppose that the potential is Keplerian, [1571 − Johannes Kepler − 1630] also known as Coulombian [1736 − Charles Augustin Coulomb − 1806] i.e., $V(r) = -\alpha/r$, where α is a positive constant. Perform the change of dependent variable

$$R(r) = \frac{1}{r}F(r),$$

and introduce two new constants $\lambda^2 = \beta E$ and $b = \beta\alpha$, to obtain the following ordinary differential equation:

$$\frac{d^2}{dr^2}F(r) + \left[\lambda^2 + \frac{b}{r} - \frac{\ell(\ell+1)}{r^2}\right]F(r) = 0.$$

(b) Introduce the change of independent variable $x = -2i\lambda r$ into the last equation to obtain the following ordinary differential equation:

$$\frac{d^2}{dx^2}F(x) + \left[-\frac{1}{4} + \frac{ib/2\lambda}{x} - \frac{\ell(\ell+1)}{x^2}\right]F(x) = 0.$$

This is called the Whittaker differential equation, as one can see in **PE 1.28**. (c) Use the Frobenius method, with an expansion around the origin, to discuss the solution of this equation.

Solution　(a) Differentiating the unknown function (dependent variable)

$$R(r) = \frac{1}{r}F(r)$$

with respect to r and introducing the result into the differential equation we obtain

$$\frac{d^2}{dr^2}F(r) + \left[-\beta V(r) + \beta E - \frac{\ell(\ell+1)}{r^2}\right]F(r) = 0.$$

Introducing the parameters λ^2 and b defined above, we finally have

$$\frac{d^2}{dr^2}F(r) + \left[\lambda^2 + \frac{b}{r} - \frac{\ell(\ell+1)}{r^2}\right]F(r) = 0,$$

which is the desired ordinary differential equation. (b) Now, for the new independent variable $x = -2i\lambda r$ we have

$$\frac{d}{dr} = \frac{dx}{dr}\frac{d}{dx} = -2i\lambda\frac{d}{dx}.$$

Substituting this result into the previous differential equation, we then get

$$\frac{d^2}{dx^2}F(x) + \left[-\frac{1}{4} + \frac{ib/2\lambda}{x} - \frac{\ell(\ell+1)}{x^2} \right] F(x) = 0,$$

which is the Whittaker differential equation. (c) To use the Frobenius method, we write $R(r)$ as

$$F(x) = \sum_{k=0}^{\infty} a_k x^{k+s},$$

where $a_0 \neq 0$ and s is a free parameter.

Differentiating this expression with respect to x, substituting into the differential equation and rearranging terms we have

$$\sum_{k=0}^{\infty} [(k+s)(k+s-1) - \ell(\ell+1)] a_k x^{k+s} + \frac{ib}{2\lambda} \sum_{k=0}^{\infty} a_k x^{k+s+1} - \frac{1}{4} \sum_{k=0}^{\infty} a_k x^{k+s+2} = 0.$$

Changing the index in the last two sums we obtain

$$\sum_{k=0}^{\infty} [(k+s)(k+s-1) - \ell(\ell+1)] a_k x^{k+s} + \frac{ib}{2\lambda} \sum_{k=1}^{\infty} a_{k-1} x^{k+s} - \frac{1}{4} \sum_{k=2}^{\infty} a_{k-2} x^{k+s} = 0,$$

whence it follows the indicial equation, $k = 0$,

$$s(s-1) - \ell(\ell+1) = 0,$$

with solutions $s_1 = \ell + 1$ and $s_2 = -\ell$. As ℓ is a nonnegative integer we have that $s_1 - s_2 = 2\ell + 1$ is always a positive integer. Then, the Frobenius series provides one solution while the other linearly independent solution must be obtained from the generalized Frobenius series, and it will not be a polynomial.

For $k = 1$, we have the following relation:

$$[s(s+1) - \ell(\ell+1)] a_1 + \frac{ib}{2\lambda} a_0 = 0;$$

and finally, the recurrence relation, valid for $k \geq 2$,

$$[(k+s)(k+s-1) - \ell(\ell+1)] a_k + \frac{ib}{2\lambda} a_{k-1} - \frac{1}{4} a_{k-2} = 0.$$

We now discuss the two cases, that is, $s = -\ell$ and $s = \ell + 1$. Let us first consider the case in which $s = -\ell$; from the relation between a_1 and a_0 we find that

$$a_1 = \frac{ib}{4\lambda\ell}a_0.$$

We thus see that we must have $\ell \neq 0$. So, $s = -\ell$ with $\ell = 1, 2, \ldots$ does not furnish a Frobenius series. On the other hand, the case $s = \ell + 1$ furnishes

$$a_1 = -\frac{i\lambda}{4(\ell+1)\lambda}a_0,$$

that is, given a_0 we obtain a_1 and using the recurrence relation

$$a_k = \frac{1}{k(k+2\ell+1)}\left(-\frac{ib}{\lambda}a_{k-1} + \frac{1}{2}a_{k-2}\right)$$

we obtain the remaining terms. Note that $s_1 - s_2 = 2\ell+1$ and the Frobenius method furnishes only one solution. To obtain another linearly independent solution, we must use the generalized Frobenius series.

SE 2.6 The classical Langevin's [1872 – Paul Langevin – 1946] theory of paramagnetism leads to the following expression for magnetic polarization:

$$P(x) = C\left(\frac{\cosh x}{\sinh x} - \frac{1}{x}\right),$$

where C is a positive constant. Expand $P(x)$ in a power series for small x, i.e., high temperatures and weak fields.

Solution Knowing that the expansions for $\sinh x$ and $\cosh x$ are given, respectively, by

$$\sinh x = x + \frac{x^3}{3!} + \frac{x^5}{5!} + \cdots$$

and

$$\cosh x = 1 + \frac{x^2}{2!} + \frac{x^4}{4!} + \cdots$$

we can calculate the quotient $x\,\cosh x/\sinh x$, obtaining

$$\frac{x\,\cosh x}{\sinh x} = 1 + \frac{1}{3}x^2 - \frac{1}{45}x^4 - \cdots$$

Then, we may write for $P(x)$:

$$P(x) = \frac{C}{x}\left(1 + \frac{1}{3}x^2 - \frac{1}{45}x^4 - \cdots - 1\right),$$

or, finally,

$$P(x) = C\left(\frac{1}{3}x - \frac{1}{45}x^3 + \cdots\right).$$

SE 2.7 The displacement x of a particle with rest mass m_0, under the action of a constant force $m_0 g$, where g is the gravitational acceleration, along the x axis, including relativistic effects, is given by

$$x = \frac{c^2}{g}\left\{\left[1 + \left(\frac{gt}{c}\right)^2\right]^{1/2} - 1\right\},$$

where c is the speed of light in vacuum. Find the displacement x as a power series in t. Compare it with the classical result $x = \frac{1}{2}gt^2$.

Solution Using the result [4]

$$\sqrt{1+x} = 1 + \frac{1}{2}x - \frac{1}{2}\cdot\frac{1}{4}x^2 + \frac{1}{2}\cdot\frac{1}{4}\cdot\frac{3}{6}x^3 - \cdots$$

valid for $|x| < 1$, which is obtained through the Maclaurin series, we can write

$$x(t) = \frac{c^2}{g}\left(1 + \frac{1}{2}\beta - \frac{1}{8}\beta^2 + \frac{1}{16}\beta^3 - \cdots - 1\right)$$

where $\beta = \left(\dfrac{gt}{c}\right)^2$. Then, simplifying, we obtain for the displacement the following expansion:

$$x(t) = \frac{1}{2}gt^2\left\{1 - \frac{1}{4}\left(\frac{gt}{c}\right)^2 + \frac{1}{8}\left(\frac{gt}{c}\right)^4 - \cdots\right\}.$$

Thus, the classical result is recovered at the limit $gt \ll c$ where only the first term in the series contributes, i.e.

$$x(t) = \frac{1}{2}gt^2.$$

SE 2.8 Find the general solution, in the neighborhood of $x = 0$, of the following homogeneous linear ordinary differential equation:

$$x(1-x)\frac{d^2}{dx^2}y(x) - 2x\frac{d}{dx}y(x) + 2y(x) = 0, \tag{2.19}$$

with $y = y(x)$.

Solution It is not difficult to see that $y(x) = x$ is a solution (inspection) of this differential equation. Once we realized that, we might use the method of reduction of order to determine the second linearly independent solution and then its general solution. However, we will use the Frobenius method—in this case, the generalized Frobenius method, because the roots of the indicial equation are integer and their difference is also an integer. Suppose a solution given by the following series:

$$y(x) = \sum_{n=0}^{\infty} c_n x^{n+s}, \tag{2.20}$$

where $c_0 \neq 0$ and s is a free parameter.

Then, differentiating with respect to x and substituting the results into the differential equation above we obtain

$$\sum_{n=0}^{\infty}(n+s)(n+s-1)c_n x^{n+s-1} - \sum_{n=0}^{\infty}(n+s)(n+s-1)c_n x^{n+s} -$$
$$-2\sum_{n=0}^{\infty}(n+s)c_n x^{n+s} + 2\sum_{n=0}^{\infty} c_n x^{n+s} = 0$$

or, rearranging terms,

$$\sum_{n=0}^{\infty}(n+s)(n+s-1)c_n x^{n+s-1} + \sum_{n=0}^{\infty}[2-(n+s)(n+s+1)]c_n x^{n+s} = 0. \tag{2.21}$$

Changing the index in the second sum, $n \to n-1$, we have

$$\sum_{n=0}^{\infty}(n+s)(n+s-1)c_n x^{n+s-1} + \sum_{n=1}^{\infty}[2-(n+s-1)(n+s)]c_{n-1} x^{n+s-1} = 0,$$

$$\tag{2.22}$$

from which we may write the indicial equation

$$s(s-1)c_0 = 0 \tag{2.23}$$

and also the following recurrence relation:

$$c_n = \frac{(n+s-1)(n+s)-2}{(n+s)(n+s-1)}c_{n-1}, \quad n \geq 1. \tag{2.24}$$

The indicial equation has roots $s = 0$ and $s = 1$. Note that these roots differ by an integer. In such cases, the usual method consists of searching for the second solution through the generalized Frobenius method, i.e, a solution with the form

$$y_2(x) = y_1(x) \ln x + \sum_{k=0}^{\infty} b_k x^k .$$

For $s = 0$, we obtain

$$c_n = \frac{(n+1)(n-2)}{n(n-1)} c_{n-1}, \qquad n \geq 1, \tag{2.25}$$

which is not defined for $n = 1$ and thus does not provide a solution of Frobenius type.

On the other hand, for $s = 1$ we obtain

$$c_n = \frac{(n-1)(n+2)}{n(n+1)} c_{n-1}, \qquad n \geq 1. \tag{2.26}$$

As $c_1 = 0$, all coefficients except c_0 are null, and it follows that

$$y_1(x) = Ax$$

with A a constant, is a solution of the differential equation.

We may search for the other linearly solution using the generalized Frobenius method, i.e., as $y(x) = x$ is a solution, we consider the expression

$$y_2(x) = x \ln x + \sum_{k=0}^{\infty} b_k x^k, \, b_0 \neq 0.$$

Then, differentiating this expression and substituting the result into the homogeneous linear ordinary differential equation, we find that $b_0 = -1/2$, b_1 is arbitrary, $b_2 = 3/2$ and the recurrence relation is given by

$$b_{n+1} = \frac{(n-1)(n+2)}{n(n+1)} b_n,$$

with $n \geq 2$, whence it follows, choosing $b_1 = 1$, the second linearly independent solution

$$y_2(x) = x \ln x - \frac{1}{2} + x + \frac{3}{2}x^2 + x^3 + \frac{5}{6}x^4 + \cdots$$

The general solution is then given by the expression

$$y(x) = A y_1(x) + B y_2(x),$$

where A and B are constants to be determined by the respective conditions of a specific problem and $y_1(x)$ and $y_2(x)$ are as aforementioned.

SE 2.9 Let $x \in \mathbb{R}$. Consider the homogeneous linear ordinary differential equation

$$x\frac{\mathrm{d}^2}{\mathrm{d}x^2}y(x) + (1-x)x\frac{\mathrm{d}}{\mathrm{d}x}y(x) - y(x) = 0, \tag{2.27}$$

known as a particular confluent hypergeometric function. (a) Introduce the change $x = 1/\xi$ to obtain another homogeneous linear ordinary differential equation for $y = y(\xi)$ and (b) using the equation obtained in (a) justify if it has a Frobenius series arround $\xi = 0$.

Solution (a) First, using the chain rule, we have for the first and second derivative operators

$$\frac{\mathrm{d}}{\mathrm{d}x} = -\xi^2\frac{\mathrm{d}}{\mathrm{d}\xi} \qquad \text{and} \qquad \frac{\mathrm{d}^2}{\mathrm{d}x^2} = \xi^4\frac{\mathrm{d}^2}{\mathrm{d}\xi^2} + 2\xi^3\frac{\mathrm{d}}{\mathrm{d}\xi}.$$

Substituting into Eq. (2.27) and simplifying, we get the homogeneous linear ordinary differential equation in variable ξ,

$$\xi^3 y'' + \xi(1+\xi)y' - y = 0, \tag{2.28}$$

with $y = y(\xi)$. (b) We look for a solution of the last ordinary differential equation in terms of a Frobenius series

$$y(\xi) = \sum_{k=0}^{\infty} a_k \xi^{k+s},$$

where $a_0 \neq 0$ and s is a free parameter. Evaluating the derivatives, substituting into Eq. (2.28) and rearranging we get

$$\sum_{k=1}^{\infty} a_{k-1}(k+s-1)^2 \xi^{k+s} + \sum_{k=0}^{\infty} a_k(k+s-1)\xi^{k+s} = 0.$$

Taking $k = 0$ in the second sum, we have the indicial equation and for $k \geq 1$, the corresponding recurrence relation

$$a_0(s-1) = 0 \qquad \text{and} \qquad a_k = -ka_{k-1},$$

respectively. The indicial equation furnishes $s = 1$. Thus, the solution in terms of a Frobenius series can be written as

$$y(\xi) = a_0 \sum_{k=0}^{\infty} (-1)^k k! \xi^{k+1}.$$

Since we made a change of variable $x = 1/\xi$ and did the analysis at point $\xi = 0$ (the same as analyzing around point $x = \infty$), we say that the equation in variable

x admits a regular solution at infinity, besides the regular at origin which, as can be verified by substitution, is given by $y(x) = \exp(x)$. In short, this is a second-order ordinary differential equation whose two linearly independent solutions are, one regular at the origin and the other regular at the infinity.

2.5   Proposed Exercises

PE 2.1 Show that $x_0 = \infty$ is another regular singular point for the hypergeometric differential equation.

PE 2.2 Let $x \in \mathbb{R}$ and let a, b be two nonnull real constants. Obtain the singular points and classify them for the Euler differential equation

$$x^2 \frac{d^2}{dx^2} y(x) + ax \frac{d}{dx} y(x) + by(x) = 0.$$

PE 2.3 Let $-1 \leq x \leq 1$. Consider the zero-order Legendre differential equation

$$(1 - x^2) \frac{d^2}{dx^2} y(x) - 2x \frac{d}{dx} y(x) = 0.$$

Does this equation have singular points? If it does, what are they?

PE 2.4 (a) Determine the radius of convergence of the geometric series

$$\sum_{n=0}^{\infty} x^n.$$

(b) Evaluate its sum.

PE 2.5

(a) Determine the convergence radius of the series

$$\sum_{n=0}^{\infty} (-1)^n \frac{x^{2n+1}}{(2n+1)!} \qquad \text{and} \qquad \sum_{n=0}^{\infty} (-1)^n \frac{x^{2n}}{(2n)!}.$$

(b) Identify them with elementary functions.

PE 2.6 Let $x \in \mathbb{R}$. Find the singular points of the differential equation

$$x^2(1 - x^2) \frac{d^2}{dx^2} y(x) + 2x \frac{d}{dx} y(x) - 4y(x) = 0$$

and classify them.

PE 2.7 What are the singular points of the ordinary differential equation

$$\frac{d^2}{dx^2}y(x) + x\frac{d}{dx}y(x) - y(x) = 0\,,$$

if any?

PE 2.8 Let $x \in \mathbb{R}$. Consider the linear, second-order ordinary differential equation

$$(x^2 - x - 2)\frac{d^2}{dx^2}y(x) - (x+1)\frac{d}{dx}y(x) + y(x) = 2022\,.$$

(a) Classify its singular points. (b) Obtain a solution for the corresponding homogeneous differential equation and a particular solution of the nonhomogeneous differential equation.

PE 2.9 Let $x \in \mathbb{R}$. Consider the so-called Airy [1801 – George Biddel Airy – 1892] ordinary differential equation

$$\frac{d^2}{dx^2}y(x) - xy(x) = 0\,.$$

Show that a solution of the Airy equation is given by

$$y_1(x) = C_1\left[1 + \frac{(x-1)^2}{2} + \frac{(x-1)^3}{6} + \frac{(x-1)^4}{24} + \cdots\right],$$

with C_1 an arbitrary constant.

PE 2.10 Let $x \in \mathbb{R}$ and $y = y(x)$. Obtain the solution of the initial value problem

$$\begin{cases} \dfrac{d^2y}{dx^2} - y = 0 \\ \qquad y(0) = 2 \\ \qquad y'(0) = 1 \end{cases}$$

using an expansion in Maclaurin series, justifying the procedure.

PE 2.11 Expand $f(x) = \cos x$ in a Maclaurin series. Obtain, for this expansion, its interval of convergence.

PE 2.12 (a) Introduce the change of variable $x = \frac{t}{b}$ into the hypergeometric differential equation given in **SE 2.1** and take the limit $b \to 0$ to obtain the confluent hypergeometric equation

$$t\frac{d^2w}{dt^2} + (c - t)\frac{dw}{dt} - a\,w = 0,$$

with $w = w(t)$. (b) Show that $t_0 = 0$ and $t_0 = \infty$ are respectively regular and irregular singular points of this equation.

PE 2.13 Show that

$$\frac{1}{(1+x)^2} = 1 - 2x + 3x^2 - \ldots = \sum_{k=1}^{\infty} (-1)^{k-1} k x^{k-1}$$

for $|x| < 1$.

PE 2.14 Assuming that $x^2 < 1$, show that

$$\frac{x}{(1-x)^2} = \sum_{k=1}^{\infty} k x^k \ .$$

PE 2.15 Expand $f(x) = e^x$ in a Maclaurin series.

PE 2.16 Expand $f(x) = \sinh x$ in a Maclaurin series.

PE 2.17 Expand $f(x) = \cosh x$ in a Maclaurin series.

PE 2.18 Using the results of the three previous exercises, verify that

$$\sinh x + \cosh x = e^x \ .$$

PE 2.19 Solve **PE 1.25** using the method of power series.

PE 2.20 Solve **PE 1.26** with the method of power series.

PE 2.21 Let $y = y(x)$. Solve, using power series, the differential equation

$$(1 - x^2)y'' + xy' - y = 0 \ .$$

PE 2.22 Solve, using the Frobenius method, the Bessel equation of order 1. Verify that it is necessary to use the generalized series.

PE 2.23 Solve, using the Frobenius method, the Bessel equation of order 0. Verify that in this case we have a series of the power type. Obtain the second linearly independent solution.

PE 2.24 Using the Frobenius method, solve the Euler equation of **PE 1.14**.

PE 2.25 Let $y = y(x)$. Solve, using the Frobenius method, the differential equation

$$xy'' + y' = 0 \ .$$

PE 2.26 Solve the Legendre equation of order zero with the Frobenius method.

PE 2.27 Let $y = y(x)$. Using the Frobenius method, solve the differential equation

$$y'' + (1 - x^2)y = 0,$$

called Hermite [1822 – Charles Hermite – 1901] differential equation.

PE 2.28 Development around the point at infinity. Use the Frobenius method to solve the differential equation (**PE 1.13**)

$$x^4 y'' + 2x^3 y' - \omega^2 y = 0,$$

where ω^2 is a positive constant and $y = y(x)$.

PE 2.29 Confluent hypergeometric equation. Let $y = y(x)$. Use the Frobenius method to obtain a solution of the differential equation

$$xy'' + (1 - x)y' - y = 0.$$

PE 2.30 Let $y = y(x)$. Solve the Airy equation

$$y'' - xy = 0,$$

using the Frobenius method.

PE 2.31 Show that the equation $x^4 y'' - y = 0$, $y = y(x)$, does not possess solutions of Frobenius type. Using the suggestion given in **PE 2.28**, the expansion around the point at infinity, what can you conclude?

PE 2.32 Let $y = y(x)$. Solve the ordinary differential equation

$$y'' + xy' + y = 0.$$

in the neighborhood of the point $x = 0$.

PE 2.33 Let $y = y(x)$. Give a complete solution for the differential equation

$$y'' + 5x^3 y = 0.$$

PE 2.34 Let $y = y(x)$. Find the power series solution of the differential equation

$$4xy'' + 2(1 - x)y' - y = 0.$$

PE 2.35 Let $y = y(x)$. Obtain a solution in power series for

$$x^2 y'' + xy' + \left(x^2 - \frac{1}{9} \right) y = 0.$$

PE 2.36 Let $y = y(x)$. Solve the ordinary differential equation

$$x^2 y'' + xy' + (x^3 - 2)y = 0$$

around the point $x = 0$.

PE 2.37 Let $y = y(x)$. Solve the ordinary differential equation

$$3(x^2 + x)y'' + (x + 2)y' - y = 0$$

in the neighborhood of the point $x = 0$.

PE 2.38 Let $\omega = \omega(x)$. Solve the ordinary differential equation

$$2x\omega'' + \omega' - \omega = 0,$$

in the neighborhood of the point $x = 0$.

PE 2.39 Show that

$$y \equiv y(x) = x^2 + x^{1/2}\,e^{-x^2/4} + \sum_{k=1}^{\infty}(-1)^k \frac{x^{2k+2}}{7.11.15\ldots(4k + 3)}$$

is a solution of

$$2x^2 y'' + (x^3 - 3x)y' + 2y = 0.$$

PE 2.40 Let $y = y(x)$. Solve the ordinary differential equation

$$8x^2 y'' + 2xy' + (1 - x)y = 0$$

around the point $x = 0$.

PE 2.41 Let $y = y(x)$. Obtain a solution for the ordinary differential equation

$$x^2 y'' + (x^3 - x)y' + (1 - x^2)y = 0.$$

PE 2.42 Let $y = y(x)$. Obtain solutions in the neighborhood of $x = 1$ for the ordinary differential equation

$$(x - 1)y'' - xy' + y = 0.$$

PE 2.43 Let $y = y(x)$. Solve the ordinary differential equation

$$xy'' + y' - y = 0.$$

PE 2.44 Let $y = y(x)$. Obtain a solution of the ordinary differential equation

$$x^2(1 + x)y'' + x(1 + x)y' - y = 0.$$

PE 2.45 Let $y = y(x)$. Solve the ordinary differential equation

$$xy'' + (x - 1)y' - y = 0.$$

PE 2.46 Let $y = y(x)$. Solve the ordinary differential equation

$$9x^2 y'' + 3x(x^2 + 2)y' + (x^2 - 2)y = 0.$$

PE 2.47 Let $y = y(x)$. Obtain a solution of the ordinary differential equation

$$x(1 + x)y'' - y' - 2y = 0.$$

PE 2.48 Show that

$$y_1(x) = \sum_{k=0}^{\infty} (-1)^k \frac{x^{k+1/2}}{2^{2k} k!(k+1)!}$$

and

$$y_2(x) = y_1(x) \ln x - 4x^{-1/2} - \sum_{k=0}^{\infty} (-1)^k \frac{H_{k+1} + H_k}{2^{2k} k!(k+1)!} x^{k+1/2}$$

are solutions of

$$4x^2 y'' + 4xy' + (x - 1)y = 0,$$

with $y = y(x)$ and where

$$H_k = \sum_{j=1}^{k} \frac{1}{j}.$$

PE 2.49 Solve the ordinary differential equation

$$2xy'' + y' - (x + x^2)y = 0.$$

PE 2.50 Let $y = y(x)$. Obtain a solution of the ordinary differential equation

$$xy'' - y' + (1 - x)y = 0.$$

References

1. W.E. Boyce, R.C. Diprima, *Elementary Differential Equations and Boundary Value Problems* (John Wiley, New York, 2007)
2. M. Braun, *Differential Equations and their Applications* (Springer, New York, 1983)
3. E.A. Coddington, N. Levison, *Theory of Ordinary Differential Equations* (McGraw-Hill, New York, 1955)
4. I.S. Gradshteyn, I.M. Ryzhik, *Table of Integrals, Series and Products* (Academic Press, New York, 1980)

Chapter 3
Laurent Series and Residues

Life is good for only two things, discovering mathematics and teaching mathematics.

1781 – Siméon Poisson – 1840

In this chapter, we present a revision of some basic facts about functions of a complex variable. This revision includes the main properties of such functions in what concerns their differentiation and integration and which will be useful, for instance, when we study integral transforms, in which the obtention of the final results will often require the calculation of integrals using the method of residues. The proofs of all theorems and of the lemma presented in this chapter can be found in [3, 4].

3.1 Functions of a Complex Variable

The set of complex numbers, usually denoted by $\mathbb{C}$, may be understood as an extension of the set of real numbers $\mathbb{R}$, necessary for the solution of algebraic equations of the type $z^2 = -1$, for example. The solutions of this equation are $z = \pm i$, where the number i, called *imaginary unit*, is given by $i = \sqrt{-1}$. The set of complex numbers is formed by all the elements of the form $z = x + iy$, where x and y are real.

To each complex number z, there corresponds a *complex conjugate* (or simply conjugate), given by $z^* = x - iy$, also denoted by $\bar{z}$. The product of two complex numbers $z_1 = x_1 + iy_1$ and $z_2 = x_2 + iy_2$, with x_1, y_1, x_2 and y_2 real numbers, is given by $z_1 z_2 = (x_1^2 - y_1^2) + i(x_1 y_2 + x_2 y_1)$, where $ix = xi$ for every real number x.

We can represent complex numbers by points on the *complex plane*, also called *Argand-Gauss* [1768 – Jean Robert Argand – 1822] *plane*, associating to $z = x + iy$ the point (x, y) of the plane. With the help of polar coordinates we then construct a *trigonometric representation* of z,

© The Author(s), under exclusive license to Springer Nature Switzerland AG 2025
E. Capelas de Oliveira, J. E. Maiorino, *Analytical Methods in Applied Mathematics*, Problem Books in Mathematics, https://doi.org/10.1007/978-3-031-74794-6_3

$$z = x + iy = |z|(\cos\theta + i\sin\theta),$$

where $|z| = \sqrt{x^2 + y^2} = \sqrt{zz^*}$ is the absolute value or module of z, and θ, known as argument, is defined by the relations $\cos\theta = x/|z|$ and $\sin\theta = y/|z|$. Due to the periodicity of the sine and cosine functions, the values θ and $\theta + 2n\pi$, with n integer, define the same complex number z. However, calculus with complex variables requires that θ be defined uniquely, and because of this we adopt the convention that $-\pi < \theta \leq \pi$. With the help of the *Euler relation*,

$$e^{i\theta} = \cos\theta + i\sin\theta,$$

we finally obtain the *polar representation* of complex numbers:

$$z = r\,e^{i\theta}, \quad \text{where } r = |z|.$$

The complex conjugate of z then takes on the form $z^* = r\,e^{-i\theta}$ and it becomes easy to verify that $|z|^2 = zz^* = r\,e^{i\theta}\,r\,e^{-i\theta} = r^2$, a real number. As with real variables, a function of a complex variable may be understood as a rule, which associates to each complex number $z = x + iy$ another complex number $w = f(z) = u(x, y) + iv(x, y)$, where $u(x, y)$ and $v(x, y)$ are real functions of x and y. An important group of complex functions is formed by the so-called *multiple-valued functions* or *multivalued functions*, i.e., those functions that associate to a given value of the variable two or more *distinct* numbers $w = f(z)$. One should not confuse this fact with the arbitrariness of the value of θ in the polar representation: $r\,e^{i\theta}$ and $r\,e^{i(\theta+2n\pi)}$ are different *representations* of the *same* number. Such arbitrariness is eliminated by the restriction imposed on the possible values of θ. Such functions are generally considered as being composed of *branches*, each of which is a *single-valued* function of the complex number z, as we will see further.

Example 3.1 Let $z \in \mathbb{C}$ and $n = 0, 1, 2, \ldots$ Consider, for example, the function $f(z) = \sqrt[n]{z}$. Using the polar representation $z = r\,e^{i\theta}$, we can easily see that $f(z)$ can be divided in n branches, according to the usual formula for roots:

$$
\begin{aligned}
\text{Main branch: } f_1(z) &= \sqrt[n]{r}\,e^{i\theta/n}; \\
\text{Second branch: } f_2(z) &= \sqrt[n]{r}\,e^{i(\theta+2\pi)/n}; \\
\text{Third branch: } f_3(z) &= \sqrt[n]{r}\,e^{i(\theta+4\pi)/n}; \\
&\ \ \vdots \qquad\qquad \vdots \\
n-\text{th branch: } f_n(z) &= \sqrt[n]{r}\,e^{i(\theta+2(n-1)\pi)/n}\,.
\end{aligned}
$$

Rigorously speaking, $f_1(z)$, $f_2(z)$, $\ldots$, $f_n(z)$ are n distinct functions, as they associate to the same complex z different values $f_i(z)$, $i = 1, \ldots, n$. Nevertheless, they share the common property that, for any z, $[f_i(z)]^n = z$, and for this reason, they are treated together as branches of a unique (n-valued) function $f(z) = \sqrt[n]{z}$.

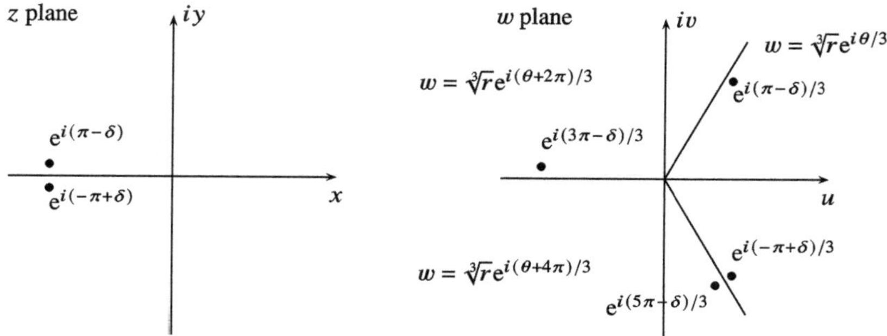

Fig. 3.1 Branch point and branches of the cubic root of z

Another important property common to each branch of $f(z) = \sqrt[n]{z}$ is that all such branches are *discontinuous* along the line uniting the origin $z = 0$ to negative infinity, i.e., the real negative half-axis.

In order to see that, let us consider the main branch of function $w = f(z) = \sqrt[3]{z}$. A point on the z plane, $z = r\,e^{i\theta}$, $-\pi \le \theta \le \pi$, will be mapped by the main branch of $f(z)$ into the point $w = f_1(z) = \sqrt[3]{r}\,e^{i\theta/3} = \sqrt[3]{r}\,e^{i\alpha}$, where $-\pi/3 \le \alpha \le \pi/3$. This means that the entire z plane will be mapped into just one-third of the w plane, as can be seen in the right-hand side of Fig. 3.1. Accordingly, the second branch will map the z plane onto the region $w = f_2(z) = \sqrt[3]{r}\,e^{i(\theta+2\pi)/3} = \sqrt[3]{r}\,e^{i\alpha}$ with $\pi/3 \le \alpha \le \pi$. For the third branch, we will have $w = f_3(z) = \sqrt[3]{r}\,e^{i(\theta+4\pi)/3} = \sqrt[3]{r}\,e^{i\alpha}$ with $-\pi \le \alpha \le \pi/3$.

A consequence of this property is that the points $e^{i(\pi-\delta)}$ and $e^{i(-\pi+\delta)}$, which are close to each other on the z plane—for a small value of δ—are taken to the points $e^{i(\pi-\delta)/3}$ and $e^{i(-\pi+\delta)/3}$ by the first branch of $f(z) = \sqrt[3]{z}$. As these points are far from each other on the w plane, this means that $f(z)$ is *discontinuous* along the negative x axis of the z plane. The same happens with the other two branches. Figure 3.1 shows the image of $e^{i(-\pi+\delta)}$ for the first branch of $f(z)$ and the three images of $e^{i(\pi-\delta)}$ for all three branches.

The point $z = 0$ is called the *branch point* of $f(z)$, and the real negative half-axis is the *cut line* (or *branch line*) of $f(z)$. The cut line of a function is not unique, and in fact, it does not even need to be a straight line: it is determined by the division of the function in branches. Whatever the division chosen, however, there will always be the same number of branches, and the branch point will always be the same as this is a characteristic of the given function. □

The concept of continuity for functions of one complex variable is formally the same as the one for functions of a real variable. We say that a complex function $f(z)$ is *continuous* at $z = z_0$ if, and only if,

$$\lim_{z \to z_0} f(z) = f(z_0).$$

However, it must be noticed that this limit may be taken along any direction or path on the z plane. This difference between complex and real functions entails other differences in the properties of such functions concerning their derivatives and integrals, as we will see in the sequence.

As for real variables, the *derivative* of a complex function $f(z)$ at $z = z_0$ is given by the limit

$$f'(z_0) = \lim_{z \to z_0} \frac{f(z) - f(z_0)}{z - z_0},$$

where, once more, the limit must be the same along any path on the complex plane.

It is easy to verify that the derivatives of complex functions obey the usual differentiation rules. If $f_1 = f_1(z)$ and $f_2 = f_2(z)$ are differentiable functions of z, then:

$$\frac{d}{dz}(f_1 + f_2) = \frac{df_1}{dz} + \frac{df_2}{dz}; \tag{3.1}$$

$$\frac{d}{dz}(f_1 f_2) = \frac{df_1}{dz} f_2 + f_1 \frac{df_2}{dz}; \tag{3.2}$$

$$\frac{d}{dz}\left(\frac{f_1}{f_2}\right) = \frac{1}{f_2^2}\left(\frac{df_1}{dz} f_2 - f_1 \frac{df_2}{dz}\right). \tag{3.3}$$

Let $f = f(w)$ and $w = w(z)$ be two differentiable complex functions. The chain rule takes on the well-known form

$$\frac{df}{dz} = \frac{df}{dw}\frac{dw}{dz}.$$

The possession of a derivative is a property of a function *at a point,* just like being continuous. In complex calculus, we work primarily with functions possessing derivatives at least on a finite region of the complex plane. Such functions have special properties, thus deserving a particular name and a more detailed study of their characteristics.

A complex function $f(z)$ is called *analytic, regular* (an older term) or *holomorphic* (a modern term) on a region R of the complex plane if, and only if, $f(z)$ possesses its first derivative $f'(z)$ on all points of R. It is possible to show that the existence of the first derivative $f'(z)$ on a region R of the complex plane entails the existence and continuity of the derivatives of all orders n, $f^{(n)}(z)$, on that region. Besides, such functions have other properties which we state as theorems.

Theorem 3.1 (Cauchy–Riemann [1826 – Bernhard Riemann – 1866] Conditions) *Let $f(z)$ be an analytic function on a given domain R on the complex plane. If $z \equiv x + iy$ and $f(z) \equiv u(x, y) + iv(x, y)$, then $u(x, y)$ and $v(x, y)$ are continuous functions of x and y and satisfy the Cauchy–Riemann conditions, also called Cauchy–Riemann equations,*

$$\frac{\partial u}{\partial x} = \frac{\partial v}{\partial y} \qquad and \qquad \frac{\partial u}{\partial y} = -\frac{\partial v}{\partial x}.$$

The reciproque of this theorem is also true and may be stated as follows:

Theorem 3.2 (Morera's [1856 – Giacinto Morera – 1909] Theorem) *Let $x, y \in \mathbb{R}$ and $z \in \mathbb{C}$ with $z = x + iy$. If $f(z) = u(x, y) + iv(x, y)$ is such that $u(x, y)$ and $v(x, y)$ are continuous and have continuous first partial derivatives on a domain R of the complex plane, then $f(z)$ is analytic on R.*

The two theorems aforementioned imply that the derivative $f'(z)$ of an analytic function $f(z) = u(x, y) + iv(x, y)$ is given, in terms of u and v, by

$$f'(z) = \frac{\partial u}{\partial x} + i\frac{\partial v}{\partial x} = \frac{\partial v}{\partial y} - i\frac{\partial u}{\partial y}.$$

Example 3.2 Let $x, y \in \mathbb{R}$ and $z \in \mathbb{C}$ with $z = x + iy$. Let $u = u(x, y) = x^2 - y^2 - 2x$ be the real part of a complex function $f(z) = u(x, y) + iv(x, y)$. Determine $f(z)$ satisfying the condition $f(0) = 0$.

We first determine $v = v(x, y)$ by means of the Cauchy-Riemann conditions,

$$\frac{\partial u}{\partial x} = 2x - 2 = \frac{\partial v}{\partial y}.$$

Integration of this expression in variable y furnishes

$$v(x, y) = 2xy - 2y + \phi(x),$$

with $\phi(x)$ a function depending on variable x only. Using the second Cauchy-Riemann condition, we have

$$-\frac{\partial v}{\partial x} = 2y + \phi'(x) \quad \Longrightarrow \quad -2y - \phi'(x) = \frac{\partial u}{\partial y} = -2y.$$

Solving the ordinary differential equation in variable x, we get

$$\phi(x) = C,$$

where C is an arbitrary constant. Thus, we can write

$$f(z) = x^2 - y^2 - 2x + i(2xy - 2y + C) = z^2 - 2z + iC.$$

Using the condition $f(0) = 0$, we find that $C = 0$ and we finally obtain

$$f(z) = z(z - 2)$$

which is the sought function. $\qquad\qquad\qquad\qquad\qquad\qquad\qquad\qquad\qquad\qquad\quad \square$

It is possible to show that all n-th degree polynomials in one complex variable $z \in \mathbb{C}$, $P_n(z) = a_0 + a_1 z + \cdots + a_n z^n$, with $a_0, a_1, \ldots, a_n$ constants, are analytic functions whose derivatives can be easily calculated by means of the formula

$$\frac{\mathrm{d}}{\mathrm{d}z}(z^n) = n z^{n-1}, \quad n \text{ integer.} \tag{3.4}$$

With this result and the differentiation rule Eq. (3.3) it becomes clear that all algebraic rational functions, i.e., functions of the form $P(z)/Q(z)$, where $P(z)$ and $Q(z)$ are polynomials, are also analytic except, of course, at the points in which $Q(z)$ is null.

A function of fundamental importance in the theory of complex variables is the *exponential function*. Let $x, y \in \mathbb{R}$ and $z \in \mathbb{C}$ with $z = x + iy$; we define

$$e^z = e^x \, e^{iy} = e^x (\cos y + i \sin y).$$

It is an analytic function and has the following properties:

1. The product of two exponential functions is given by $e^{z_1 + z_2} = e^{z_1} \, e^{z_2}$.
2. For any complex number z, $e^z \neq 0$.
3. The inverse of an exponential is given by $\frac{1}{e^z} = e^{-z}$.
4. For every z, $e^{z + 2\pi i} = e^z$.
5. For every z and for every real number μ, $(e^z)^\mu = e^{\mu z}$.

Using the definition of the exponential function, we can define the fundamental trigonometric and hyperbolic functions:

$$\cos z = \frac{e^{iz} + e^{-iz}}{2}; \qquad\qquad \sin z = \frac{e^{iz} - e^{-iz}}{2i};$$

$$\cosh z = \frac{e^z + e^{-z}}{2}; \qquad\qquad \sinh z = \frac{e^z - e^{-z}}{2}.$$

From these functions, we construct the remaining trigonometric and hyperbolic functions, using the same definitions of the corresponding functions of a real variable. All such functions are analytic on their domains, and their derivatives bear the same relationships to each other as their real counterparts.

Example 3.3 Let $z \in \mathbb{C}$. Evaluating the derivatives of trigonometric and hyperbolic functions, we can find that

$$\frac{\mathrm{d}}{\mathrm{d}z} \sin z = \cos z; \qquad\qquad \frac{\mathrm{d}}{\mathrm{d}z} \cos z = -\sin z;$$

and

$$\frac{\mathrm{d}}{\mathrm{d}z} \sinh z = \cosh z; \qquad\qquad \frac{\mathrm{d}}{\mathrm{d}z} \cosh z = \sinh z.$$

Note the similarity with the real functions. $\square$

Finally, we must mention that these functions, because they are analytic, can be expanded in Taylor series identical to their real counterparts. On the other hand, logarithmic functions and inverse trigonometric and hyperbolic functions require greater care, as they are multivalued functions [3].

3.2   Laurent Series, Zeros, and Singularities

As we already said, in the study of linear ordinary differential equations, a point can be classified as regular or singular, and a singular point can be further classified as a regular singular point or an irregular singular point. Here we extend this classification to the complex plane in order to incorporate all possibilities.

Let C_1 and C_2 be two concentric circumferences with center at a point $z = a$ of the complex plane and radii R_1 and R_2, $R_1 < R_2$, respectively. Assume that $f(z)$ is a single-valued analytic function on C_1 and C_2, and also on the region R limited by them, the annular region. Let $z = a + h$ be an arbitrary point of R.

Then we may write $f(a + h)$ in terms of a series that generalizes the Taylor series, the so-called *Laurent* [1813 – Pierre Alphonse Laurent – 1854] *series*:

$$f(a + h) = a_0 + a_1 h + a_2 h^2 + \cdots + \frac{a_{-1}}{h} + \frac{a_{-2}}{h^2} + \cdots$$

The coefficients in this series are given by

$$a_n = \frac{1}{2\pi i} \oint_C \frac{f(z)}{(z - a)^{n+1}} \mathrm{d}z , \quad n = 0,\ \pm 1,\ \pm 2,\ \ldots$$

where C is an arbitrary circumference completely contained within region R, and integration is performed in the counterclockwise sense.

The expression $a_0 + a_1(z - a) + a_2(z - a)^2 + \cdots$ is called the *analytic part* of the Laurent series and the remaining part, composed of powers of $(z - a)^{-1}$, is its *principal part*. If the principal part is null the series is reduced to a Taylor series.

The point $z = a$ is called a *zero* (or *root*) of function $f(z)$, if $f(a) = 0$. If $f(z)$ is analytic at $z = a$, its Laurent series around this point reduces to a Taylor series,

$$f(z) = \sum_{n=0}^{\infty} a_n (z - a)^n,$$

with $a_0 = 0$. If $a_1 \neq 0$, the point $z = a$ is called a *simple zero* (or a zero of order one). It may happen that in the expansion of $f(z)$ several coefficients are null. Let

a_m be the first nonnull coefficient in the Taylor series. In such a case we say that $z = a$ is a *zero of order m*. The order of a zero can be determined without using the Taylor series, by calculating the limit

$$\lim_{z \to a} \frac{f(z)}{(z-a)^n}, \qquad \text{for } n = 1, 2, 3, \ldots$$

The least value of n for which this limit is nonnull is the order of the zero.

Example 3.4 Obtain the Laurent series for $f(z) = (1 - z^2)^{-1}$ which converges in the annular region $1/4 < |z - 1| < 2$ and determine the exact convergence region.

First, we note that the annular region is centered at $z = 1$; thus, we should expand $f(z)$ in powers of $z - 1$. As $f(z)$ can be written in the form

$$f(z) = -\frac{1}{z-1} \cdot \frac{1}{z+1}$$

we obtain

$$f(z) = -\frac{1}{z-1} \cdot \frac{1}{2 + (z-1)} = \frac{-1/2}{z-1} \left[1 - \left(-\frac{z-1}{2} \right) \right]^{-1}.$$

Using the geometric series,

$$\sum_{i=1}^{\infty} q^i = \frac{1}{1-q}, \qquad \text{for } |q| < 1$$

which converges, in this case, for $|z - 1| < 2$, we get

$$f(z) = \frac{-1/2}{z-1} \sum_{n=0}^{\infty} \left(-\frac{z-1}{2} \right)^n = \sum_{n=0}^{\infty} \frac{(-1)^{n+1}}{2^{n+1}} (z-1)^{n-1}$$

whose exact convergence region is $0 < |z - 1| < 2$ ☐

The points on the complex plane on which a function $f(z)$ ceases to be analytic are called *singularities* of the function. We have seen an example of singularity when we dealt with multivalued functions: the cut lines along each branch are entirely composed of singular points. As we have already said, even though it is always possible to choose different cut lines for one and the same multivalued function, the *branch point* will be the same for all possible choices, being itself a singularity. In the case in which $f(z)$ is analytic on a neighbourhood of a point a, but not at $z = a$, we say that $f(z)$ has an *isolated singularity* at this point.

Isolated singularities may be classified in four great groups, depending on the behavior of function $f(z)$ when $z \to a$ in an arbitrary way:

1. Removable Singularity: If a function $f(z)$ is not defined at $z = a$, but there exists $\lim_{z \to a} f(z)$, then $z = a$ is a removable singularity. In this case we define $f(a) = \lim_{z \to a} f(z)$, so the function becomes analytic at a and on the neighborhood of this point. The functions $f(z) = \sin z/z$ and $g(z) = 1/(z - \cot g\, z)$ have removable singularities at $z = 0$.

2. Poles: If $f(z)$ has in the principal part of its Laurent series a finite number of terms given by

$$\frac{a_{-1}}{z - a} + \frac{a_{-2}}{(z - a)^2} + \cdots + \frac{a_{-n}}{(z - a)^n} ,$$

where $a_{-n} \neq 0$, then $z = a$ is a pole of order n. If $f(z)$ has a pole at $z = a$, then $\lim_{z \to a} |f(z)| = \infty$. Examples of this kind of function are $f(z) = 1/\sin z$ and $g(z) = 1/z$ at $z = 0$.

3. Essential Singularity: Any isolated singularity of a function $f(z)$, which is not a pole nor a removable singularity will be an essential singularity. In such cases $f(z)$ does not have a limit nor does its module tend to infinity, but it *oscillates fastly* when $z \to a$. The Laurent series will then have an *infinite* number of terms in its principal part. As an example, a singular point of this kind is $z = 0$ for the function $f(z) = e^{1/z}$.

4. Branch Point: A point $z = z_0$ is a branch point (algebraic branch point) if the multivalued function $f(z)$ is discontinuous upon transversing a small circumference arround this point. This is equivalent to saying that z_0 is the point of convergence of the cut lines of the branches of $f(z)$, as we have seen before. In the case studied, the branch point of function $f(z) = \sqrt[n]{z}$ was the origin $z = 0$.

In the theory of functions of complex variables, it is useful to introduce the so-called *point at infinity*, for which one uses the usual notation "∞". We will not explain here how this is done. What is important to us now is to notice that in the same way as for any common point on the complex plane, we can define the properties and the behavior of a complex function "at infinity," which corresponds here to the point at infinity, ∞. Thus, we say that a function $f(z)$ is *analytic at infinity* if the function

$$g(z) = f(1/z)$$

is *analytic at* $z = 0$. Besides, it is possible to introduce in the same way the concepts of *pole at infinity*, *branch point at infinity*, etc., that is, by considering the behavior of $g(z)$ at the origin. An examination of the most common functions will show that all functions that are analytic at infinity possess at least one singularity at some (finite) point z on the plane. We might ask ourselves whether there are complex functions which are analytic on the entire complex plane. The answer is given by the following theorem:

Theorem 3.3 (Liouville's [1809 − Joseph Liouville − 1882] Theorem) *The only function $f(z)$ that is analytic on the whole complex plane and at the point at infiniy is the constant function, $f(z) = constant$.*

Example 3.5 Let $z \in \mathbb{C}$. Obtain the Laurent series for

$$f(z) = \frac{\cos z - 1}{z^2}$$

and classify the singularity.

Note that $z = 0$ is a regular point for $\cos z$. Then, using the Maclaurin series for $\cos z$,

$$\cos z = \sum_{n=0}^{\infty} (-1)^n \frac{z^{2n}}{(2n)!} = 1 + \sum_{n=1}^{\infty} (-1)^n \frac{z^{2n}}{(2n)!} \,,$$

substituting it into $f(z)$ and rearranging, we have

$$f(z) = \sum_{n=1}^{\infty} (-1)^n \frac{z^{2n-2}}{(2n)!} \,,$$

which is the Laurent series for $f(z)$. Now, let us write explicitly its first three terms:

$$f(z) = -\frac{1}{2} + \frac{z^2}{4!} - \frac{z^4}{6!} + \cdots$$

We see that this is a Maclaurin series. As the Laurent series is a Maclaurin series, singularity $z = 0$ is a removable singularity. We can verify this explicitly by evaluating the limit

$$\lim_{z \to 0} \frac{\cos z - 1}{z^2} \qquad \to \left(\frac{0}{0} \right) .$$

As this is an undetermined quotient, we can use the l'Hôpital rule to get

$$\lim_{z \to 0} \frac{-\sin z}{2z} = -\frac{1}{2} \lim_{z \to 0} \frac{\sin z}{z} = -\frac{1}{2} .$$

Note that, this is indeed the first term of the Laurent series, i.e., the limit of $f(z)$ when $z \to 0$. $\qquad\qquad\qquad\qquad\qquad\qquad\qquad\qquad\qquad\qquad\qquad\qquad$ $\square$

After presenting the Laurent series, we now introduce the concept of *residue*, a particular coefficient of this series expansion, one that plays an important role in calculating several real integrals as we will see when we discuss the residue theorem. Remark that, from now on, except where explicitly mentioned, we will always consider counterclockwise orientation when we perform path integrals on the complex plane.

3.3 Residues

Let C be a circumference centered at $z = a$, R the region limited by C and $f(z)$ an analytic function on $R \cup C$ except at $z = a$. Then $f(z)$ has associated with it a Laurent series around $z = a$ given by

$$f(z) = \sum_{n=-\infty}^{\infty} a_n(z-a)^n = \underbrace{a_0 + a_1(z-a) + \cdots}_{\text{analytic part}} + \underbrace{\frac{a_{-1}}{z-a} + \frac{a_{-2}}{(z-a)^2} + \cdots}_{\text{principal part}}$$

where the coefficients are given by

$$a_n = \frac{1}{2\pi i} \oint_\Gamma \frac{f(z)}{(z-a)^{n+1}} dz, \quad n = 0, \pm 1, \pm 2, \ldots$$

with Γ a closed, counterclockwise oriented curve on the complex plane.
   In the special case $n = -1$ we have

$$\oint_\Gamma f(z)dz = 2\pi i a_{-1}.$$

This expression involves only the a_{-1} coefficient and is called the *residue* of $f(z)$ at $z = a$. Thus, in order to calculate the residue of a function at a point, it is enough to expand it in a Laurent series around that point and take the coefficient a_{-1} of the power z^{-1}. In the case in which $z = a$ is a pole of order k we have

$$a_{-1} = \lim_{z \to a} \frac{1}{(k-1)!} \frac{d^{k-1}}{dz^{k-1}} \left[(z-a)^k f(z) \right].$$

Example 3.6 Let $z \in \mathbb{C}$. Obtain the residue for $f(z) = \dfrac{1/z}{\sin^2 z}$.
   We must find the coefficient of (z^{-1}). Using the same argument for the $\cos z$, we can expand the $\sin z$ in a Maclaurin series

$$\sin z = \sum_{n=0}^{\infty} (-1)^n \frac{z^{2n+1}}{(2n+1)!} = z - \frac{z^3}{3!} + \frac{z^5}{5!} - \cdots$$

This is a convergent series; its square is given by

$$\sin^2 z = \left(z - \frac{z^3}{3!} + \frac{z^5}{5!} - \cdots \right) \left(z - \frac{z^3}{3!} + \frac{z^5}{5!} - \cdots \right)$$

or, considering only the first three terms, we have

$$\sin^2 z = z^2 - \frac{z^4}{3} + \frac{23}{360}z^6 - \cdots$$

Using long division we obtain the Laurent series

$$f(z) = \underbrace{\frac{1}{z^3} + \frac{1/3}{z}}_{\text{principal part}} + \underbrace{\frac{17}{360}z + \cdots}_{\text{analitic part}}$$

This expansion shows that $f(z)$ has a pole of order three, the highest negative power of z, and that its residue, the coefficient of z^{-1}, is equal to $1/3$. Note that this residue can also be evaluated using the expression for residues,

$$a_{-1} = \lim_{z \to 0} \frac{1}{2!}\frac{\mathrm{d}^2}{\mathrm{d}z^2}\left(z^3\frac{1}{z\sin^2 z}\right)$$

because at $z = 0$ we have a pole of order three. $\square$

3.4   Residues and the Evaluation of Real Integrals

We present here an important tool for evaluating real integrals with the help of path integrals on the complex plane, the so-called residue theorem.

Theorem 3.4 (Residue Theorem) *Let C be a simple and closed curve on the complex plane, counterclockwise oriented. Let R be the region limited by C and $f(z)$ an analytic function on $C \cup R$ except on singularities a, b, c, $\ldots$ in R; let us assume that on these points $f(z)$ has residues given respectively by a_{-1}, b_{-1}, c_{-1} $\ldots$ Then*

$$\oint_C f(z)\mathrm{d}z = 2\pi i(a_{-1} + b_{-1} + \cdots).$$

The residue theorem is used to evaluate definite real integrals, usually by extending the integrals along an infinite or semi-infinite domain. It is also important in the calculation of Laplace and Fourier transforms, as we will see in Chap. 6. The main difficulty for this application lies in the choice of an adequate function $f(z)$—as it must somehow reproduce the function under the integral sign—and in the choice of the contour C along which the integral is carried. These are a few simple cases:

(a) Integrals of the kind

$$\int_{-\infty}^{\infty} F(x)\mathrm{d}x,$$

where $F(x)$ is a rational function.

In this case we evaluate $\oint_C F(z)\mathrm{d}z$ along a contour C formed by the straight line segment on the real axis from $-R$ to R and by the semicircunference Γ above the real axis as shown in Fig. 3.2. Once the calculation is done, we take the limit $R \to \infty$ and we then show that the integral along Γ goes to zero, while the integral along the real axis becomes the definite integral sought for.

(b) Integrals of the kind

$$\int_0^{2\pi} G(\sin\theta, \cos\theta)\mathrm{d}\theta, \tag{3.5}$$

where G is a rational function.

Employing the polar representation of complex numbers we write $z = \mathrm{e}^{i\theta}$, whence it follows that $2i\sin\theta = z - z^{-1}$ and $2\cos\theta = z + z^{-1}$. It's easy to see that the integral in Eq. (3.5) is equivalent to an integral $\oint_C F(z)\mathrm{d}z$, where the curve C is a circumference centered at the origin and unit radius.

(c) Integrals of the kind

$$\int_{-\infty}^{\infty} F(x)\cos mx\,\mathrm{d}x \quad \text{or} \quad \int_{-\infty}^{\infty} F(x)\sin mx\,\mathrm{d}x,$$

where $F(x)$ is rational.

For this kind of integral we take $\oint_C F(z)\,\mathrm{e}^{imz}\,\mathrm{d}z$ where C is the contour presented in (a) as shown in Fig. 3.2.

Example 3.7 Let $x \in \mathbb{R}$. Evaluate the integral

$$\Lambda = \int_0^{\infty} \frac{\mathrm{d}x}{(1+x^2)^2}.$$

In order to evaluate this integral, we consider the integral on the complex plane

$$\oint_C \frac{\mathrm{d}z}{(1+z^2)^2},$$

with $z = x + iy$ and where C is a contour as in Fig. 3.2. The singularities of this function are poles at $z_1 = i$ and $z_2 = -i$, both of order two. However, only z_1 contributes because z_2 is out of the region limited by C. Going through the contour

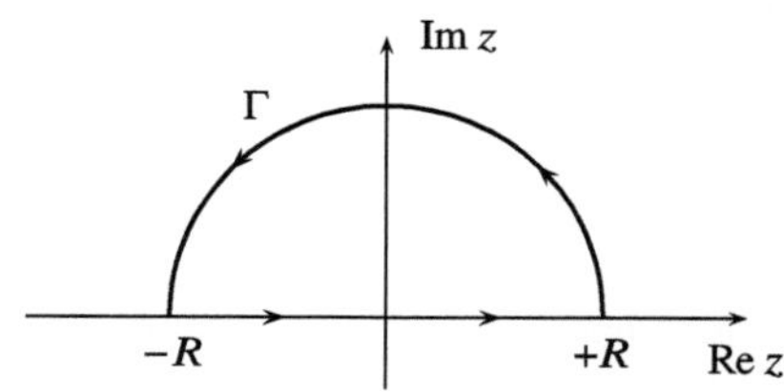

Fig. 3.2 Contour of integration for the application of the residue theorem to calculate real integrals in cases (a) and (c)

in the counterclockwise sense, we can write

$$\oint_C \frac{dz}{(1+z^2)^2} = \int_{-R}^{R} \frac{dx}{(1+x^2)^2} + \oint_\Gamma \frac{dz}{(1+z^2)^2} \, .$$

Taking the limit $R \to \infty$, using Jordan's lemma, shown in **SE 3.4**, and the residue theorem, we get

$$\int_{-\infty}^{\infty} \frac{dx}{(1+x^2)^2} = 2\pi i \, \mathrm{Res}(z = i).$$

As we have a pole of order two at $z = i$ we obtain

$$2 \int_0^\infty \frac{dx}{(1+x^2)^2} = 2\pi i \lim_{z \to i} \frac{1}{1!} \frac{d}{dz} \left\{ (z-i)^2 \frac{1}{(z-i)^2(z+i)^2} \right\}.$$

Evaluating the derivatives and calculating the limit we have, after simplification,

$$\int_0^\infty \frac{dx}{(1+x^2)^2} = \frac{\pi}{4}$$

which is the desired result. Note that, in this particular case the substitution $x = \tan \xi$ would lead to the same result. $\qquad\qquad\square$

3.5 Solved Exercises

SE 3.1 Use the Cauchy–Riemann conditions to show that $f(z) = |z|^2$ is not an analytic function.
Solution: If $z = x + iy$, then $f(z) = |z|^2 = zz^* = x^2 + y^2 \equiv u(x, y) + iv(x, y)$. We therefore have $u(x, y) = x^2 + y^2$ and $v(x, y) = 0$. Thus

$$\frac{\partial u}{\partial x} = 2x \neq 0 = \frac{\partial v}{\partial y};$$

$$\frac{\partial u}{\partial y} = 2y \neq 0 = -\frac{\partial v}{\partial x}.$$

From Theorem 3.1 it follows that $f(z)$ is not analytic on any region of the complex plane.

SE 3.2 Expand the function $f(z) = \sin z / z$ in a Laurent series around the origin.
Solution: As we have seen in the previous section, the function $\sin z$ is analytic and can be represented by means of a Taylor series identical to the real series for $\sin x$, which has already been presented in the previous chapter:

$$\sin z = z - \frac{z^3}{3!} + \frac{z^5}{5!} - \cdots$$

Thus,

$$f(z) = \frac{\sin z}{z} = 1 - \frac{z^2}{3!} + \frac{z^4}{5!} - \cdots$$

We can see that in this case, the expansion does not have a principal part and the Laurent series obtained coincides with the Taylor series.

SE 3.3 What is the residue of $f(z) = \sin z / z$ at $z = 0$?
Solution: Using the result of the previous solved exercise, we realize that for this function $a_{-1} = 0$. Therefore, $\mathrm{Res}[\sin z / z] = 0$ at $z = 0$.

Solving the next exercise requires the use of an important result known as *Jordan's* [1838 – Camille Marie–Ennemond Jordan – 1922] *lemma*.

Lemma 3.5 (Jordan's Lemma) *Consider the integral*

$$I_R = \int_{\Gamma_R} e^{iRz} f(z) dz ,$$

where $\Gamma_R = \{z = R e^{i\theta}, 0 \leq \theta \leq \pi\}$ is a semicircumference of radius R with center at the origin, passing through the upper half of the complex plane. Suppose that $f(z)$ is analytic on this half-plane and that the greatest value $f(R)$ of $|f(z)|$ for any $z \in \Gamma_R$ tends to zero when $R \to \infty$. Then, the integral aforementioned will also become null at this limit, i.e., $\lim_{R \to \infty} I_R = 0$.

SE 3.4 Calculate the integral

$$\int_{-\infty}^{\infty} \frac{x^2 + 2}{(x^2 + 1)(x^2 + 4)} dx .$$

Solution: Let us choose the integral

$$I = \oint_{\Gamma} \frac{z^2 + 2}{(z^2 + 1)(z^2 + 4)} dz \tag{3.6}$$

and the oriented contour shown in Fig. 3.3.

From Eq. (3.6) we see that the integrand is regular on the upper half plane, except at the simple poles $z = i$ and $z = 2i$.

Using the residue theorem and taking $R > 2$, we see that the contour chosen comprehends both poles. Hence,

$$\oint_{\Gamma} \frac{z^2 + 2}{(z^2 + 1)(z^2 + 4)} dz = 2\pi i \left(\frac{1}{6i} + \frac{1}{6i} \right) = \frac{4\pi}{3} .$$

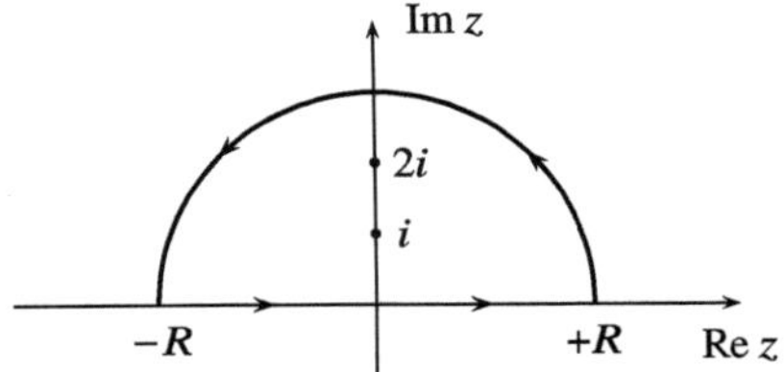

Fig. 3.3 Contour on the complex plane for the integral of **SE 3.4**

We can then write, going through the contour,

$$\oint_\Gamma \frac{z^2 + 2}{(z^2 + 1)(z^2 + 4)}\, dz = \int_{-R}^{R} \frac{x^2 + 2}{(x^2 + 1)(x^2 + 4)}\, dx +$$

$$+ \oint_{\Gamma_R} \frac{z^2 + 2}{(z^2 + 1)(z^2 + 4)}\, dz = \frac{4\pi}{3}\, .$$

We must prove that the integral $\oint_{\Gamma_R}$ is equal to zero when $R \to \infty$. For this sake we use the following identities

$$|z^2 + 2| \le |z|^2 + 2 = R^2 + 2$$

$$|z^2 + 1| \ge |z|^2 - 1 = R^2 - 1$$

$$|z^2 + 4| \ge |z|^2 - 4 = R^2 - 4.$$

Thus, for $|z| = R$ the integral in Γ_R does not exceed

$$\pi R \frac{R^2 + 2}{(R^2 - 1)(R^2 - 4)} = \pi R \frac{R^2}{R^4} \frac{1 + 2/R^2}{(1 - 1/R^2)(1 - 4/R^2)}$$

which goes to zero when $R \to \infty$. Therefore,

$$\int_{-\infty}^{\infty} \frac{x^2 + 2}{(x^2 + 1)(x^2 + 4)}\, dx = \frac{4\pi}{3}\, .$$

SE 3.5   Unit circumference Let $a > b > 0$. Evaluate the integral

$$\int_0^{2\pi} \frac{d\theta}{a + b\cos\theta}\, .$$

Solution: In problems like this, in which the denominator is never null and the limits of integration differ by a complete turn of the trigonometric circumference, we introduce the so-called polar form, i.e., $z = \exp(i\theta)$. With this change, the integral in z becomes a path integral along a circumference centered at the origin, denoted by C, with unit radius and counterclockwise orientation. Then

$$\int_0^{2\pi} \frac{d\theta}{a + b\cos\theta} = \oint_C \frac{dz/iz}{a + b\left(\dfrac{1+z^2}{2z}\right)} = \frac{2}{i}\oint_C \frac{dz}{bz^2 + 2az + b}\,.$$

The roots of the denominator (the function's poles) are given by

$$z_1 = -\frac{a}{b} + \frac{1}{b}\sqrt{a^2 - b^2} \quad \text{and} \quad z_2 = -\frac{a}{b} - \frac{1}{b}\sqrt{a^2 - b^2}$$

and are both real, as $a > b$, but only z_1 lies inside the unit circumference because $a > b$. This means that z_1 is the only singularity contributing for the integral.

Then, using the residue theorem we have

$$\oint_C \frac{dz}{bz^2 + 2az + b} = 2\pi i \lim_{z \to z_1}\left[(z - z_1)\frac{1}{b(z - z_1)(z - z_2)}\right] = \frac{2\pi i}{b}\frac{1}{z_1 - z_2},$$

whence we can write

$$\int_0^{2\pi} \frac{d\theta}{a + b\cos\theta} = \frac{2}{i}\frac{\pi i}{\sqrt{a^2 - b^2}} = \frac{2\pi}{\sqrt{a^2 - b^2}}.$$

SE 3.6 Branch Point Let $0 < \alpha < 1$. Evaluate the real integral

$$\int_0^\infty \frac{x^\alpha}{(1 + x)^2}dx\,.$$

Solution: Consider a function of the form

$$f(z) = \frac{z^\alpha}{(1 + z)^2},$$

with $0 < |z| < 1$, which has a branch point at $z = 0$ and a pole of second order at $z = -1$. Let us then consider a contour C such that $z = 0$ lies outside the region enclosed by C while $z = -1$ lies inside it, as shown in Fig. 3.4. This contour is oriented in the counterclockwise sense and is composed of two concentric arcs of circumference C_1 and C_2 of radii ϵ and R, respectively, and of two line segments L_1 and L_2. Using the residue theorem we can write

$$\oint_C \frac{z^\alpha}{(1 + z)^2}dz = \int_{C_2} \frac{z^\alpha}{(1 + z)^2}dz + \int_R^\epsilon \frac{(x\,e^{2i\pi})^\alpha}{(1 + x)^2}dx + \int_{2\pi}^0 \frac{(\epsilon\,e^{i\theta})^\alpha}{(1 + \epsilon\,e^{i\theta})^2}i\epsilon\,e^{i\theta}\,d\theta +$$

$$+ \int_\epsilon^R \frac{x^\alpha}{(1 + x)^2}dx = 2\pi i \lim_{z \to -1}\frac{d}{dz}\left[(1 + z)^2\frac{z^\alpha}{(1 + z)^2}\right] = -2\pi i\alpha\,e^{i\pi\alpha}\,.$$

Take the limits $\epsilon \to 0$ and $R \to \infty$. From the first limit, it follows that the integral in θ goes to zero; and by Jordan's lemma, the integral on C_2 is also null.

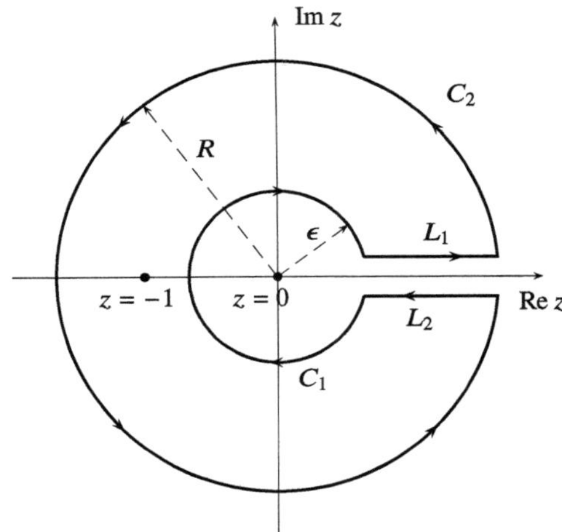

Fig. 3.4 Contour of integration for **SE 3.6**

Thus

$$e^{2i\pi\alpha}\int_{\infty}^{0}\frac{x^{\alpha}}{(1+x)^{2}}dx + \int_{0}^{\infty}\frac{x^{\alpha}}{(1+x)^{2}}dx = 2\pi i\,e^{i\pi\alpha},$$

whence we conclude that

$$\int_{0}^{\infty}\frac{x^{\alpha}}{(1+x)^{2}}dx = -\frac{\pi}{\sin\pi\alpha}.$$

SE 3.7 Obtain the Laurent series around the singularity $z = 0$ for the function

$$f(z) = \frac{1}{z}\cosh z^{-1}.$$

What kind of singularity is this?

Solution: Consider the change of independent variable $1/z = x$, which allows us to write

$$\frac{1}{z}\cosh z^{-1} = x\cosh x.$$

Then, using the series expansion for the hyperbolic cosine we have

$$x\cosh x = x\left(1 + \frac{x^{2}}{2!} + \frac{x^{4}}{4!} + \cdots\right)$$

whence, finally

$$\frac{1}{z}\cosh z^{-1} = \frac{1}{z}\left(1 + \frac{1}{z^2 2!} + \frac{1}{z^4 4!} + \cdots\right)$$

which is the desired expression. As we have infinite terms with negative powers of z, we have an essential singularity.

SE 3.8 Let $x, y \in \mathbb{R}$ and $z \in \mathbb{C}$ with $z = x + iy$. Consider the complex function

$$f(z) = \frac{e^{ikz} - e^{imz}}{z^2},$$

with $k, m \in \mathbb{N}$. This function has a simple pole at the origin and using the principal value, it is possible to show that [1]

$$\int_0^\infty \frac{\cos kx - \cos mx}{x^2}\, dx = -\frac{\pi}{2}(|k| - |m|). \tag{3.7}$$

Evaluate the real integral

$$\int_0^\infty \frac{\sin^2 x}{x^2}\, dx.$$

Solution: This integral can be considered a particular case of the result mentioned in the statement of the exercise. To see this, we take $k = 2$ and $m = 0$ in Eq. (3.7), obtaining

$$\int_0^\infty \frac{\cos 2x - 1}{x^2}\, dx = -\pi.$$

Using the well-known relation $\cos 2\theta = 1 - 2\sin^2\theta$ and rearranging, we have

$$\int_0^\infty \frac{\sin^2 x}{x^2}\, dx = \frac{\pi}{2},$$

which is the desired result.

Interistingly, this result is the same result obtained in **PE 3.27.**, for a different integral, namely

$$\int_0^\infty \frac{\sin x}{x}\, dx = \frac{\pi}{2}.$$

3.6 Proposed Exercises

PE 3.1 Let $z \in \mathbb{C}$. Solve the quadratic equation

$$z^2 + 8(i - 1)z + 63 - 16i = 0.$$

PE 3.2 Let $z \in \mathbb{C}$ with $z \neq 0$ and suppose that

$$\Omega = \frac{1 + z + z^2}{1 - z + z^2}$$

is a real number. Then, show that $|z| = 1$.

PE 3.3 (Putnam, 1989) Show that: If $11z^{10} + 10iz^9 + 10iz - 11 = 0$, then $|z| = 1$.

PE 3.4 Let $z \in \mathbb{C}$. Evaluate the integral

$$\frac{1}{i} \int_\Gamma \frac{dz}{z}$$

where $\Gamma : z(t) = \exp(it)$ and $0 \leq t \leq \pi/2$.

PE 3.5 Let $z \in \mathbb{C}$. Evaluate the integral

$$\frac{1}{2\pi i} \oint_\Gamma \frac{\tan z}{z^2 - 1} \, dz,$$

where Γ is a counterclockwise oriented circumference centered at the origin, with radius $3/2$.

PE 3.6 Let $z \in \mathbb{C}$ and

$$f(z) = \frac{6 \sin z}{z(z^2 + 4)\sqrt{z + 4}},$$

with $f(0) = 1$. Classify its singular points.

PE 3.7 Let $z \in \mathbb{C}$. Evaluate the integral

$$i \oint_\Gamma \frac{z}{(z + 3)(z - 1)^2} \, dz,$$

where Γ is a counterclockwise oriented circumference such that $z = 1$ is inside the region enclosed by Γ and $z = -3$ is outside it.

PE 3.8 Let $x, y \in \mathbb{R}$, $z \in \mathbb{C}$ and let Γ be an ellipse with equation $9x^2 + y^2 = 9$, oriented in the counterclockwise sense. Evaluate the integral

$$i \oint_\Gamma \left[\frac{z \exp(\pi z)}{z^4 - 16} + z \exp(\pi/z) \right] dz.$$

PE 3.9 Let $x, y \in \mathbb{R}$ and $z \in \mathbb{C}$ with $z = x + iy$. Evaluate the integral

$$\int_{\Gamma} \mathrm{Im}(z^2)\, dz$$

from the point $z = 0$ up to $z = 2 + 4i$, along the parable of equation $y = x^2$.

PE 3.10 Let Γ be a unitary circumference oriented in the counterclockwise sense. Use the Laurent series to evaluate the integral

$$\int_{\Gamma} z\, e^{2/z}\, dz\,.$$

PE 3.11 Show that the function $f(z) = \sin z$ is analytic. Do the same for $f(z) = \cos z$.

PE 3.12 Show that the function $f(z) = \mathrm{tg}\, z/z$ is *meromorphic*, i.e., that it can be written as the quotient of two analytic functions.

PE 3.13 Consider an analytic function $f(z) = u(x, y) + i v(x, y)$ and suppose that the second order partial derivatives of u and v with respect to x and y exist and are continuous. Show that the real and imaginary parts of $f(z)$ satisfy the Laplace equation,

$$\frac{\partial^2 f}{\partial x^2} + \frac{\partial^2 f}{\partial y^2} = 0.$$

Functions with this property are called *harmonic functions*.

PE 3.14 Expand the functions below in Laurent series around $z = 0$:

$$\text{(a)}\quad f(z) = \exp(1/z)$$
$$\text{(b)}\quad f(z) = z^{1/2}$$

Classify $z = 0$ for these functions.

PE 3.15 Calculate the residues of the functions of **PP 3.14**.

PE 3.16 Obtain the Laurent series around the singularity $z = 0$ for the function

$$f(z) = \frac{1}{z^3}(z - \sin z)\,.$$

What kind of singularity is this?

PE 3.17 Do the same as in the previous exercise for the function

$$f(z) = \frac{z}{(z + 1)(z + 2)}$$

around the point $z = -2$.

PE 3.18 Analogously to **PE 3.16** for $1/(z \cosh z^{-1})$ around $z = 0$.

PE 3.19 Find the residue of $F(z) = (\cotg z \coth z)/z^3$ around $z = 0$.

PE 3.20 Let $t \in \mathbb{R}$. Calculate

$$\frac{1}{2\pi i} \oint_C \frac{e^{zt}}{z^2(z^2 + 2z + 2)} dz$$

along the circumferences whose equations are: (a) $|z| = 1$ and (b) $|z| = 3$.

PE 3.21 Calculate

$$\oint_C \frac{2 + 3\sin \pi z}{z(z - 1)^2} dz$$

where C is a square with vertices at $3 + 3i$, $3 - 3i$, $-3 + 3i$ and $-3 - 3i$.

PE 3.22 Prove that

$$\int_0^\infty \frac{\ln(x^2 + 1)}{x^2 + 1} dx = \pi \ln 2.$$

PE 3.23 Calculate the integral

$$\int_0^\infty \frac{dx}{1 + x^2}$$

using residues.

PE 3.24 Let $0 < a < 1$. Calculate the integral

$$\int_{-\infty}^\infty \frac{e^{ax}}{1 + e^x} dx \; .$$

PE 3.25 Using the residue theorem, show that:

$$\text{(a)} \quad \int_0^{2\pi} \frac{\cos 3\theta}{5 - 4\cos \theta} d\theta = \frac{\pi}{12} \; ;$$

$$\text{(b)} \quad \int_0^\pi \frac{d\theta}{1 + \sin^2 \theta} = \frac{\pi}{2\sqrt{2}} \; .$$

PE 3.26 Consider the function

$$f(z) = (y^3 - 3x^2 y) + i(x^3 - 3xy^2).$$

(a) Are the Cauchy–Riemann conditions satisfied on all points of the real axis?

(b) Are the partial derivatives $\dfrac{\partial u}{\partial x}$, $\dfrac{\partial u}{\partial y}$, $\dfrac{\partial v}{\partial x}$, and $\dfrac{\partial v}{\partial y}$, where $u = y^3 - 3x^2 y$ and

$v = x^3 - 3xy^2$, continuous on all points of the real axis? (c) Is the function $f(z)$ analytic on all points of the real axis?

PE 3.27 Using the residue theorem, show that

$$\int_0^\infty \frac{\sin x}{x}\,dx = \frac{\pi}{2}.$$

An interesting generalization of this result can be found in [2].

PE 3.28 Given the function

$$f(z) = \frac{3z^4 - 2z^3 + 8z^2 - 2z + 5}{z - i},$$

show that $f(z)$ is not continuous at $z = i$, but that this is a removable discontinuity.

PE 3.29 If $\operatorname{tg} z = \operatorname{tg}(x + iy) = u(x, y) + iv(x, y)$, prove that

$$u(x, y) = \frac{\sin 2x}{\cos 2x + \cosh 2y},$$

$$v(x, y) = \frac{\sinh 2y}{\cos 2x + \cosh 2y}$$

and

$$\operatorname{Re}\left\{\frac{1 + i\,\operatorname{tg}(\theta/2)}{1 - i\,\operatorname{tg}(\theta/2)}\right\} = \cos\theta,$$

where θ is a real number.

PE 3.30 Consider the limit

$$I = \lim_{R\to\infty}\int_{-R}^{R} f(x)\,dx,$$

with $f(x)$ real. This limit is called the *principal value*, denoted by **P**, of the integral of $f(x)$ and is given by

$$\mathbf{P}\int_{-\infty}^{\infty} f(x)\,dx = \lim_{\substack{R\to\infty \\ \epsilon\to 0}}\left(\int_{-R}^{x_0-\epsilon} f(x)\,dx + \int_{x_0+\epsilon}^{R} f(x)\,dx\right),$$

if this integral exists.

Show that even though the integral

$$\int_{-1}^{1}\frac{dx}{x},$$

does not exist, its principal value exists and is identically null, that is,

$$\mathbf{P}\int_{-1}^{1}\frac{\mathrm{d}x}{x}=0.$$

PE 3.31 For $a>0$, show that

$$\mathbf{P}\int_{-\infty}^{\infty}\frac{\cos x}{a^{2}-x^{2}}\mathrm{d}x=\pi\,\frac{\sin a}{a}.$$

PE 3.32 Use polar coordinates on the plane, $x=r\cos\theta$ and $y=r\sin\theta$, in order to obtain the Cauchy–Riemann conditions in polar form.

PE 3.33 Show that $f(z)=e^{y}(\cos x+i\sin x)$ is not analytic at any point of the complex plane.

PE 3.34 Assuming that $\omega_1=u+iv$ is an analytic function, show that the function $\omega_2=-v+iu$ is also analytic.

PE 3.35 Show that an analytic function $f(z)$, whose derivative $\mathrm{d}f/\mathrm{d}z$ is null, is constant.

PE 3.36 Find the order of each pole of

$$f(z)=\frac{z+2}{z^{2}(z-1)(z^{2}+16)^{3}}.$$

PE 3.37 Determine and classify the singularities of the following functions:

$$\text{(a)}\quad\frac{4z^{3}+i}{z^{3}-1};$$

$$\text{(b)}\quad\sec z;$$

$$\text{(c)}\quad z\operatorname{cosec}z.$$

PE 3.38 Calculate the residue of functions (a) $\pi\cotg\pi z$ and (b) $\pi\operatorname{cosec}\pi z$.

PE 3.39 Show that, for $|z|>1$,

$$\frac{1}{z-1}=\sum_{k=-\infty}^{-1}z^{k}.$$

PE 3.40 Show that, for $|z|<2$,

$$\frac{2}{2-z}=\sum_{k=0}^{\infty}\left(\frac{z}{2}\right)^{k}.$$

PE 3.41 Show that, for $0 < a < 1$ and $0 < b < 1$,

$$\int_0^\infty \frac{x^{a-1} - x^{b-1}}{1 - x}\,dx = \pi(\cotg a\pi - \cotg b\pi).$$

PE 3.42 Show that

$$\frac{1}{\pi} \int_0^\infty \frac{\ln(1 + x^2 y^2)}{x^2 + 1}\,dx = \ln(1 + y).$$

For this relation to be valid, what restriction must be imposed on the possible values of y?

PE 3.43 Show that

$$\int_0^\infty \frac{\ln x}{(x^2 + 1)^4}\,dx = -\frac{23\pi}{96}.$$

PE 3.44 Show that

$$\int_0^\infty \frac{\sin x}{x(x^2 + 1)^2}\,dx = \frac{\pi}{2}\left(1 - \frac{3}{2e}\right).$$

PE 3.45 Use residues to calculate

$$\int_0^\infty \frac{dt}{\sqrt{t}}(1 + t)^{-3}.$$

PE 3.46 Show that

$$\int_0^\infty e^{-it}\, t^{\mu-1} dt = \frac{\Gamma(\mu)}{i^\mu},$$

where $\Gamma(\mu)$ is the gamma function given in **PE 4.28**. Use the contour of integration shown in Fig. 3.5.

Fig. 3.5 Contour of integration for **PE 3.46**

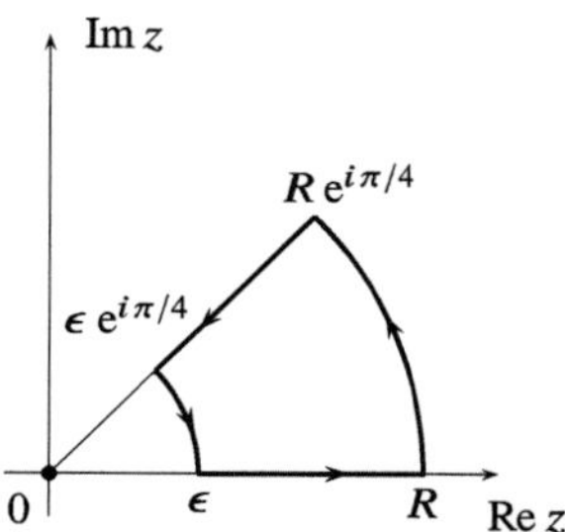

PE 3.47 Using the result of the previous exercises, show that:

$$(a) \int_0^\infty \cos x \; x^{\mu-1} dx = \cos \frac{\pi \mu}{2} \Gamma(\mu);$$

$$(b) \int_0^\infty \sin x \; x^{\mu-1} dx = \sin \frac{\pi \mu}{2} \Gamma(\mu).$$

PE 3.48 Show that, for $a > 1$,

$$\int_0^{2\pi} \frac{d\theta}{a^2 \cos^2 \theta + \sin^2 \theta} = \frac{2\pi}{a}.$$

PE 3.49 For p a positive integer such that $p > 2$, show that

$$\int_0^\infty \frac{x}{x^p + 1} dx = \frac{\pi}{p} \operatorname{cosec} \left(\frac{2\pi}{p} \right).$$

PE 3.50 Integrate the function $f(z) = z/(1 - e^{-iz})$ on a rectangular contour with vertices at $\pm \pi$ and $\pm \pi + iR$ to show that

$$\frac{1}{2\pi} \int_0^\pi \frac{x \sin x}{1 - \cos x} dx = \ln 2.$$

References

1. M.J. Ablowitz, A.S. Fokas, *Complex Variables: Introduction and Applications*. Cambridge Texts in Applied Mathematics (Cambridge University Press, Cambridge, 1997)
2. E. Capelas de Oliveira, Residue theorem and related integrals. Int. J. Math. Educ. Sci. and Tech. **32**, 156–160 (2001)
3. E. Capelas de Oliveira, W.A. Rodrigues Jr., *Funções Analíticas com Aplicações* (Analytic Functions with Applications) (Editora Livraria da Física, São Paulo, 2006)
4. R.V. Churchill, *Complex Variables with Applications* (McGraw-Hill, New York, 1960)

Chapter 4
Special Functions

Besides the elementary functions usually studied in basic calculus courses—rational, trigonometric, logarithmic and exponential functions—there exists another (huge) class of functions called *special functions* [2], the knowledge of which is essential for solving many real problems arising in exact sciences. These families of functions appear almost always as sets of solutions of some particular family of ordinary differential equations. For this reason we begin this chapter investigating the solutions of a general linear second-order ordinary differential equation with three regular singular points, which will be transformed into a *hypergeometric equation*, whose solutions are the so-called *hypergeometric functions*. In the sequence, by considering a certain limit process applied to the hypergeometric equation, we will obtain the *confluent hypergeometric equation*, whose solutions are (of course) called *confluent hypergeometric functions*. As a special case of the first group of functions we will study the *Legendre functions*, and in the second group we will study the *Bessel functions*.

4.1 Differential Equation with Three Singular Points

Let $p(x)$ and $q(x)$ be real functions. Consider the homogeneous linear second-order ordinary differential equation

$$\frac{\mathrm{d}^2}{\mathrm{d}z^2}u(z) + p(z)\frac{\mathrm{d}}{\mathrm{d}z}u(z) + q(z)u(z) = 0. \tag{4.1}$$

There exists a close relation between the properties of functions $p(z)$ and $q(z)$ appearing in this equation and the properties of its *solutions*. In fact, one may prove

© The Author(s), under exclusive license to Springer Nature Switzerland AG 2025
E. Capelas de Oliveira, J. E. Maiorino, *Analytical Methods in Applied Mathematics*,
Problem Books in Mathematics, https://doi.org/10.1007/978-3-031-74794-6_4

that the general solution of the differential equation above will have singularities at the points of the complex plane in which $p(z)$ and $q(z)$ have their *poles*. The points in which $p(z)$ and $q(z)$ are *analytic* functions are called *ordinary points* of the equation. Correspondingly, the points in which $p(z)$ and/or $q(z)$ present singularities are called *singular points* of the differential equation. We have the following definitions:

Definition 4.1.1 (Regular Singular Point) Every point of the complex plane in which the general solution of the ordinary differential equation has a pole or a branch point.

Definition 4.1.2 (Irregular Singular Point) Every point in which the general solution presents an essential singularity.

We shall study here a linear second-order ordinary differential equation with three regular singular points z_1, z_2 and z_3. It may be shown that for this to be possible, it is necessary and sufficient that $p(z)$ has simple poles at the three points z_1, z_2 and z_3, and that $q(z)$ has poles of order smaller than or equal to two at those points. Besides, we will require that the point at infinity be an ordinary point of the differential equation. To ensure that this condition is satisfied, we introduce the change of independent variable $z = 1/\omega$ into Eq. (4.1), obtaining

$$\frac{\mathrm{d}^2}{\mathrm{d}\omega^2} u(\omega) + P(\omega)\frac{\mathrm{d}}{\mathrm{d}\omega} u(\omega) + Q(\omega)u(\omega) = 0,$$

where

$$P(\omega) = \frac{2}{\omega} - \frac{1}{\omega^2}\, p\left(\frac{1}{\omega}\right) \tag{4.2}$$

and

$$Q(\omega) = \frac{1}{\omega^4}\, q\left(\frac{1}{\omega}\right).$$

Infinity will be an ordinary point if, and only if, $P(\omega)$ and $Q(\omega)$ are analytic at $\omega = 0$. The requirement that $p(z)$ has simple poles and that $q(z)$ has poles of order smaller than or equal to two at z_1, z_2 and z_3 is satisfied by supposing that

$$p(z) = \frac{A}{z - z_1} + \frac{B}{z - z_2} + \frac{C}{z - z_3} + D \tag{4.3}$$

and

$$q(z) = \frac{1}{(z - z_1)(z - z_2)(z - z_3)}\left(\frac{E}{z - z_1} + \frac{F}{z - z_2} + \frac{G}{z - z_3}\right),$$

where A, B, C, D, E, F and G are constants. Replacing Eq. (4.3) into Eq. (4.2) we find

$$P(\omega) = \frac{2}{\omega} - \frac{1}{\omega}\left(\frac{A}{1 - \omega z_1} + \frac{B}{1 - \omega z_2} + \frac{C}{1 - \omega z_3} + \frac{D}{\omega}\right),$$

which must be analytic at $\omega = 0$. This condition is satisfied if $D = 0$ and $A + B + C = 2$.

Once $p(z)$ and $q(z)$ are defined, we employ the Frobenius method in order to find the solutions of the differential equation in the form of a Taylor series around the singularities z_1, z_2 and z_3, i.e., we will suppose three solutions of the form

$$u_i(z) = (z - z_i)^r \sum_{n=0}^{\infty} a_n(z - z_i)^n, \qquad (4.4)$$

for $i = 1,\ 2,\ 3$ and where r is a free parameter. However, before we replace this formula into the differential equation, we will rewrite function $p(z)$ and $q(z)$ as

$$p(z) = \frac{F_i(z)}{(z - z_i)} \quad \text{and} \quad q(z) = \frac{G_i(z)}{(z - z_i)^2}\ ,$$

for $i = 1, 2, 3$.

From this definition and the form of $p(z)$ and $q(z)$ it follows, for instance, that

$$F_1(z) = A + \frac{B(z - z_1)}{(z - z_2)} + \frac{C(z - z_1)}{(z - z_3)} = p(z)(z - z_1)$$

and

$$G_1(z) = \frac{1}{(z - z_2)(z - z_3)}\left[E + \frac{F(z - z_1)}{(z - z_2)} + \frac{G(z - z_1)}{(z - z_3)}\right]$$

$$= q(z)(z - z_1)^2$$

are *analytic* functions around z_1, the same being true of $F_2(z)$, $G_2(z)$, $F_3(z)$ and $G_3(z)$ around z_2 and z_3, respectively. For this reason, we may also expand them in Taylor series around $z = z_i$, for $i = 1, 2, 3$:

$$F_i(z) = \sum_{n=0}^{\infty} \frac{1}{n!}\left[\frac{\mathrm{d}^n}{\mathrm{d}z^n}F_i(z)\right]_{z=z_i}(z - z_i)^n$$

$$= F_i(z_i) + F_i'(z_i)(z - z_i) + \frac{F_i''(z_i)}{2!}(z - z_i)^2 + \cdots$$

$$
G_i(z) = \sum_{n=0}^{\infty} \frac{1}{n!} \left[\frac{\mathrm{d}^n}{\mathrm{d}z^n} G_i(z) \right]_{z=z_i} (z - z_i)^n
$$

$$
= G_i(z_i) + G_i'(z_i)(z - z_i) + \frac{G_i''(z_i)}{2!}(z - z_i)^2 + \cdots
$$

At this point it should be clear that the calculations for one of the singularities will be valid for the other two. So, we will restrict our attention to the series around z_1, generalizing the results obtained for the other two cases.

Our aim here is to obtain the indicial equation for the series, so we will write only the first few terms of the sums. Calculating $u_1'(z)$ and $u_1''(z)$ in the form of Eq. (4.4) and substituting the results into Eq. (4.1) we obtain:

$$
\sum_{n=0}^{\infty} a_n (n + r)(n + r - 1)(z - z_1)^{n+r-2} +
$$

$$
+ \frac{\left[F_1(z_1) + F_1'(z_1)(z - z_1) + \frac{F_1''(z_1)}{2!}(z - z_1)^2 + \cdots \right]}{(z - z_1)} \sum_{n=0}^{\infty} a_n (n+r)(z-z_1)^{n+r-1} +
$$

$$
\frac{\left[G_1(z_1) + G_1'(z_1)(z - z_1) + \frac{G_1''(z_1)}{2!}(z - z_1)^2 + \cdots \right]}{(z - z_1)^2} \sum_{n=0}^{\infty} a_n (z - z_1)^{n+r} = 0,
$$

or in the following form:

$$
a_0[r(r - 1) + F_1(z_1)r + G_1(z_1)](z - z_1)^{r-2} +
$$

$$
\{a_1[(1+r)r + F_1(z_1)(1+r) + G_1(z_1)] + a_0[F_1'(z_1)r + G_1'(z_1)]\}(z-z_1)^{r-1} + \cdots = 0.
$$

The indicial equation is obtained by setting equal to 0 the factor multiplying the term with the smallest exponent. Given that $a_0 \neq 0$ by hypothesis, we have

$$
r(r - 1) + F_1(z_1)r + G_1(z_1) = 0,
$$

and for the other two singularities we will similarly obtain

$$
r(r - 1) + F_2(z_2)r + G_2(z_2) = 0;
$$

$$
r(r - 1) + F_3(z_3)r + G_3(z_3) = 0.
$$

We can rewrite these equations as

$$
r^2 + [F_i(z_i) - 1]r + G_i(z_i) = 0,
$$

for $i = 1$, 2, 3. Then, if α and α' are the roots of the first indicial equation around z_1, this equation implies that

$$\begin{cases} \alpha + \alpha' = -[F_1(z_1) - 1] = 1 - A; \\ \alpha\alpha' = G_1(z_1) = \dfrac{E}{(z_1 - z_2)(z_1 - z_3)}. \end{cases}$$

In the same way, if β and β' are roots of the indicial equation around z_2 and if γ and γ' are the roots associated with the third singularity, we will have respectively:

$$\begin{cases} \beta + \beta' = 1 - B; \\ \beta\beta' = \dfrac{F}{(z_2 - z_1)(z_2 - z_3)}; \end{cases}$$

$$\begin{cases} \gamma + \gamma' = 1 - C; \\ \gamma\gamma' = \dfrac{G}{(z_3 - z_1)(z_3 - z_2)}. \end{cases}$$

With these results we may rewrite the original differential equation as

$$\frac{\mathrm{d}^2 u}{\mathrm{d}z^2} + \left(\frac{1 - \alpha - \alpha'}{z - z_1} + \frac{1 - \beta - \beta'}{z - z_2} + \frac{1 - \gamma - \gamma'}{z - z_3} \right) \frac{\mathrm{d}u}{\mathrm{d}z} +$$

$$+ \left[\frac{(z_1 - z_2)(z_1 - z_3)\alpha\alpha'}{z - z_1} + \frac{(z_2 - z_1)(z_2 - z_3)\beta\beta'}{z - z_2} + \frac{(z_3 - z_1)(z_3 - z_2)\gamma\gamma'}{z - z_3} \right] \times$$

$$\times \frac{u}{(z - z_1)(z - z_2)(z - z_3)} = 0, \tag{4.5}$$

with the restriction $\alpha + \alpha' + \beta + \beta' + \gamma + \gamma' = 1$, which arises from the requirement that the point $z = \infty$ be an ordinary point of the differential equation being solved.

This ordinary differential equation is the so-called *Riemann equation*. Any solution $u(z)$ of this differential equation can be represented by the Riemann–Papperitz [1857 – Johannes Erwin Papperitz – 1938] symbol, which is given by

$$u(z) = \mathbf{P} \left\{ \begin{matrix} z_1 & z_2 & z_3 & \\ \alpha & \beta & \gamma & z \\ \alpha' & \beta' & \gamma' & \end{matrix} \right\},$$

where the first three columns show the singular points of the differential equation, together with the roots of the corresponding indicial equations, while the last column shows the independent variable. The equality aforementioned is equivalent to saying that $u(z)$ satisfies the ordinary differential equation Eq. (4.5).

Example 4.1 The associated Legendre differential equation, as we will see in this chapter, has three singular points at $z = -1$, $z = 1$ and $z = \infty$. Write the corresponding Riemann–Papperitz symbol.

Let ν and μ be two constants. The associated Legendre differential equation is

$$(1 - z^2)\frac{d^2}{dz^2}u(z) - 2z\frac{d}{dz}u(z) + \left[\nu(\nu + 1) - \frac{\mu^2}{1 - z^2}\right]u(z) = 0,$$

which can be written in the form

$$(1 - z)^2(1 + z)^2\frac{d^2}{dz^2}u(z) - 2z(1 - z)(1 + z)\frac{d}{dz}u(z)$$

$$+[\nu(\nu + 1)(1 - z)(1 + z) - \mu^2]u(z) = 0.$$

Here we consider only the singularity at $z = -1$, as the treatment of the singularity at $z = 1$ is analogous to it; in order to deal with the point $z = \infty$ we have to make a change of variable $z \to 1/(z + 1)$ and then discuss the resulting equation at $z = -1$. Thus, consider the Frobenius series around $z = -1$,

$$u(z) = \sum_{n=0}^{\infty} a_n(z + 1)^{k+s}$$

with $a_0 \neq 0$ and s a free parameter. Evaluating the derivatives, substituting into the associated Legendre differential equation and rearranging, we can write

$$\sum_{n=0}^{\infty} a_n\left[(n + s)(n + s - 1) + 2(n + s) - \nu(\nu + 1)\right](z + 1)^{n+s+2} +$$

$$\sum_{n=0}^{\infty} a_n\left[-4(n + s)(n + s - 1) - 6(n + s) + 2\nu(\nu + 1)\right](z + 1)^{n+s+1} +$$

$$\sum_{n=0}^{\infty} a_n\left[4(n + s)(n + s - 1) + (n + s) - \mu^2\right](z + 1)^{n+s} = 0.$$

In order to obtain the recurrence relation, we introduce a change of index $n \to n-2$ in the first sum and $n \to n - 1$ in the second sum. On the other hand, since we are interested only in the Riemann–Papperitz symbol, we need only the roots of the auxiliary equation. To this end, we consider only the case $n = 0$ in the third sum,

$$[4s(s - 1) + 4s - \mu^2]a_0(z + 1)^s = 0,$$

whose roots are $s_1 = \mu/2$ and $s_2 = -\mu/2$. With the same procedure, we obtain the same roots for the auxiliary equation relatively to the point $z = 1$, i.e., $s_1 = \mu/2$ and $s_2 = -\mu/2$. Also, at $z = \infty$ we obtain $s_1 = -\nu$ and $s_2 = \nu + 1$. With these results we can finally write the Riemann–Papperitz symbol

$$u(z) = \mathbf{P} \left\{ \begin{array}{ccc} -1 & \infty & 1 \\ \mu/2 & -\nu & \mu/2 \\ -\mu/2 & \nu+1 & \mu/2 \end{array} \; z \right\},$$

which is the desired result $\square$

4.2 Hypergeometric Equation

The solution of the Riemann equation contains nine different parameters, the three singular points and the six roots of the indicial equations related by the restriction $\alpha + \alpha' + \beta + \beta' + \gamma + \gamma' = 1$. We can reduce these parameters to only *three* independent parameters by means of a change of dependent variable of the form

$$u(z) = (z - z_1)^{-r}(z - z_2)^{-s}(z - z_3)^{-t}v(z),$$

with $r + s + t = 0$, and a change of independent variable,

$$z' = \frac{\bar{A}z + \bar{B}}{\bar{C}z + \bar{D}}, \tag{4.6}$$

where $\bar{A}$, $\bar{B}$, $\bar{C}$ and $\bar{D}$ are constants which are to be determined.

The first transformation modifies the roots of the indicial equation at the three singularities, in such a way that

$$v(z) = (z - z_1)^r (z - z_2)^s (z - z_3)^t u(z)$$

$$= (z - z_1)^r (z - z_2)^s (z - z_3)^t \mathbf{P} \left\{ \begin{array}{ccc} z_1 & z_2 & z_3 \\ \alpha & \beta & \gamma \\ \alpha' & \beta' & \gamma' \end{array} \; z \right\}$$

$$= \mathbf{P} \left\{ \begin{array}{ccc} z_1 & z_2 & z_3 \\ \alpha+r & \beta+s & \gamma+t \\ \alpha'+r & \beta'+s & \gamma'+t \end{array} \; z \right\}.$$

Note that the singular points of the equation are still the same. With the change of independent variable Eq. (4.6) it is possible to shift the three singularities z_1, z_2 and z_3 to the standard points $z_1' = 0$, $z_2' = \infty$ and $z_3' = 1$. Choosing $r = -\alpha$, $s = \alpha + \gamma$ and $t = -\gamma$, we finally obtain

$$u(z) = \mathbf{P}\left\{\begin{matrix} z_1 & z_2 & z_3 & \\ \alpha & \beta & \gamma & z \\ \alpha' & \beta' & \gamma' & \end{matrix}\right\}$$

$$= \left(\frac{z-z_1}{z-z_2}\right)^{\alpha}\left(\frac{z-z_3}{z-z_2}\right)^{\gamma}\mathbf{P}\left\{\begin{matrix} 0 & \infty & 1 & \\ 0 & \alpha+\beta+\gamma & 0 & z' \\ \alpha'-\alpha & \alpha+\beta'+\gamma & \gamma'-\gamma & \end{matrix}\right\}$$

$$= \left(\frac{z-z_1}{z-z_2}\right)^{\alpha}\left(\frac{z-z_3}{z-z_2}\right)^{\gamma}v(z')\,.$$

Introducing parameters a, b and c defined by

$$\alpha+\beta+\gamma = a, \qquad \alpha+\beta'+\gamma = b, \qquad 1+\alpha-\alpha' = c,$$

and recalling that $\alpha+\alpha'+\beta+\beta'+\gamma+\gamma' = 1$ we get

$$\mathbf{P}\left\{\begin{matrix} z_1 & z_2 & z_3 & \\ \alpha & \beta & \gamma & z \\ \alpha' & \beta' & \gamma' & \end{matrix}\right\} =$$

$$= \left(\frac{z-z_1}{z-z_2}\right)^{\alpha}\left(\frac{z-z_3}{z-z_2}\right)^{\gamma}\mathbf{P}\left\{\begin{matrix} 0 & \infty & 1 & \\ 0 & a & 0 & z' \\ 1-c & b & c-a-b & \end{matrix}\right\}\,.$$

The differential equation corresponding to the Riemann–Papperitz symbol appearing at the end of this equality, for $v(z')$, is

$$\frac{\mathrm{d}^2 v}{\mathrm{d}z'^2} + \left\{\frac{c}{z'-0} + \lim_{z_2'\to\infty}\left(\frac{1-a-b}{z'-z_2'}\right) + \frac{1-c+a+b}{z'-1}\right\}\frac{\mathrm{d}v}{\mathrm{d}z'} +$$

$$+ \lim_{z_2'\to\infty}\left\{\frac{(0-z_2')(0-1)0}{(z'-0)^2(z'-z_2')(z'-1)} + \frac{(z_2'-0)(z_2'-1)ab}{(z'-0)(z'-z_2')^2(z'-1)} + \right.$$

$$\left. + \frac{(1-0)(1-z_2')0}{(z'-0)(z'-z_2')(z'-1)^2}\right\}v = 0,$$

that is

$$z'(1-z')\frac{\mathrm{d}^2 v}{\mathrm{d}z'^2} + [c-(a+b+1)z']\frac{\mathrm{d}v}{\mathrm{d}z'} - abv = 0.$$

This result shows that any linear second-order ordinary differential equation with three regular singular points may be put in the form of a *hypergeometric equation*,

$$z(1 - z)\frac{d^2u}{dz^2} + [c - (a + b + 1)z]\frac{du}{dz} - abu = 0, \tag{4.7}$$

whose solution $u(z)$ is denoted by

$$u(z) = {}_2F_1(a, b; c; z).$$

We will now search for a solution with the form of a power series (Frobenius series) around the origin, where it is analytic since the corresponding indicial equation has roots 0 and $1 - c$. Suppose that

$${}_2F_1(a, b; c; z) = \sum_{n=0}^{\infty} c_n z^n.$$

Introducing this expression into Eq. (4.7) and imposing the normalization condition $c_0 = 1$ we obtain

$${}_2F_1(a, b; c; z) = \frac{\Gamma(c)}{\Gamma(a)\Gamma(b)} \sum_{n=0}^{\infty} \frac{\Gamma(a + n)\Gamma(b + n)}{\Gamma(c + n)\Gamma(n + 1)} z^n = \sum_{n=0}^{\infty} \frac{(a)_n (b)_n}{(c)_n} \frac{z^n}{n!},$$
$$\tag{4.8}$$

where $\Gamma(\cdot)$ is the gamma function (whose relation with the beta function can be seen in PE 4.28) and $(\cdot)_k$ is the so-called Pochhammer [1841 – Leo August Pochhammer – 1920] symbol, defined in **PE 4.31**. Notice that if we put $a = 1$ and $b = c$ we find

$${}_2F_1(1, b; b; z) = \sum_{n=0}^{\infty} z^n,$$

which, for $|z| < 1$, is the geometric series. For this reason the series for ${}_2F_1(a, b; c; z)$ is called *hypergeometric*. It is worth observing that if we interchange the places of z_1, z_2 and z_3 in the Riemann equation, Eq. (4.5), or if we exchange α with α', or β with β', the Riemann equation remains unaltered, but we are led to a different hypergeometric equation. The first permutation may take place in $3! = 6$ different ways, and the second one gives rise to $2 \times 2 = 4$ combinations of αs and βs, providing a total of 24 different equations (and solutions), known as *Kummer* [1810 – Ernst Eduard Kummer – 1893] *solutions*. Given that a linear second-order ordinary differential equation can have only two linearly independent solutions, there exists among these 24 solutions a series of relations, known as *Kummer relations,*, which make it possible to obtain one solution from other known solutions. For a complete table, see [4].

Finally, we should mention that if the root $1 - c$ is not an integer, a second linearly independent solution of the hypergeometric equation is given by

$$u_2(z) = z^{1-c} \, {}_2F_1(b - c + 1, a - c + 1; 2 - c; z).$$

Example 4.2 Consider a simple pendulum of mass m and length ℓ. Using Newton's second law, we obtain the following nonlinear ordinary differential equation

$$\frac{\mathrm{d}^2}{\mathrm{d}t^2}\varphi(t) + \frac{g}{\ell}\sin\varphi(t) = 0$$

where g is a constant, the gravitational acceleration, and $\varphi(t)$ is the angle, varying with time t, formed by the vertical axis and the pendulum rod. Note that this ordinary differential equation is nonlinear because the dependent variable, $\varphi(t)$, is the argument of the sine function. Obtain the period of oscillation.

To solve this nonlinear ordinary differential equation we first write it as

$$\frac{\mathrm{d}}{\mathrm{d}t}\left\{\frac{1}{2}\left[\frac{\mathrm{d}}{\mathrm{d}t}\varphi(t)\right]^2 - \frac{g}{\ell}\cos\varphi(t)\right\} = 0,$$

which implies that the quantity between braces is constant. Thus, to obtain the constant we must impose an initial condition, $\varphi(0) = \varphi_0$. Manipulating the last equation and using the initial condition, we obtain

$$\frac{\mathrm{d}}{\mathrm{d}t}\varphi(t) = \sqrt{\frac{2g}{\ell}\left[\cos\varphi(t) - \cos\varphi_0\right]}$$

which is a separable nonlinear first-order ordinary differential equation. Using the trigonometric identity (double arc) and integrating, we can write

$$2\sqrt{\frac{g}{\ell}}\int \mathrm{d}t = \int \frac{\mathrm{d}\varphi}{\sqrt{k^2 - \sin^2(\varphi/2)}},$$

where we introduced a constant $k = \sin(\varphi_0/2)$ and simplified the expression by writing $\varphi(t) = \varphi$. To evaluate the integral in the second member, we introduce an adequate change of variable $\sin\theta = \frac{1}{k}\sin(\varphi/2)$; remembering that $\varphi \leq \varphi_0$, we get

$$\sqrt{\frac{g}{\ell}}\int \mathrm{d}t = \int \frac{\mathrm{d}\theta}{\sqrt{1 - k^2\sin^2\theta}}$$

which can be identified with an elliptic integral of the first kind. We solve only the integral on the right. The time taken for the angle to change from zero to φ_0 is a quarter of the period T. Thus, θ goes from zero to $\pi/2$ and we can write for the integral

$$\Omega(k) = \int_0^{\pi/2} \frac{\mathrm{d}\theta}{\sqrt{1 - k^2\sin^2\theta}},$$

which is the so-called complete elliptic integral of the first kind [1].

To evaluate $\Omega(k)$, we first introduce an expansion for the integrand,

$$\frac{1}{\sqrt{1 - k^2 \sin^2 \theta}} = \sum_{n=0}^{\infty} \frac{(\frac{1}{2})_n}{n!} k^{2n} \sin^{2n} \theta.$$

We then exchange the order of integral and sum and use the known relation [5, 6]

$$\int_0^{\pi/2} \sin^n x \, dx = \begin{cases} \dfrac{\pi}{2} \dfrac{(\frac{1}{2})_{n/2}}{(\frac{n}{2})!}, & n = 0, 2, 4, \ldots \\[2ex] \dfrac{(\frac{n-1}{2})!}{(\frac{3}{2})_{(n-1)/2}}, & n = 1, 3, 5, \ldots \end{cases}$$

We thus obtain $\Omega(k)$ expressed in terms of a series

$$\Omega(k) = \frac{\pi}{2} \sum_{n=0}^{\infty} \frac{(\frac{1}{2})_n (\frac{1}{2})_n}{(1)_n \, n!} k^{2n}.$$

Identifying this result with Eq. (4.8), we obtain

$$T = 2\pi \sqrt{\frac{\ell}{g}} \, {}_2F_1\left[\frac{1}{2}, \frac{1}{2}; 1; \sin^2\left(\frac{\varphi_0}{2}\right)\right]$$

which is the period of oscillation; it depends on the amplitude and does not depend on the mass. We remark that for the small-angle approximation, $\varphi_0 \ll 1$, we have $\sin 0 = 0$ and ${}_2F_1(a, b; c; 0) = 1$; then

$$T = 2\pi \sqrt{\frac{\ell}{g}},$$

which is the period for small amplitudes.   □

4.3 Confluent Hypergeometric Equation

Introducing the change of independent variable

$$z = \frac{x}{b}$$

into the hypergeometric equation Eq. (4.7) and taking its limit as $b \to \infty$, we obtain an ordinary differential equation with the form

$$x\frac{d^2u}{dx^2} + (c - x)\frac{du}{dx} - au = 0,$$

for $u = u(x)$, where a and c are new parameters, which should not be confused with the original parameters of the hypergeometric equation Eq. (4.7). This equation is called a *confluent hypergeometric equation* or *Kummer equation*. The name stands from the fact that departing from the three singularities of the original hypergeometric equation Eq. (4.7) we arrive at an ordinary differential equation with only two singular points, because two of the original singularities have merged into a unique singularity at infinity. After this *confluence* we are left with a regular singularity at $z = 0$ and an irregular singularity at infinity.

We might use the Frobenius method to obtain the solutions of this new equation (**PE 2.29**), but we prefer to proceed in the same way as we did to arrive at this differential equation, that is, we will start from the solutions of the hypergeometric equation Eq. (4.8), perform the same change of variable, and then take the same limit. For $c \neq 0, -1, -2,\dots$ two linearly independent solutions of the Kummer equation, also called confluent hypergeometric functions, are:

$$u_1 = \lim_{b\to\infty} \, {}_2F_1\left(a, b; c; \frac{z}{b}\right) \equiv {}_1F_1(a; c; z)$$

$$= \frac{\Gamma(c)}{\Gamma(a)}\sum_{n=0}^{\infty}\frac{\Gamma(a + n)}{\Gamma(c + n)}\frac{z^n}{n!} = \sum_{n=0}^{\infty}\frac{(a)_n}{(b)_n}\frac{z^n}{n!},$$

as

$$\lim_{b\to\infty}\frac{\Gamma(b + n)}{\Gamma(b)}b^{-n} = 1,$$

and

$$u_2 = z^{1-c}\,{}_1F_1(a - c + 1; 2 - c; z).$$

Example 4.3 Let $z \in \mathbb{C}$ and $\mu \in \mathbb{C}$, with $\text{Re}(\mu) > 0$. The incomplete gamma function is defined by the following integral:

$$\gamma(\mu, z) = \int_0^z e^{-t}t^{\mu-1}\,dt\,.$$

Express the gamma function in terms of a confluent hypergeometric function.

Using the series expansion formula for the exponential function, exchanging the order of the integral with the sum and integrating, we obtain

$$\int_0^z e^{-t}t^{\mu-1}dt = \sum_{k=0}^{\infty}\frac{(-1)^k}{k!}\int_0^z t^{k+\mu-1}\,dt = z^{\mu}\sum_{k=0}^{\infty}\frac{(-1)^k}{k+\mu}\frac{z^k}{k!} \equiv \gamma(\mu, z),$$

which can be written as

$$\gamma(\mu, z) = z^{\mu} \sum_{k=0}^{\infty} \frac{\Gamma(k + \mu)}{\Gamma(k + \mu + 1)} (-z)^k .$$

Identifying this expression with the confluent hypergeometric function we get

$$\gamma(\mu, z) = \frac{z^{\mu}}{\mu} {}_1F_1(\mu; \mu + 1; -z),$$

which is the desired result $\qquad\qquad\square$

In the next two sections, we introduce two particular cases of hypergeometric and confluent hypergeometric functions, the so-called Legendre functions and Bessel functions, respectively.

4.4 Legendre Functions

As we aforementioned, Legendre functions are a particular case of hypergeometric functions. The differential equation for the Legendre functions is

$$(1 - z^2)\frac{d^2u}{dz^2} - 2z\frac{du}{dz} + \left[\nu(\nu + 1) - \frac{\mu^2}{1 - z^2}\right] u = 0, \quad u = u(z), \qquad (4.9)$$

where μ and ν are usually unrestricted complex numbers. The most frequent representations, given in terms of the hypergeometric function, are:

$$\Gamma(1 - \mu)\mathscr{B}_{\nu}^{\mu}(z) = \frac{(z + 1)^{\mu/2}}{(z - 1)^{-\mu/2}} {}_2F_1\left(-\nu, \nu + 1; 1 - \mu; \frac{1 - z}{2}\right)$$

and

$$\Gamma\left(\frac{3}{2} + \nu\right)\mathscr{D}_{\nu}^{\mu}(z) = \frac{e^{i\pi\mu}}{2^{-\nu-1}}\pi^{1/2}\Gamma(\nu + \mu + 1)\frac{(z^2 - 1)^{\mu/2}}{z^{-\nu-\mu-1}} \times$$

$$\times {}_2F_1\left(1 + \frac{\nu}{2} + \frac{\mu}{2}, \frac{1}{2} + \frac{\nu}{2} + \frac{\mu}{2}; \frac{3}{2} + \nu; \frac{1}{z^2}\right).$$

The first function is called a *Legendre function of the first kind*, and the other is the *Legendre function of the second kind*; the parameters μ and ν are arbitrary complex numbers.

For μ and ν integers and for z real, with $-1 < x < 1$, these functions are called *associated Legendre functions* and are denoted by $Q_n^m(x)$ and $P_n^m(x)$, respectively.

In the case in which $\mu = m = 1, 2, 3 \ldots$ is an integer, the following relations are valid:

$$\mathscr{B}_v^m(z) = (z^2 - 1)^{m/2} \frac{d^m}{dz^m} \mathscr{B}_v(z);$$

$$\mathscr{D}_v^m(z) = (z^2 - 1)^{m/2} \frac{d^m}{dz^m} \mathscr{D}_v(z);$$

$$P_v^m(x) = (-1)^m (1 - x^2)^{m/2} \frac{d^m}{dx^m} P_v(x);$$

$$Q_v^m(x) = (-1)^m (1 - x^2)^{m/2} \frac{d^m}{dx^m} Q_v(x);$$

where $\mathscr{B}_v(z)$, $\mathscr{D}_v(z)$, $P_\mu(x)$ and $Q_v(x)$ are the solutions of Eq. (4.9) when $\mu = 0$. From these expressions we see that we just need to know the solution for $\mu = 0$ in order to obtain the other solutions by differentiation.

Let us then assume $\mu = 0$ and $v = n =$ an integer in Eq. (4.9):

$$(1 - z^2)\frac{d^2\omega}{dz^2} - 2z\frac{d\omega}{dz} + n(n + 1)\omega = 0, \qquad \omega = \omega(z).$$

Its general solution, for $z = x \in \mathbb{R}$, is given by

$$\omega(x) = AP_n(x) + BQ_n(x),$$

where $P_n(x)$ are the classical well-known Legendre polynomials, $Q_n(x)$ are Legendre functions of the second kind and A and B are two arbitrary constants.

The first three Legendre polynomials are:

$$P_0(x) = 1, \quad P_1(x) = x, \quad P_2(x) = \frac{1}{2}(3x^2 - 1),$$

and in the general case (obtained, for instance, through the Frobenius method)

$$P_n(x) = \sum_{m=0}^{[n/2]} (-1)^m \frac{(2n - 2m)!}{2^n m!(n - m)!(n - 2m)!} x^{n-2m}$$
$$= {}_2F_1\left(-n, n + 1; 1; \frac{1 - x}{2}\right),$$

where $[n/2]$ represents the greatest integer smaller than or equal to $n/2$ and ${}_2F_1(a, b; c; z)$ is the hypergeometric function.

For the Legendre functions of the second kind, we have

$$Q_0(x) = \frac{1}{2} \ln \left(\frac{x+1}{x-1} \right), \qquad Q_1(x) = \frac{x}{2} \ln \left(\frac{x+1}{x-1} \right) - 1,$$

whence it is clear that $Q_n(x)$ is not defined at the extremes of the interval. In the general case $Q_n(x)$ is related to $P_n(x)$ by

$$Q_n(x) = \frac{1}{2} \int_{-1}^{1} (x-t)^{-1} P_n(t)dt .$$

It is important to notice that once $P_0(x)$ and $P_1(x)$ are known, all the remaining polynomials get determined by means of a pure recurrence relation, as in the first relation in **PE 4.23**. This is also true for $Q_n(x)$.

Finally, the orthogonality relation for Legendre polynomials is

$$\int_{-1}^{1} P_n(x) P_m(x)dx = \begin{cases} 0 & n \neq m \\ \dfrac{2}{2n+1} & n = m \end{cases}$$

where the so-called weight function is equal to 1.

Example 4.4 Let $x, t \in \mathbb{R}$. We define the generating function for Legendre polynomials as the two-variable function whose power series expansion has these polynomials as coefficients of the series:

$$\mathsf{G}(x,t) \equiv (1 - 2xt + t^2)^{-1/2} = \sum_{k=0}^{\infty} P_k(x)\, t^k,$$

where $P_k(x)$ are the Legendre polynomials. Using the generating function for the Legendre polynomials, show the recurrence relation involving a derivative,

$$P_k(x) = P'_{k+1}(x) - 2x P'_k(x) + P'_{k-1}(x),$$

where the prime $'$ denotes the derivative with respect to x and $k \geq 1$.

Differentiating the expression for the generating function with respect to x, we can write

$$\frac{\partial}{\partial x} \mathsf{G}(x,t) = t(1 - 2xt + t^2)^{-3/2} = \sum_{k=0}^{\infty} P'_k(x)\, t^k .$$

Using the first equality and the generating function, we can write the equation

$$\frac{\partial}{\partial x} \mathsf{G}(x,t) = \frac{t}{1 - 2xt + t^2} \mathsf{G}(x,t),$$

which is a first order differential equation.

Using the series expansions in powers of t, we can obtain the corresponding power series expansion for the left member, involving the polynomials. Then, by identifying the coefficients with the same degree we get the desired result. $\qquad\square$

4.5 Bessel Functions

As we have also mentioned, Bessel functions are particular cases of confluent hypergeometric functions and are given by

$$
{}_1F_1(v + 1/2, 2v + 1; 2iz) = \Gamma(1 + v)\, e^{iz} \left(\frac{z}{2}\right)^{-v} J_v(z),
$$

where $J_v(z)$ is the Bessel function of order v, which has the following representation as a Frobenius series:

$$
J_v(z) = \sum_{m=0}^{\infty} \frac{(-1)^m (z/2)^{v+2m}}{m!\,\Gamma(v + m + 1)} .
$$

The second linearly independent solution of the Bessel differential equation of order v is given by

$$
Y_v(z) = [\sin(\pi v)]^{-1}[J_v(z)\cos(\pi v) - J_{-v}(z)],
$$

called Bessel function of order v of the second kind.

Bessel functions are solutions of the linear ordinary differential equation

$$
z^2 \frac{d^2\omega}{dz^2} + z\frac{d\omega}{dz} + (z^2 - v^2)\omega = 0, \quad \omega = \omega(z),
$$

where v is the parameter denoting its order. Such functions belong to a wider class of functions called *cylindrical functions*, denoted by $C_v(z)$, which are defined from the following recurrence relations:

$$
C_{v-1}(z) + C_{v+1}(z) = \frac{2v}{z}C_v(z),
$$

$$
C_{v-1}(z) - C_{v+1}(z) = 2\frac{d}{dz}C_v(z).
$$

The first one is called *pure recurrence relation* because it does not involve the derivative.

An analogous treatment can be given to the Bessel function with an imaginary argument, called *modified Bessel function* and denoted by

$$I_\nu(z) = \exp\left(\frac{-i\pi}{2}\nu\right) J_\nu\left(z\, e^{i\frac{\pi}{2}}\right),$$

while for the second linearly independent solution we have

$$K_\nu(z) = \frac{\pi}{2}[\sin(\pi\nu)]^{-1}[I_{-\nu}(z) - I_\nu(z)].$$

We close this chapter presenting the so-called *generating function* for the Bessel functions, analogous to the Legendre polynomials. Let $z \in \mathbb{C}$ and $t \in \mathbb{R}$; the generating function for the Bessel function is given by

$$\exp\left[\frac{z}{2}\left(t - \frac{\alpha^2}{t}\right)\right] = \sum_{n=-\infty}^{\infty} \left(\frac{t}{\alpha}\right)^n J_n(\alpha z),$$

where α is a parameter.

Example 4.5 Let $n \in \mathbb{N}$. Using the recurrence relation for the Bessel function

$$J_{n-1}(x) = \frac{n}{x} J_n(x) + J_n'(x),$$

where the prime $'$ denotes de derivative with respect to x, show the result

$$\int_0^1 x^{n+1} J_n(\mu x)\, dx = \frac{J_{n+1}(\mu)}{\mu},$$

where μ is a nonzero constant and $n = 0, 1, 2, \ldots$

Multiplying the recurrence relation by x^n we have

$$x^n J_{n-1}(x) = nx^{n-1} J_n(x) + x^n J_n'(x) = \frac{d}{dx}\left[x^n J_n(x)\right].$$

Introducing the change of variable $x \to \mu x$ in this expression and integrating we have

$$\mu \int x^n J_{n-1}(\mu x)\, dx = x^n J_n(\mu x).$$

Taking the integration on the interval $[0, 1]$ and writing $n \to n+1$ we have

$$\int_0^1 x^{n+1} J_n(\mu x)\, dx = \frac{J_{n+1}(\mu)}{\mu}$$

which is the desired result. $\qquad\qquad\square$

4.6 Solved Exercises

SE 4.1 A particle P with variable mass m is at rest at a distance ℓ from the origin O, when it begins to be atracted toward the origin by a force proportional to the product mz, where z is the distance from P to the origin. The mass m of the particle decreases with time t according to the formula

$$m = (a + b\,t)^{-1},$$

where a and b are constants. Obtain the horary equation for this particle.
Solution: Using Newton's second law we may write the following linear ordinary differential equation:

$$F = -k^2 mz = \frac{\mathrm{d}p}{\mathrm{d}t} = \frac{\mathrm{d}}{\mathrm{d}t}\left(m\frac{\mathrm{d}z}{\mathrm{d}t}\right),$$

where k^2 is a proportionality constant.

Introducing the expression for the mass we obtain

$$\frac{\mathrm{d}^2}{\mathrm{d}u^2}z(u) - \frac{1}{u}\frac{\mathrm{d}}{\mathrm{d}u}z(u) + \lambda^2 z(u) = 0,$$

where we have introduced the parameter $\lambda = k/b$, together with a new independent variable $a + b\,t = u$.

We now introduce a change of dependent variable with the form

$$z(u) = uF(u),$$

whence we obtain the ordinary differential equation

$$F'' + \frac{1}{u}F' + \left(\lambda^2 - \frac{1}{u^2}\right)F = 0,$$

which is easily identified as a Bessel equation of order one, whose solution is given by

$$F = A\,J_1(\lambda u) + B\,Y_1(\lambda u),$$

where A and B are constants. Hence, the solution of the initial linear ordinary differential equation is given by

$$z(t) = (a + b\,t)\left\{A\,J_1\left[\lambda(a + b\,t)\right] + B\,Y_1\left[\lambda(a + b\,t)\right]\right\},$$

where the constants are determined by the conditions of the problem, namely $z(0) = \ell$, that is, the particle is initially at a distance ℓ from the origin, and $\dot{z}(0) = 0$, i.e., the particle is initially at rest, null initial velocity.

Imposing these conditions on the solution, we obtain the following linear system for A and B:

$$\begin{cases} A\,J_1'(\lambda a) + B\,Y_1'(\lambda a) = -\ell/\lambda a^2 \\ A J_1(\lambda a) + B\,Y_1(\lambda a) = \ell/a. \end{cases}$$

Employing the relation

$$C_1(x) + x\frac{\mathrm{d}}{\mathrm{d}x}C_1(x) = xC_0(x),$$

which is valid for Bessel functions $J_1(x)$ and $Y_1(x)$, we obtain the solution of the system, given by

$$A = \frac{\lambda L\pi}{2}Y_0(\lambda a) \quad \text{and} \quad B = -\frac{\lambda L\pi}{2}J_0(\lambda a).$$

From this we finally obtain the solution of the problem, namely

$$z(t) = \frac{\lambda L\pi}{2}(a + bt)\left\{Y_0(\lambda a)J_1[\lambda(a + bt)] - J_0(\lambda a)Y_1[\lambda(a + bt)]\right\}.$$

SE 4.2 Laplace's projective equation in spherical coordinates (r, θ, ϕ), after using the method of separation of variables, which we will present in Chap. 9, gives rise to the following three ordinary differential equations:

$$\frac{\mathrm{d}^2}{\mathrm{d}\phi^2}\Phi(\phi) + m^2\Phi(\phi) = 0;$$

$$\sin^2\theta\frac{\mathrm{d}^2}{\mathrm{d}\theta^2}\Theta(\theta) + \sin\theta\,\cos\theta\frac{\mathrm{d}}{\mathrm{d}\theta}\Theta(\theta) + \left[\ell(\ell+1) - \frac{m^2}{\sin^2\theta}\right]\Theta(\theta) = 0;$$

and

$$\frac{\mathrm{d}^2}{\mathrm{d}r^2}R(r) + \frac{2}{r}\frac{\mathrm{d}}{\mathrm{d}r}R(r) + \left[\frac{n(n+2)}{(1+r^2)^2} - \frac{\ell(\ell+1)}{r^2(1+r^2)}\right]R(r) = 0;$$

where $\ell = 0, 1, 2, \cdots, m = -\ell, -\ell+1, \cdots, l-1, l, n = 0, 1, 2, \cdots$ and $n \geq \ell$.

The equation involving the ϕ variable has already been discussed in the first chapter. Now, (a) solve the equation in variable θ, (b) solve the equation in the r variable, considering only one polynomial solution, and (c) discuss the polynomial case when $\ell = 0$.

Solution:

(a) Consider the equation in θ; introducing the change of independent variable

$$\cos\theta = x,$$

we obtain the equation

$$(1 - x^2)\frac{d^2}{dx^2}\Theta(x) - 2x\frac{d}{dx}\Theta(x) + \left[\ell(\ell+1) - \frac{m^2}{1-x^2}\right]\Theta(x) = 0$$

which is the Legendre associated differential equation, whose general solution, written in terms of θ, is given by

$$\Theta(\theta) = A\, P_\ell^m(\cos\theta) + B\, Q_\ell^m(\cos\theta),$$

where A and B are constants and $P_\ell^m(x)$ and $Q_\ell^m(x)$ are the associated Legendre functions of the first and second kinds, respectively.

(b) Here we introduce, into the differential equation involving the r variable, the following change of dependent variable:

$$R(r) = \left(\frac{r^2}{1+r^2}\right)^{\ell/2} F(r).$$

With this change we obtain the differential equation

$$\frac{d^2}{dr^2}F(r) + 2\frac{\ell+1+r^2}{r(1+r^2)}\frac{d}{dr}F(r) + \left[\frac{n(n+2)-\ell(\ell+2)}{(1+r^2)^2}\right]F(r) = 0.$$

Now, a change of independent variable $1 + r^2 = \frac{1}{x^2}$ leads to the differential equation

$$(1 - x^2)\frac{d^2}{dx^2}F(x) - (2\ell+3)x\frac{d}{dx}F(x) + [n(n+2)-\ell(\ell+2)]\,F(x) = 0,$$

which is a Gegenbauer [1849 – Leopold Bernhard Gegenbauer – 1903] equation. As we are only interested in its polynomial solution, we obtain

$$F(x) = A\, C_{n-\ell}^{\ell+1}(x),$$

where A is a constant and $C_\mu^\nu(x)$ are the Gegenbauer polynomials. The solution of the differential equation in the r variable is then given by

$$R(r) = A\,\left(\frac{r^2}{1+r^2}\right)^{\ell/2} C_{n-\ell}^{\ell+1}\left(\frac{1}{\sqrt{1+r^2}}\right).$$

(c) For the polynomial case, when $\ell = 0$ we get

$$R(r) = A\, C_n^1\left(\frac{1}{\sqrt{1+r^2}}\right) = A\, U_n\left(\frac{1}{\sqrt{1+r^2}}\right),$$

where $U_n(x)$ are the Chebyshev [1821 – Pafnuty Lvovich Chebyshev – 1894] polynomials of the second kind.

SE 4.3 Let $a > 0$ and $|z| < 1$. Use the integral representation of the hypergeometric function given in **PE 4.41**,

$$
{}_2F_1(a, b; c; z) = \frac{\Gamma(c)}{\Gamma(c - b)\Gamma(b)} \int_0^1 t^{b-1}(1 - t)^{c-b-1}(1 - z t)^{-a}\, dt
$$

with $\mathrm{Re}(c) > \mathrm{Re}(b) > 0$, to prove the following relation:

$$
{}_2F_1(a, b; c; z) = (1 - z)^{-a}\, {}_2F_1\left(a, c - b; c; \frac{z}{z - 1}\right).
$$

Solution: Introducing the change of variable

$$
t = 1 - \xi
$$

into the integral representation we obtain

$$
\begin{aligned}
{}_2F_1(a, b; c; z) &= \frac{\Gamma(c)}{\Gamma(c - b)\Gamma(b)} \int_0^1 \xi^{c-b-1}(1 - \xi)^{b-1}(1 - z + z\xi)^{-a}\, d\xi \\
&= \frac{\Gamma(c)(1 - z)^{-a}}{\Gamma(c - b)\Gamma(b)} \int_0^1 \xi^{c-b-1}(1 - \xi)^{b-1}\left(1 - \frac{z}{z - 1}\xi\right)^{-a}\, d\xi .
\end{aligned}
$$

Comparing this expression with the integral representation aforementioned we may write

$$
{}_2F_1(a, b; c; z) = (1 - z)^{-a}\, {}_2F_1\left(a, c - b; c; \frac{z}{z - 1}\right).
$$

SE 4.4 Let $-1 < x < 1$ and $m, n \in \mathbb{N}$ with $m \neq n$. Show that

$$
\int_{-1}^1 P_n(x) P_m(x)\, dx = 0,
$$

where $P_n(\cdot)$ is the Legendre polynomial. This relation is known as orthogonality relation.

Solution: As we know, the Legendre polynomial $P_n(x)$ satisfies a linear ordinary differential equation, known as the Legendre differential equation, which can be written in the form

$$
\frac{d}{dx}\left[(1 - x^2)\frac{d}{dx}P_n(x)\right] + n(n + 1)P_n(x) = 0 .
$$

We are interested in obtaining an integral involving the product of two Legendre polynomials. Then, we write an analog to the above ordinary differential equation for $P_m(x)$, multiply the first equation by $P_m(x)$ and the second one by $P_n(x)$. Subtracting the expressions from each other and rearranging, we can write the result in the following appropriate way

$$(n + m + 1)(n - m)P_n(x)P_m(x)$$
$$\frac{\mathrm{d}}{\mathrm{d}x}\left\{(1 - x^2)\left[P_n(x)\frac{\mathrm{d}}{\mathrm{d}x}P_m(x) - P_m(x)\frac{\mathrm{d}}{\mathrm{d}x}P_n(x)\right]\right\} .$$

Integrating this equation with respect to x on $-1 \le x \le 1$ we obtain

$$(n + m + 1)(n - m)\int_{-1}^{1} P_n(x)P_m(x)\,\mathrm{d}x$$
$$= \left\{(1 - x^2)\left[P_n(x)\frac{\mathrm{d}}{\mathrm{d}x}P_m(x) - P_m(x)\frac{\mathrm{d}}{\mathrm{d}x}P_n(x)\right]\right\}\Bigg|_{-1}^{1},$$

which can be rewritten in the form

$$(n + m + 1)(n - m)\int_{-1}^{1} P_n(x)P_m(x)\,\mathrm{d}x = 0$$

because $(1 - x^2) = 0$ at $x = \pm 1$. Then, since m and n are nonnegative integers, we have $m + n + 1 \ne 0$. Also, by hypothesis, $n \ne m$, and we conclude that

$$\int_{-1}^{1} P_n(x)P_m(x)\,\mathrm{d}x = 0, \qquad n \ne m .$$

This expression says that the set of Legendre polynomials is orthogonal with respect to the unity weight function on the interval $-1 < x < 1$.

SE 4.5 The generating function of Bessel functions may be used, for example, to prove the so-called Schläfli [1814 – Ludwig Schläfli – 1895] integral representation,

$$J_n(z) = \frac{1}{2\pi i}\oint_C \frac{\mathrm{d}t}{t^{n+1}}\exp\left[\frac{z}{2}\left(t - \frac{1}{t}\right)\right],$$

where contour C circles the origin in counterclockwise sense. From the integral representation aforementioned, prove the so-called *Bessel integral*:

$$J_n(z) = \frac{1}{\pi}\int_0^{\pi} \cos(n\theta - z\sin\theta)\mathrm{d}\theta .$$

Solution: Introducing the change of variable $t = \exp(i\theta)$, the parametrization of a circumference with unit radius centered at the origin, we can write

$$J_n(z) = \frac{1}{2\pi i} \int_0^{2\pi} \frac{i\,e^{i\theta}\,d\theta}{(e^{i\theta})^{n+1}} \exp\left[\frac{z}{2}\left(e^{i\theta} - e^{-i\theta}\right)\right].$$

Now, using Euler's relation [3] and simplifying terms we obtain

$$J_n(z) = \frac{1}{2\pi} \int_0^{2\pi} e^{-in\theta}\, e^{iz\sin\theta}\, d\theta\ .$$

This expression can be rewritten as

$$J_n(z) = \frac{1}{2\pi} \int_0^{2\pi} [\cos(z\sin\theta - n\theta) + i\sin(z\sin\theta - n\theta)]d\theta\ .$$

Let us introduce another change of variable, $\theta = x + \pi$. Then we may write

$$J_n(z) = \frac{(-1)^n}{2\pi} \int_{-\pi}^{\pi} [\cos(z\sin x + nx) - i\sin(z\sin x + nx)]dx\ .$$

As the sine function is an odd function, its integral over a symmetric interval is null, and we then get

$$J_n(z) = \frac{(-1)^n}{2\pi} \int_0^{\pi} \cos(z\sin x + nx)dx\ .$$

Finally, a new change of variable given by $x = -\theta + \pi$ brings us to the integral

$$J_n(z) = \int_0^{\pi} \cos(z\sin\theta - n\theta)d\theta\ ,$$

which is the desired result.

SE 4.6 Calculate, using the hypergeometric function, the integral

$$\int_0^{\mu} \frac{dx}{\sqrt{1 - x^8}}\ ,$$

where $\mu > 0$.

Solution: Introducing the change of variable $x^8 = u$ we can rewrite the integral as

$$\int_0^{\mu} \frac{dx}{\sqrt{1 - x^8}} = \frac{1}{8} \int_0^{\mu^8} du\, u^{-7/8}(1 - u)^{-1/2}\ .$$

Now, calling $u = \mu^8 t$ we obtain

$$\int_0^\mu \frac{dx}{\sqrt{1-x^8}} = \frac{1}{8}\mu^{1/8}\int_0^1 dt\, t^{-7/8}(1-\mu^8 t)^{-1/2}\,,$$

which can be identified as a hypergeometric function, i.e.,

$$\int_0^\mu \frac{dx}{\sqrt{1-x^8}} = \frac{\mu}{8}\frac{\Gamma(9/8-1/8)\Gamma(1/8)}{\Gamma(9/8)}\,{}_2F_1\left(\frac{1}{8},\frac{9}{8};\frac{1}{2};\mu^8\right)$$

or finally, after simplifications,

$$\int_0^\mu \frac{dx}{\sqrt{1-x^8}} = \mu\,{}_2F_1\left(\frac{1}{8},\frac{9}{8};\frac{1}{2};\mu^8\right)\,,$$

which is the desired result.

SE 4.7 Find a solution for the linear ordinary differential equation

$$x^2(x^2-1)\frac{d^2}{dx^2}y(x) + x(x^2-1)\frac{d}{dx}y(x) + \frac{1}{16}y(x) = 0$$

and express it in terms of a hypergeometric function.
Solution: Introducing the change of variable $x^2 = t$, we may write

$$t^2(t-1)\frac{d^2}{dt^2}y(t) + t(t-1)\frac{d}{dt}y(t) + \frac{1}{64}y(t) = 0\,.$$

Now, making the change of dependent variable

$$y(t) = t^{1/8}H(x),$$

we arrive at the ordinary differential equation

$$t(1-t)\frac{d^2}{dt^2}H(x) + \frac{5}{4}(1-t)\frac{d}{dt}H(t) - \frac{1}{64}H(t) = 0\,,$$

which is a hypergeometric equation, with a solution given by

$$H(t) = {}_2F_1\left(\frac{1}{8},\frac{1}{8};\frac{5}{4};t\right)\,,$$

from which we obtain, reintroducing the original variables,

$$y(x) = A\,x^{1/8}\,{}_2F_1\left(\frac{1}{8},\frac{1}{8};\frac{5}{4};x^2\right)\,,$$

where A is an arbitrary real constant.

SE 4.8 Show that

$$
\mathbf{P} \left\{ \begin{array}{ccc} 0 & \infty & 1 \\ \alpha & \beta & \gamma \\ \alpha' & \beta' & \gamma' \end{array} \; z \right\} = \mathbf{P} \left\{ \begin{array}{ccc} 1 & 0 & \infty \\ \alpha & \beta & \gamma \\ \alpha' & \beta' & \gamma' \end{array} \; \frac{1}{1-z} \right\} .
$$

Solution: The ordinary differential equation for the first scheme is given by

$$
\frac{\mathrm{d}^2 y}{\mathrm{d}z^2} + \left(\frac{1 - \alpha - \alpha'}{z} + \frac{1 - \gamma - \gamma'}{z - 1} \right) \frac{\mathrm{d}y}{\mathrm{d}z} + \left(-\frac{\alpha\alpha'}{z} + \beta\beta' + \frac{\gamma\gamma'}{z - 1} \right) \frac{y}{z(z - 1)} = 0.
$$

Then, inverting the expression $t = (1 - z)^{-1}$ we obtain $z = (t - 1)t^{-1}$, whence we find for the parameters

$$
\begin{aligned}
z \to 0 \quad t &= 1 \implies 1 \text{ with exponents } \alpha \text{ and } \alpha' \\
z \to 1 \quad t &= \infty \implies \infty \text{ with exponents } \gamma \text{ and } \gamma' \\
z \to \infty \quad t &= 0 \implies 0 \text{ with exponents } \beta \text{ and } \beta',
\end{aligned}
$$

which is the desired result.

4.7 Proposed Exercises

PE 4.1 Show that

$$
\int_0^\infty e^{-x} \frac{\mathrm{d}x}{\sqrt{x}} = \sqrt{\pi} .
$$

PE 4.2 Show that

$$
\int_{-\infty}^\infty e^{-x^2} \, \mathrm{d}x = \sqrt{\pi} .
$$

PE 4.3 Use the definition of the beta function to calculate

$$
B \left(\frac{1}{2}, \frac{1}{2} \right) .
$$

PE 4.4 Let $a, b \in \mathbb{R}_+$. Evaluate the integral

$$
\int_{-1}^0 {}_2F_1(-a, b; b; -x),
$$

where ${}_2F_1(-a, b; b; -x)$ is a hypergeometric function.

PE 4.5 Let $a, b, c \in \mathbb{R}$ be arbitrary parameters with $c \neq 0, -1, -2, \ldots$ What must be the relations between parameters a, b and c so that $_2F_1(a, b; c; 1)$ is defined?

PE 4.6 Let $|x| < 1$. Evaluate $_2F_1(1, 1; 2; -x)$.

PE 4.7 Using the duplication formula for the gamma function, evaluate $\Gamma(3/2)$.

PE 4.8 Let $z \in \mathbb{C}$. The so-called Mittag–Leffler [1846 – Magnus Gustaf (Gösta) Mittag-Leffler – 1927] function, which can be considered a generalization of the exponential function, is given by

$$E_\alpha(z) = \sum_{k=0}^{\infty} \frac{z^k}{\Gamma(\alpha k + 1)},$$

where α is a complex parameter with $\mathrm{Re}(\alpha) > 0$. Let $z = x \in \mathbb{R}$. Evaluate (a) $E_1(x)$ and (b) $E_2(x^2)$.

PE 4.9 Let $x \in \mathbb{R}$, $\alpha > 0$ and $\beta > 0$. The Mittag-Leffler function with two parameters is defined by the series

$$E_{\alpha,\beta}(x) = \sum_{k=0}^{\infty} \frac{x^k}{\Gamma(\alpha k + \beta)}.$$

Show that $E_{1,2}(x) = {}_1F_1(1; 2; x)$; this is a relation between the Mittag–Leffler function with two parameters and the confluent hypergeometric function.

PE 4.10 Consider the case of a hypergeometric differential equation, Eq. (4.7), for which none of the numbers c, $a - b$, $c - a - b$ is an integer. Knowing that one of its solutions is given by Eq. (4.8), show that

$$\omega_1(z) = {}_2F_1(a, b; a + b - c + 1; 1 - z)$$

and

$$\omega_2(z) = (1 - z)^{c-a-b} {}_2F_1(c - b, c - a; c - a - b + 1; 1 - z)$$

are also two linearly independent solutions of the same differential equation.

PE 4.11 Do as in the preceding exercise and suppose that $a = -n$ (integer) in order to show that

$$\omega(z) = z^\lambda (1 - z)^\mu P_n(z),$$

where λ and μ are parameters and $P_n(z)$ is a polynomial of degree n in z, and that in this case the series is finite.

PE 4.12 Show that (quadratic transformation)

$$_2F_1(a, b; a + b + 1/2; \sin^2 \theta) = {}_2F_1(2a, 2b; a + b + 1/2; \sin^2 \theta/2).$$

PE 4.13 Show that

$$(a) \quad \ln(1 \pm z) = \pm z \, _2F_1(1, 1; 2; \mp z),$$

$$(b) \quad (1 + z)^a = {}_2F_1(-a, b; b; -z).$$

PE 4.14 Show that, formally,

$$(a) \quad \frac{d}{dz} \, _1F_1(a, c; z) = \frac{a}{c} \, _1F_1(a + 1, c + 1; z),$$

$$(b) \quad \frac{d}{dz} U(a; c; z) = -aU(a + 1; c + 1; z),$$

where

$$U(a; c; z) = \frac{\pi}{\sin \pi c} \left[\frac{_1F_1(a; c; z)}{\Gamma(c)\Gamma(1 + a - c)} - z^{1-c} \frac{_1F_1(a + 1 - c; 2 - c; z)}{\Gamma(a)\Gamma(2 - c)} \right]$$

is also a solution of the confluent hypergeometric equation.

PE 4.15 Show that

$$(a) \; _1F_1 \left(v + \frac{1}{2}; 2v + \frac{1}{2}; 2z \right) = \Gamma(1 + v) \, e^z \left(\frac{z}{2} \right)^{-v} I_v(z),$$

$$(b) \; U \left(v + \frac{1}{2}; 2v + \frac{1}{2}; 2z \right) = \pi^{-1/2} \, e^z (2z)^{-v} K_v(z).$$

PE 4.16 Given that the Whittaker function is defined as

$$M_{x;\mu}(z) = e^{-z/2} z^{\mu+1/2} \, _1F_1(\frac{1}{2} + \mu - x; 1 + 2\mu; z),$$

find the ordinary differential equation satisfied by $M_{x;\mu}(z)$.

PE 4.17 As in the previous problem, find the linear ordinary differential equations satisfied by the following functions:

(a) Laguerre [1834 – Edmond Nicolas Laguerre – 1886]

$$L_n^\alpha(z) = \frac{(\alpha + 1)n}{n!} \, _1F_1(-n; \alpha + 1; z);$$

(b) Hermite

$$H_n(z) = 2^n U\left(\frac{-n}{2}; \frac{1}{2}; z^2\right);$$

(c) Weber [1842 – Heinrich Weber – 1913]

$$D_\nu(z) = 2^{\nu/2}\, e^{-z^2/4}\, U\left(\frac{-\nu}{2}; \frac{1}{2}; \frac{z^2}{2}\right).$$

PE 4.18 Let $P_\nu^\mu(x)$ be an associated Legendre function of the first kind, ν and μ real paremeters. Show that

(a) $P_{-\nu-1}^\mu(x) = P_\nu^\mu(x)$;

(b) $P_\nu^{-m}(x) = \dfrac{\Gamma(1+\nu-m)}{\Gamma(1+\nu+m)}(-1)^m P_\nu^m(x), \quad m = 1, 2, 3, \ldots$;

(c) $W[P_\nu^\mu(x); Q_\nu^\mu(x)] = \dfrac{\Gamma(1+\nu+m)}{\Gamma(1+\nu-m)}(1-x^2)^{-1}$;

where W is the Wronskian and $Q_\nu^\mu(\cdot)$ is the associated Legendre function of second kind.

PE 4.19 Gegenbauer functions are the solutions of the Gegenbauer equation

$$(z^2 - 1)\frac{d^2\omega}{dz^2} + (2\nu + 1)z\frac{d\omega}{dz} - \alpha(\alpha + 2\nu)\omega = 0,$$

and can be written in terms of the hypergeometric function as

$$C_\alpha^\nu(z) = \Gamma(\alpha + 2\nu)[\Gamma(\alpha+1)\Gamma(2\nu)]^{-1}\, {}_2F_1\left(\alpha + 2\nu, -\alpha; \nu + \frac{1}{2}; \frac{1-z}{2}\right).$$

Express $C_\alpha^\nu(z)$ in terms of the Legendre function of the first kind $\mathscr{B}_\mu^\nu(z)$.

PE 4.20 Let $\alpha, \beta \in \mathbb{R}$ and $n \in \mathbb{N}$. Jacobi [1805 – Carl Gustav Jacob Jacobi – 1851] polynomials $P_n^{(\alpha,\beta)}(x)$ are related to the hypergeometric function by

$$P_n^{(\alpha,\beta)}(x) = \frac{\Gamma(n+\alpha+1)}{n!\Gamma(\alpha+1)}\, {}_2F_1\left(-n, \alpha + \beta + n + 1; \alpha + 1; \frac{1-x}{2}\right).$$

Find the ordinary differential equation for $P_n^{(\alpha,\beta)}(x)$. What are the resulting equations when $\alpha = \beta = 0$ and when $\alpha = \beta = \lambda - 1/2$, with $\lambda > 1/2$?

PE 4.21 Let $n \in \mathbb{N}$. Show the Rodrigues [1795 – Benjamin-Olinde Rodrigues – 1851] formula,

$$P_n(x) = \frac{1}{2^n n!} \frac{d^n}{dx^n} [(x^2 - 1)^n],$$

where $P_n(x)$ are Legendre polynomials.

PE 4.22 Show that the generating function for the Legendre polynomials is

$$(1 - 2xz + z^2)^{-1/2} = \sum_{n=0}^{\infty} P_n(x) z^n,$$

where $-1 < x < 1$ and $|z| < 1$ and $P_n(x)$ are Legendre polynomials.

PE 4.23 Let $n = 1, 2, 3, \ldots$ Prove the recurrence relations

(a) $(n + 1) P_{n+1}(x) = (2n + 1) x P_n(x) - n P_{n-1}(x)$;

(b) $(n + 1) P_n(x) = P'_{n+1}(x) - x P'_n(x)$;

where the prime denotes differentiation with respect to x.

PE 4.24 Let $n \in \mathbb{N}$. Show that

(a) $T_n(x) = {}_2F_1\left(-n, n; \frac{1}{2}; \frac{1-x}{2}\right)$;

(b) $T_n(x) = U_n(x) - x U_{n-1}(x)$;

where $U_n(x) = (n + 1) {}_2F_1\left(-n, n + 2; \frac{3}{2}; \frac{1-x}{2}\right)$ and $T_n(x)$ and $U_n(x)$ are Chebyshev (Tchebichef) polynomials of the first kind and second kind, respectively.

PE 4.25 Let $v \in \mathbb{R}$. Denote by $I_v(x)$ the modified Bessel function of order v.

(a) Show that

$$I_{v-1}(z) - I_{v+1}(z) = 2\frac{v}{z} I_v(z)$$

and

$$I_{v-1}(z) + I_{v+1}(z) = 2\frac{d}{dz} I_v(z).$$

(b) Obtain the second-order ordinary differential equation for $I_v(z)$.

PE 4.26 Let $J_v(\cdot)$, $I_v(\cdot)$ and $K_v(\cdot)$ be Bessel functions. Show the following Wronskians:

$$\text{(a) } W[J_\nu(z), J_{-\nu}(z)] = -\frac{2}{\pi z} \sin \pi \nu, \ \nu \neq \text{integer};$$

$$\text{(b) } W[I_\nu(z), K_\nu(z)] = -\frac{1}{z}.$$

PE 4.27 Show the following special relations:

$$\text{(a) } J_{1/2}(z) = Y_{-1/2}(z) = \left(\frac{\pi z}{2}\right)^{-1/2} \sin z;$$

$$\text{(b) } Y_{1/2}(z) = -J_{-1/2}(z) = -\left(\frac{\pi z}{2}\right)^{-1/2} \cos z;$$

$$\text{(c) } K_{1/2}(z) = K_{-1/2}(z) = \left(\frac{\pi}{2z}\right)^{1/2} e^{-z};$$

$$\text{(d) } e^{iz\cos\alpha} = \sum_{n=-\infty}^{\infty} i^n e^{in\alpha} J_n(z);$$

with $\alpha \in \mathbb{R}$.

PE 4.28 The so-called *beta function* is defined by

$$B(m, n) = \int_0^1 (1 - x)^{m-1} x^{n-1} dx = \frac{\Gamma(m)\Gamma(n)}{\Gamma(m + n)},$$

where

$$\Gamma(m) = \int_0^\infty e^{-t} t^{m-1} dt$$

is the *gamma function*, with m and n real numbers different from $0, -1, -2\ldots$, which generalizes the concept of factorial.

Show that, if $0 < m < 1$,

$$\Gamma(m)\Gamma(1 - m) = \frac{\pi}{\sin \pi m}.$$

PE 4.29 Show that, for $0 < m < 1$,

$$\Gamma(m)\Gamma(1 - m) = \int_0^\infty \frac{u^{m-1}}{1 + u} du.$$

PE 4.30 Show that

$$\Gamma(z) = \int_0^1 \left(\ln \frac{1}{x}\right)^{z-1} dx.$$

PE 4.31 The Pochhammer symbol $(a)_k$ is defined as

$$(a)_k = a(a+1)\ldots(a+k-1),$$

for k integer, with $(a)_0 = 1$. Express $(a)_k$ in terms of factorials.

PE 4.32 Let $z > t$. Using the beta function, show that, for $0 < \alpha < 1$,

$$\int_t^z \frac{dx}{(z-x)^{1-\alpha}(x-t)^{\alpha}} = \frac{\pi}{\sin \pi \alpha}.$$

PE 4.33 Let $p, q \in \mathbb{R}^*$. Show that

$$B(p+1, q+1) = 2 \int_0^{\pi/2} \cos^{2p+1} \theta \, \sin^{2q+1} \theta \, d\theta .$$

PE 4.34 Let $\mathrm{Re}(s) > 0$. Show that

$$\sqrt{\pi}\, \Gamma(2s) = 2^{2s-1} \Gamma(s)\Gamma(s+1/2).$$

This is the so-called *duplication formula* for the gamma function.

PE 4.35 Show that

$$\int_0^{\pi/2} \sqrt{\cos \theta} \, d\theta = \frac{(2\pi)^{3/2}}{16\left[\left(\frac{1}{4}\right)!\right]^2},$$

where $\left(\frac{1}{4}\right)! = \Gamma(5/4)$.

PE 4.36 We define the *double factorial* for a positive integer k as

$$(2k)!! = 2k(2k-2)\ldots 6.4.2 \,,$$

$$(2k+1)!! = (2k+1)(2k-1)\ldots 5.3.1 \,.$$

Show that $(2n)!! = 2^n n!$.

PE 4.37 Show that $(2n+1)!! = (2n+1)!/(2^n n!)$.

PE 4.38 Show that

$$\int_{-1}^1 (1-x^2)^{1/2} x^{2n} dx = \begin{cases} \dfrac{\pi/2} & \text{if } n = 0; \\[2mm] \pi \dfrac{(2n-1)!!}{(2n+2)!!} & \text{if } n = 1, 2, \ldots \end{cases}$$

PE 4.39 Show that

$$\frac{1}{\pi} \int_{-1}^{1} (1 - x^2)^{-1/2} x^{2n} \, dx = \begin{cases} 1 & \text{if } n = 0; \\ \dfrac{(2n - 1)!!}{(2n)!!} & \text{if } n = 1, 2, \ldots \end{cases}$$

PE 4.40 The so-called associated Legendre polynomials can be given by

$$P_m^m(x) = (2m - 1)!!(1 - x^2)^{m/2},$$

where $m = 0, 1, 2, \ldots$ and, by definition, $(-1)!! = 1$. Show that

$$\int_{-1}^{1} [P_m^m(x)]^2 \, dx = \frac{2}{2m + 1}(2m)!, \quad m = 0, \ 1 \ldots$$

PE 4.41 Let $a \in \mathbb{R}$. Use the beta function to show that

$$_2F_1(a, b; c; x) = \frac{1}{B(b, c - b)} \int_0^1 (1 - t)^{c-b-1} t^{b-1} (1 - tx)^{-a} \, dt,$$

for $|x| < 1$ and $c > b > 0$.

PE 4.42 Let $a, b, c \in \mathbb{R}$. Show that

$$\text{(a)} \quad _2F_1(a, b; c; z) = (1 - z)^{-a} \, _2F_1\left(a, c - b; c; \frac{z}{z - 1}\right);$$

$$\text{(b)} \quad _2F_1(a, b; c; z) = (1 - z)^{c-a-b} \, _2F_1(c - a, c - b; c; z).$$

PE 4.43 Using the integral representation shown in **PE 4.41**, calculate $_2F_1(a, b; c; 1)$.

PE 4.44 Let $a, b \in \mathbb{R}$ with $1 + b - a$ a nonnegative integer. Show that

$$_2F_1(a, b; b - a - 1; -1) = \frac{\Gamma(1 + b - a)\Gamma(1 + \frac{b}{2})}{\Gamma(1 + b)\Gamma(1 + \frac{b}{2} - a)}.$$

PE 4.45 Show that

$$\int_0^1 \frac{x^{m-1}(1 - x)^{n-1}}{(x + r)^{m+n}} \, dx = \frac{B(m, n)}{r^n(1 + r)^m},$$

where $m, n \in \mathbb{N}$ and r is a real positive constant.

PE 4.46 Using the results of **PE 4.41**, find the integral representation of the confluent hypergeometric function,

$$_1F_1(a; c; x) = \frac{\Gamma(c)}{\Gamma(a)\Gamma(c - a)} \int_0^1 (1 - t)^{c-a-1} t^{a-1} e^{tx} \, dt,$$

for $\mathrm{Re}(c) > \mathrm{Re}(a) > 0$.

PE 4.47 Show that

$$\text{(a)} \quad _1F_1(a; c; z) = e^z \, _1F_1(c - a; c; -z);$$

$$\text{(b)} \quad U(a; c; z) = z^{1-c} U(a + 1 - c; 2 - c, z);$$

using the integral representation

$$U(a; c; x) = \frac{1}{\Gamma(a)} \int_0^\infty e^{-xt} \, t^{a-1} (1 + t)^{c-a-1} dt,$$

where $\mathrm{Re}(a) > 0$, $\mathrm{Re}(x) > 0$ and $c \neq 1, 2, 3, \ldots$.

PE 4.48 Let $a, c \in \mathbb{R}$. For $|s| > 1$, show that

$$\int_0^\infty e^{-st} \, _1F_1(a; c; t) dt = \frac{1}{s} \, _2F_1\left(a, 1; c; \frac{1}{s}\right),$$

which is interpreted as the Laplace transform of a confluent hypergeometric function.

PE 4.49 (a) Using the series representation of the Bessel function of order μ and **PE 4.33**, show that

$$J_\mu(z) = \frac{2}{\sqrt{\pi}\,\Gamma(\mu + 1/2)} \left(\frac{z}{2}\right)^\mu \int_0^{\pi/2} \sin^{2\mu}\theta \cos(z \cos \theta) d\theta,$$

for $\mathrm{Re}(\mu) > -1/2$. (b) Find an integral representation for $J_0(z)$.

PE 4.50 Consider the classical Legendre differential equation,

$$(1 - x^2)\frac{d^2}{dx^2} y(x) - 2x \frac{d}{dx} y(x) + n(n + 1)y(x) = 0,$$

with $n = 0, 1, 2 \ldots$, which has singularities at $x = \pm 1$ and $x = \infty$ and a solution that can be defined by the following scheme:

$$y(x) = P \left\{ \begin{array}{ccc} -1 & \infty & 1 \\ 0 & n+1 & 0 \ x \\ 0 & -n & 0 \end{array} \right\}.$$

Show that the scheme

$$y(x) = P \left\{ \begin{matrix} 0 & \infty & 1 \\ 0 & n+1 & 0 \\ 0 & -n & 0 \end{matrix} \; \tfrac{1}{2}(1-x) \right\}$$

is equivalent to the previous one.

References

1. ⋆ M. Abramowitz, I.A. Stegun,[1] *Handbook of Mathematical Functions* (Dover Publications, Inc., New York, 1972)
2. E. Capelas de Oliveira, *Funções Especiais com Aplicações* (Special Functions with Applications), Editora (Livraria da Física, São Paulo, 2005)
3. E. Capelas de Oliveira, W.A. Rodrigues Jr., *Funções Analíticas com Aplicações* (Analytic Functions with Applications), Editora (Livraria da Física, São Paulo, 2006)
4. A. Erdelyi, W. Magnus, F. Oberhettinger, F.G. Tricomi *Higher Transcendental Functions*, vols. I, II, and III (McGraw-Hill, New York, 1953)
5. ⋆ I.S. Gradshteyn, I.M. Ryzhik, *Table of Integrals, Series and Products* (Academic Press, New York, 1980)
6. ⋆ Z.X. Wang, D.R. Guo, *Special Functions* (World Scientific, Singapore, 1989)

---

[1] A ⋆ indicates a table.

Chapter 5
Fourier, Fourier-Bessel, and Fourier–Legendre Series

Don't worry about your difficulties in mathematics; I can assure you that mine are still greater.

1879 – Albert Einstein – 1955

There exist discontinuous functions that cannot be represented by power series (cf. Chap. 3). However, powers are not the unique type of known functions: there exist also, e.g., the trigonometric functions and hyperbolic functions, studied in basic mathematics, and the special functions, some of which were discussed in the preceding chapter. These functions can also be used to express other functions, as in the case of powers. Among the series thus generated we have the so-called Fourier [1768 – Jean Baptiste Joseph Fourier – 1830] series, Fourier–Bessel series, and Fourier–Legendre series, which are the subject of this chapter.

5.1 Fourier Series

Let $x \in \mathbb{R}$. Consider a periodic function $f(x)$ defined on a closed interval $[a, b]$. The Fourier series expansion for $f(x)$ is given by

$$\frac{f(x^{+}) + f(x^{-})}{2} = \frac{a_0}{2} + \sum_{k=1}^{\infty} a_k \cos\left\{\frac{k\pi(2x - b - a)}{(b - a)}\right\} +$$

$$+ \sum_{k=1}^{\infty} b_k \sin\left\{\frac{k\pi(2x - b - a)}{(b - a)}\right\},$$

where the coefficients a_k and b_k are respectively given by

$$a_k = \frac{2}{b - a} \int_a^b f(x) \cos\left\{\frac{k\pi(2x - b - a)}{(b - a)}\right\} dx, \qquad k = 0, 1, 2\ldots$$

© The Author(s), under exclusive license to Springer Nature Switzerland AG 2025
E. Capelas de Oliveira, J. E. Maiorino, *Analytical Methods in Applied Mathematics*,
Problem Books in Mathematics, https://doi.org/10.1007/978-3-031-74794-6_5

and

$$b_k = \frac{2}{b-a} \int_a^b f(x) \sin\left\{\frac{k\pi(2x-b-a)}{(b-a)}\right\} dx, \qquad k = 1, 2, 3\ldots$$

Due to their importance, for instance, when we solve a linear ordinary differential equation, we present two theorems involving the term-by-term differentiation and integration of a Fourier series. For the sake of simplicity, we consider the symmetric interval $[-\pi, \pi]$ instead of $[a, b]$.

Theorem 5.1 (Differentiation) *Let $f(x)$ be a continuous function on the interval $[-\pi, \pi]$, with $f(-\pi) = f(\pi)$ and assume that $f'(x)$ is smooth by parts on this interval. Then, the Fourier series for $f'(x)$ can be obtained by term-by-term differentiation of the Fourier series for $f(x)$ and the corresponding differentiated series converges pointwise to $f'(x)$.*

Example 5.1 In this example, we discuss the differentiation of the Fourier series, as in Theorem 5.1. Let $f(x)$ be a continuous function on $[-\ell, \ell]$ such that $f(\ell) = f(-\ell)$ and suppose that $f'(x)$ is continuous by parts and has lateral derivatives on this interval; then the Fourier series for $f(x)$ is differentiable. Show that

$$f'(x) = \sum_{k=1}^{\infty} \frac{k\pi}{\ell}\left[-a_k \sin\left(\frac{k\pi x}{\ell}\right) + b_k \cos\left(\frac{k\pi x}{\ell}\right)\right].$$

Consider the Fourier series for f and f',

$$f(x) = \frac{a_0}{2} + \sum_{k=1}^{\infty}\left[a_k \cos\left(\frac{k\pi x}{\ell}\right) + b_k \sin\left(\frac{k\pi x}{\ell}\right)\right]$$

and

$$f'(x) = \frac{A_0}{2} + \sum_{k=1}^{\infty}\left[A_k \cos\left(\frac{k\pi x}{\ell}\right) + B_k \sin\left(\frac{k\pi x}{\ell}\right)\right],$$

where the coefficients for the Fourier series associated with $f(x)$ are known. On the other hand, the coefficients for the Fourier series associated with $f'(x)$ will be given by the following expressions:

$$A_0 = \frac{1}{\ell}\int_{-\ell}^{\ell} f'(x)dx;$$

$$A_k = \frac{1}{\ell}\int_{-\ell}^{\ell} f'(x)\cos\left(\frac{k\pi x}{\ell}\right)dx;$$

$$B_k = \frac{1}{\ell}\int_{-\ell}^{\ell} f'(x)\sin\left(\frac{k\pi x}{\ell}\right)dx.$$

Thus, we can determine the relation between coefficients A_0, A_k and B_k and the coefficients a_0, a_k and b_k. Integrating by parts A_k, with $k = 0, 1, 2, \ldots$, and choosing $u = \cos\left(\dfrac{k\pi x}{\ell}\right)$, we have

$$A_k = \frac{1}{\ell}\left\{ f(x)\cos\left(\frac{k\pi x}{\ell}\right)\Big|_{-\ell}^{\ell} + \frac{k\pi}{\ell}\int_{-\ell}^{\ell} f(x)\sin\left(\frac{k\pi x}{\ell}\right)dx \right\}$$

$$= \frac{1}{\ell}\left\{ f(\ell)\cos(k\pi) - f(-\ell)\cos(-k\pi) + \frac{k\pi}{\ell}\int_{-\ell}^{\ell} f(x)\sin\left(\frac{k\pi x}{\ell}\right)dx \right\}.$$

By hypothesis $f(\ell) = f(-\ell)$, so that $f(-\ell)\cos(-k\pi) = f(\ell)\cos(k\pi)$; then

$$A_k = \frac{k\pi}{\ell}\left\{ \frac{1}{\ell}\int_{-\ell}^{\ell} f(x)\sin\left(\frac{k\pi x}{\ell}\right)dx \right\} = \frac{k\pi}{\ell}b_k.$$

In particular, for $k = 0$, we obtain $A_0 = 0$. Similarly, integrating by parts B_k for $k = 1, 2, \ldots$ and choosing $u = \sin\left(\dfrac{k\pi x}{\ell}\right)$, we get

$$B_k = \frac{1}{\ell}\left\{ f(x)\sin\left(\frac{k\pi x}{\ell}\right)\Big|_{-\ell}^{\ell} - \frac{k\pi}{\ell}\int_{-\ell}^{\ell} f(x)\cos\left(\frac{k\pi x}{\ell}\right)dx \right\}$$

$$= \frac{1}{\ell}\left\{ f(\ell)\sin(k\pi) - f(-\ell)\sin(-k\pi) - \frac{k\pi}{\ell}\int_{-\ell}^{\ell} f(x)\cos\left(\frac{k\pi x}{\ell}\right)dx \right\}$$

$$= \frac{1}{\ell}\left\{ \underbrace{2f(\ell)\sin(k\pi)}_{=0} - \frac{k\pi}{\ell}\int_{-\ell}^{\ell} f(x)\cos\left(\frac{k\pi x}{\ell}\right)dx \right\}$$

$$= -\frac{k\pi}{\ell}\left\{ \frac{1}{\ell}\int_{-\ell}^{\ell} f(x)\cos\left(\frac{k\pi x}{\ell}\right)dx \right\} = -\frac{k\pi}{\ell}a_k.$$

With these results we can write

$$f'(x) = \sum_{k=1}^{\infty} \frac{k\pi}{\ell}\left[-a_k\sin\left(\frac{k\pi x}{\ell}\right) + b_k\cos\left(\frac{k\pi x}{\ell}\right) \right],$$

which is the desired result, i.e., the same result we would have obtained by differentiating $f(x)$ term by term. Besides, at the points of discontinuity of $f'(x)$ we have

$$\frac{f'(x^+) + f'(x^-)}{2} = \sum_{k=1}^{\infty} \frac{k\pi}{\ell}\left[-a_k\sin\left(\frac{k\pi x}{\ell}\right) + b_k\cos\left(\frac{k\pi x}{\ell}\right) \right],$$

which is the desired result. $\qquad\square$

Theorem 5.2 (Integration) *Let $f(x)$ be a function continuous by parts on the symmetric interval $[-\pi, \pi]$ and periodic with period 2π. Then, the Fourier series of $f(x)$, convergent or not, can be integrated term by term between any limits.*

Example 5.2 Assuming that the Fourier series expansions of functions $f(x) = x$ and $f(x) = x^2$ are known, obtain by integration the Fourier series expansion of function $f(x) = x^3$, both on the interval $(-\ell, \ell)$.

The Fourier series for $f(x) = x^2$ is given by

$$x^2 = \frac{\ell^2}{3} + \frac{4\ell^2}{\pi^2} \sum_{k=1}^{\infty} \frac{(-1)^k}{k^2} \cos\left(\frac{k\pi}{\ell}x\right),$$

for $-\ell < x < \ell$. Integrating this expression from $-\ell$ to x we get, after simplification,

$$\frac{x^3}{3} = \frac{\ell^3}{3}x + \frac{4\ell^3}{\pi^3} \sum_{k=1}^{\infty} \frac{(-1)^k}{k^3} \sin\left(\frac{k\pi}{\ell}x\right),$$

for $-\ell < x < \ell$. As we know the Fourier series for $f(x) = x$, we can write

$$x^3 = \frac{2\ell^3}{\pi} \sum_{k=1}^{\infty} \frac{(-1)^{k+1}}{k} \sin\left(\frac{k\pi}{\ell}x\right) + \frac{12\ell^3}{\pi^3} \sum_{k=1}^{\infty} \frac{(-1)^k}{k^3} \sin\left(\frac{k\pi}{\ell}x\right),$$

which can be rewritten in the form

$$x^3 = \frac{2\ell^3}{\pi} \sum_{k=1}^{\infty} \left[\frac{(-1)^k}{k}\left(\frac{6}{\pi^2 k^2} - 1\right)\right] \sin\left(\frac{k\pi}{\ell}x\right)$$

for $-\ell < x < \ell$, which is the desired result. $\square$

5.1.1 *Parseval Identiy*

An interesting relation involving the Fourier coefficients is known as Parseval [1755 – Marc-Antoine Parseval des Chênes – 1836] identity. We introduce it by the following theorem.

Theorem 5.3 *If the Fourier series for $f(x)$ converges uniformly to the function $f(x)$ on the interval $(-\ell, \ell)$, then*

$$\frac{1}{\ell} \int_{-\ell}^{\ell} [f(x)]^2 \, dx = \frac{a_0^2}{2} + \sum_{k=1}^{\infty} (a_k^2 + b_k^2),$$

since the integral exists.

Proof Let $f(x)$ be a function expressed by its Fourier series,

$$f(x) = \frac{a_0}{2} + \sum_{k=1}^{\infty} \left[a_k \cos\left(\frac{k\pi}{\ell}x\right) + b_k \sin\left(\frac{k\pi}{\ell}x\right) \right].$$

Multiplying $f(x)$ by $f(x)$ and integrating term-by-term from $-\ell$ up to $+\ell$ (the series is uniformly convergent), we can write

$$\int_{-\ell}^{\ell} [f(x)]^2 \, dx = \frac{a_0}{2} \int_{-\ell}^{\ell} f(x) \, dx$$

$$+ \sum_{k=1}^{\infty} \left\{ a_k \int_{-\ell}^{\ell} f(x) \cos\left(\frac{k\pi}{\ell}x\right) \, dx + b_k \int_{-\ell}^{\ell} f(x) \sin\left(\frac{k\pi}{\ell}x\right) \, dx \right\}.$$

Then, using the results obtained for the Fourier coefficients,

$$\int_{-\ell}^{\ell} f(x) \cos\left(\frac{k\pi}{\ell}x\right) \, dx = \ell a_k, \quad \text{with} \quad k = 0, 1, 2, \ldots$$

and

$$\int_{-\ell}^{\ell} f(x) \sin\left(\frac{k\pi}{\ell}x\right) \, dx = \ell b_k, \quad \text{with} \quad k = 1, 2, 3, \ldots$$

we obtain

$$\int_{-\ell}^{\ell} [f(x)]^2 \, dx = \frac{a_0^2}{2}\ell + \ell \sum_{k=1}^{\infty} (a_k^2 + b_k^2),$$

which is the desired result. $\qquad\square$

5.2 Fourier–Bessel Series

Let $r \in \mathbb{R}^*$. We call Fourier–Bessel series of order m a series of the form

$$f(r) = \sum_{n=1}^{\infty} a_n \mathcal{J}_m(k_{mn}r),$$

where $\mathcal{J}_m(x)$ is the Bessel function of order m and the coefficients a_n are given by the relation

$$a_n = \frac{2}{a^2 [\mathcal{J}_{m+1}(k_{mn}a)]^2} \int_0^a r f(r) \mathcal{J}_m(k_{mn}r) \, dr.$$

This type of series is useful in problems involving Dirichlet [1805 – Peter Gustav Lejeune Dirichlet – 1859] conditions, that is, in which the boundary conditions provide the values of the function at two distint points. Other boundary conditions are treated in the same way.

5.3   Fourier–Legendre Series

Just as in the cases of Fourier series and Fourier–Bessel series, we introduce the so-called Fourier–Legendre series. We call Fourier–Legendre series all series of the type

$$f(x) = \sum_{l=0}^{\infty} a_l \mathcal{P}_l(x),$$

where $l = 0,\ 1,\ 2,\ \ldots, \mathcal{P}_l(x)$ are the Legendre polynomials and the coefficients a_l are given by

$$a_l = \frac{2l+1}{2} \int_{-1}^{1} f(x)\mathcal{P}_l(x)\,\mathrm{d}x,$$

In Chap. 7 we will discuss the so-called Sturm–Liouville [1803 – Jacques Charles François Sturm – 1855] problem, and we will see that the series discussed earlier are particular cases of a larger class of functions.

5.4   Solved Exercises

SE 5.1 Let $-\pi \leq x \leq \pi$. Show that the coefficients a_k in the Fourier series

$$f(x) = \frac{a_0}{2} + \sum_{k=1}^{\infty}(a_k \cos kx + b_k \sin kx)$$

are given by the relation

$$a_k = \frac{1}{\pi} \int_{-\pi}^{\pi} f(x) \cos kx\,\mathrm{d}x,$$

for all k.

Solution: We suppose that a certain function $f(x)$ is represented by the series

$$f(x) = \frac{a_0}{2} + \sum_{k=1}^{\infty} (a_k \cos kx + b_k \sin kx),$$

and that this series is uniformly convergent on the interval $-\pi \leq x \leq \pi$. Multiply the series by $\cos mx$, with m a positive integer:

$$f(x) \cos mx = \frac{a_0}{2} \cos mx + \sum_{k=1}^{\infty} (a_k \cos kx + b_k \sin kx) \cos mx.$$

This series is still uniformly convergent, and thus, we can integrate it term by term:

$$\int_{-\pi}^{\pi} f(x) \cos mx \, dx = \frac{a_0}{2} \int_{-\pi}^{\pi} \cos mx \, dx + \sum_{k=1}^{\infty} a_k \int_{-\pi}^{\pi} \cos kx \cos mx \, dx +$$

$$+ \sum_{k=1}^{\infty} b_k \int_{-\pi}^{\pi} \sin kx \cos mx \, dx.$$

Using the orthogonality properties of the sine and cosine functions, we have

$$a_k = \frac{1}{\pi} \int_{-\pi}^{\pi} f(x) \cos kx \, dx.$$

Thus, this relation allows us to calculate any coefficient a_k when the function $f(x)$ is known. The calculation of coefficients b_k is done in the same way, i.e., we multiply the original expression by $\sin mx$ and use again the orthogonality relations.

SE 5.2 Let $u(x, y)$, $0 \leq x, y \leq 1$, a continuous function of the point M with coordinates (x, y), denote the temperature of a square plate. It is constant in time and satisfies the two-dimensional Laplace equation written in cartesian coordinates, which we will see in Chap. 9. Suppose that $u(x, y)$ satisfies the conditions

$$u(0, y) = u(1, y) = u(x, 0) = 0$$

and assume also that each point with abscissa x and ordinate 1 has a temperature $F(x)$ given by

$$u(x, 1) \equiv F(x) = \sum_{k=1}^{\infty} \frac{100}{2^k} \sin(\pi kx), \qquad 0 \leq x \leq 1.$$

(a) Let $0 \leq x \leq 1$. Show that, for all x, $|F(x)| \leq 100$. (b) Calculate the numerical value of the temperature at the center of the plate. (c) Evaluate $F(x)$ and show that $0 \leq F(x) < 67$.

Solution:

(a) Taking the modulus of both sides we can write

$$|F(x)| = \left| \sum_{k=1}^{\infty} \frac{100}{2^k} \sin(\pi kx) \right| \le \sum_{k=1}^{\infty} \frac{100}{2^k} |\sin(\pi kx)| \le \sum_{k=1}^{\infty} \frac{100}{2^k},$$

which is a geometric series with the first term equal to 50 and ratio equal to $1/2$; its sum is equal to 100, so,

$$|F(x)| \le 100.$$

(b) In order to solve this item, we must use the method of separation of variables, which we shall see in Chap. 9. If we do this, we find for the temperature distribution

$$u(x, y) = \sum_{k=1}^{\infty} \frac{100}{2^k} \frac{\sinh(\pi ky)}{\sinh(\pi k)} \sin(\pi kx).$$

In our case, for the temperature at the center of the plate, we must calculate $u(\frac{1}{2}, \frac{1}{2})$, i.e.,

$$u\left(\frac{1}{2}, \frac{1}{2}\right) = 25 \sum_{k=0}^{\infty} \frac{1}{2^{2k}} \frac{(-1)^k}{\cosh\pi (2k + 1)/2}.$$

Substituting the values of the first three nonnull terms, we get

$$u\left(\frac{1}{2}, \frac{1}{2}\right) \simeq 9.8.$$

(c) In order to calculate $F(x)$, we use the Euler formula [1] to write the sine function as a sum of two exponentials, so that we can write

$$F(x) = 100 \sum_{k=1}^{\infty} \frac{\sin(\pi kx)}{2^k} = \frac{50}{i} \sum_{k=1}^{\infty} \frac{1}{2^k} \left(e^{i\pi kx} - e^{-i\pi kx} \right).$$

Manipulating the series, we have

$$F(x) = \frac{25}{i} e^{it} \sum_{k=0}^{\infty} \left(\frac{e^{it}}{2} \right)^k - \frac{25}{i} e^{-it} \sum_{k=0}^{\infty} \left(\frac{e^{-it}}{2} \right)^k,$$

where $t = \pi x$. Both series are geometric series and their sums are known. Thus, we can write, after some simplification,

$$F(x) = \frac{200\sin\pi x}{5 - 4\cos\pi x}.$$

Finally, in order to obtain the maximum value, we differentiate the above expression with respect to x and equate it to zero, obtaining

$$\frac{d}{dx}F(x) = \frac{5\cos\pi x - 4}{(5 - 4\cos\pi x)^2} = 0.$$

This implies that $\cos\pi x = 4/5$ and hence that $\sin\pi x = 3/5$, whence we conclude that

$$F_{\max}(x) = 200\frac{3/5}{5 - 4(4/5)} = \frac{600}{9} < 67.$$

Note that for a maximum we must have a negative second derivative; this can be confirmed by direct calculation. Finally, since for $0 < x < 1$ we have $F(x) \geq 0$, we can write

$$0 \leq F(x) < 67,$$

which is the desired expression.

SE 5.3 Consider the function $f(x) = x$ on the interval $-\pi < x < \pi$ expressed by the Fourier series

$$x = 2\left(\sin x - \frac{\sin 2x}{2} + \frac{\sin 3x}{3} - \cdots\right).$$

Integrating it term by term, we obtain the series (**PE 5.22**)

$$x^2 = 4\left[\frac{\pi^2}{12} + \sum_{k=1}^{\infty}(-1)^k\frac{\cos kx}{k^2}\right].$$

Integrate this last series term by term to calculate the sum

$$S = 1 - \frac{1}{3^3} + \frac{1}{5^3} - \frac{1}{7^3} + \cdots$$

Solution: Integrating both sides of the expression for the expansion of x^2 given earlier, from a, a constant, to x, we can write

$$\int_a^x t^2\, dt = \int_a^x \frac{\pi^2}{3}\, dt + \int_a^x \left\{\sum_{k=1}^{\infty}4(-1)^k\frac{\cos kt}{k^2}\right\} dt$$

or, after some simplification,

$$\frac{x^3}{3} - \frac{\pi^2 x}{3} = \Omega + \sum_{k=1}^{\infty} 4(-1)^k \frac{\sin kx}{k^3}$$

where Ω is a constant. In order to determine this constant, we integrate again both sides with respect to x on the interval $(-\pi, \pi)$, and we easily conclude that $\Omega = 0$. We then have for the integrated series

$$\frac{x}{12}(x^2 - \pi^2) = \sum_{k=1}^{\infty}(-1)^k \frac{\sin kx}{k^3}.$$

Finally, to determine the sum, we substitute $x = \pi/2$ into this expression, whence it follows that

$$\frac{\pi^3}{96} - \frac{\pi^3}{24} = \sum_{k=1}^{\infty}(-1)^k \frac{\sin \pi k/2}{k^3}$$

and changing the index $k \to 2k - 1$, we obtain the desired result, that is

$$S = \sum_{k=1}^{\infty} \frac{(-1)^k}{(2k-1)^3} = 1 - \frac{1}{3^3} + \frac{1}{5^3} - \frac{1}{7^3} + \ldots = \frac{\pi^3}{32}.$$

SE 5.4 Let $f(x) = x^3 - \pi^2 x$ be a function defined on the closed interval $[-\pi, \pi]$, periodic with period 2π. (a) Obtain the Fourier series for $f(x)$ and (b) use the Parseval identity to evaluate the sum

$$\sum_{k=1}^{\infty} \frac{1}{k^6}.$$

Solution: (a) Since $f(x)$ is an odd function, $a_k = 0$ for all $k = 0, 1, \ldots$ We must then evaluate only the integral

$$b_k = \frac{2}{\pi} \int_0^{\pi} (x^3 - \pi^2 x) \sin kx \, dx.$$

Using integration by parts and simplifying yields

$$b_k = \frac{12}{k^3}(-1)^k,$$

$k = 1, 2, \ldots$ Then, the corresponding Fourier series is given by

$$f'(x) = \sum_{k=1}^{\infty} \frac{12}{k^3}(-1)^k \sin kx \,.$$

(b) Using the Parseval identity we can write

$$144 \sum_{k=1}^{\infty} \frac{1}{k^6} = \frac{2}{\pi} \int_0^{\pi} (x^3 - \pi^2 x)^2 \, dx,$$

where we have used the fact that the integrand is an even function. The remaining integral is an immediate integral from which it follows, after integration and simplification,

$$\sum_{k=1}^{\infty} \frac{1}{k^6} = \frac{\pi^6}{945},$$

which is the desired result.

SE 5.5 (Orthogonality of Bessel functions) Let $u(x) = J_\nu(\lambda x)$ and $v(x) = J_\nu(\mu x)$ be two first kind Bessel functions of order $\nu > -1$ and λ and μ two real parameters. (a) Show the orthogonality of these Bessel functions with respect to the weight function $w(x) = x$,

$$\int_0^1 x J_\nu(\lambda x) J_\nu(\mu x) \, dx = \frac{\mu J_\nu(\lambda) J_\nu'(\mu) - \lambda J_\nu(\mu) J_\nu'(\lambda)}{\lambda^2 - \mu^2},$$

for $\lambda \neq \mu$ and where the prime denotes derivative. (b) Show that in the case $\mu = \lambda$ the relation is

$$\int_0^1 x [J_\nu(\lambda x)]^2 \, dx = \frac{1}{2} \left\{ [J_\nu'(\lambda)]^2 + \left(1 - \frac{\nu^2}{\lambda^2}\right) [J_\nu(\lambda)]^2 \right\}.$$

(c) Discuss the particular case in which λ and μ are roots of the Bessel function, i.e.,

$$J_\nu(\lambda) = 0 = J_\nu(\mu) \,.$$

Solution: (a) The functions $u = u(x)$ and $v = v(x)$ satisfy the following Bessel equations of order ν:

$$x^2 u'' + x u' + (\lambda^2 x^2 - \nu^2) u = 0$$

and

$$x^2 v'' + x v' + (\mu^2 x^2 - \nu^2) v = 0,$$

respectively. Multiplying the first equation by v, the second one by u, subtracting each other, and rearranging, we can write

$$\frac{\mathrm{d}}{\mathrm{d}x}\left[x(uv' - vu')\right] = (\lambda^2 - \mu^2)x\,u\,v\,.$$

Integrating this last expression from 0 to 1 we have

$$x(uv' - vu')|_{x=0}^{x=1} = (\lambda^2 - \mu^2)\int_0^1 x\,u\,v\,\mathrm{d}x\,.$$

With the condition $v > -1$, the first member of the precedent equation, evaluated at $x = 0$, goes to zero [3] and we get

$$\int_0^1 x J_\nu(\lambda x) J_\nu(\mu x)\,\mathrm{d}x = \frac{\mu J_\nu(\lambda) J_\nu'(\mu) - \lambda J_\nu(\mu) J_\nu'(\lambda)}{\lambda^2 - \mu^2}\,,$$

which is the desired result. $\square$

(b) In order to evaluate the integral in which $\lambda = \mu$, we must consider the limit $\lambda \to \mu$ in the last equation, which yields an indeterminacy in the right-hand side of the equation. In order to raise it we use the l'Hôpital [1661 − Guillaume François Antoine–Marquis de l'Hôpital − 1704] rule, obtaining

$$\int_0^1 x[J_\nu(\lambda x)]^2\,\mathrm{d}x = \frac{1}{2}\left\{[J_\nu'(\lambda)]^2 + \left(1 - \frac{\nu^2}{\lambda^2}\right)[J_\nu(\lambda)]^2\right\}\,,$$

which is the desired result.

(c) In the case $J_\nu(\lambda) = 0 = J_\nu(\mu)$, the two expressions obtained in the previous items are combined so that we can write

$$\int_0^1 x[J_\nu(\lambda x)]^2\,\mathrm{d}x = \begin{cases} \dfrac{1}{2}[J_\nu'(\lambda)]^2, & \text{for } \lambda = \mu; \\[2mm] 0, & \text{for } \lambda \neq \mu\,. \end{cases}$$

This relation is important when the boundary conditions are Dirichlet conditions. A similar expression can be obtained for Neumann conditions, when λ and μ are roots of the equation $J_\nu'(\lambda) = 0 = J_\nu'(\mu)$:

$$\int_0^1 x[J_\nu(\lambda x)]^2\,\mathrm{d}x = \begin{cases} \dfrac{1}{2}\left(1 - \dfrac{\nu^2}{\lambda^2}\right)[J_\nu(\lambda)]^2, & \text{for } \lambda = \mu; \\[2mm] 0, & \text{for } \lambda \neq \mu\,. \end{cases}$$

Finally, when λ and μ are roots of the equation $\lambda J_\nu'(\lambda) + h\, J_\nu(\lambda) = 0 = \mu J_\nu'(\mu) + h J_\nu(\mu)$, we obtain the following expression:

$$\int_0^1 x[J_\nu(\lambda x)]^2\, dx = \begin{cases} \dfrac{1}{2}\left(1 + \dfrac{h^2 - \nu^2}{\lambda^2}\right)[J_\nu(\lambda)]^2, & \text{for } \lambda = \mu; \\[2ex] 0, & \text{for } \lambda \neq \mu. \end{cases}$$

These three expressions are of great importance when, in solving a particular problem, expansions emerge in terms of a Fourier–Bessel series.

5.5 Proposed Exercises

PE 5.1 Evaluate the integral

$$\int_{-\pi}^{\pi} \sin x \, \cos x \, dx.$$

PE 5.2 Let $n, k \in \mathbb{N}$. Evaluate the integral

$$\int_{-\pi}^{\pi} \sin kx \, \sin nx \, dx.$$

PE 5.3 Let $n, k \in \mathbb{N}$. Evaluate the integral

$$\int_{-\pi}^{\pi} \sin kx \, \cos nx \, dx.$$

PE 5.4 Evaluate the sum

$$\sum_{k=1}^{\infty} \frac{\sin(2kx)}{k}, \qquad 0 < x < \pi.$$

PE 5.5 Let $f(x)$ be the following 2π-periodic function:

$$f(x) = \begin{cases} 0, & \text{if } -\pi < x < 0, \\ x, & \text{if } 0 < x < \pi. \end{cases}$$

Obtain the expansion of $f(x)$ in a Fourier series on $-\pi < x < \pi$.

PE 5.6 Show that

$$\sum_{k=1}^{\infty} \frac{1}{(2k-1)^2} = \frac{\pi^2}{8}.$$

PE 5.7 Show that if $f(x)$ is an odd and 2ℓ-periodic function, then the Fourier series of $f(x)$ is a sine series given by

$$f(x) = \sum_{k=1}^{\infty} b_k \sin\left(\frac{k\pi x}{\ell}\right),$$

with

$$b_k = \frac{2}{\ell} \int_0^{\ell} f(x) \sin\left(\frac{k\pi x}{\ell}\right) dx, \qquad k = 1, 2, \ldots$$

PE 5.8 The so-called signal function is given by

$$f(x) = \operatorname{sgn} x = \begin{cases} -1, \text{ if } & -\pi < x < 0, \\ 0, \text{ if } & x = 0, \\ 1, \text{ if } & 0 < x < \pi, \end{cases}$$

with $f(x \pm 2k\pi) = f(x)$. Obtain its Fourier series.

PE 5.9 Let $f(x)$ be a continuous function on $[-\ell, \ell]$ such that $f(\ell) = f(-\ell)$, and let $f'(x)$ be continuous by parts and with lateral derivatives on this interval. Show that the Fourier series of $f(x)$ is a differentiable series.

PE 5.10 Let f be a 2ℓ-periodic function. Using the usual Fourier series, obtain the corresponding complex Fourier series.

PE 5.11 Show that $f(x) = \cos x$ is (a) an even function and (b) periodic with period 2π.

PE 5.12 Express the function $f(x) = \cosh x$ as a Fourier series to show that

$$\cosh x = \frac{\sinh \pi}{\pi} + \frac{2 \sinh \pi}{\pi} \sum_{n=1}^{\infty} \frac{(-1)^n \cos nx}{1 + n^2}, \quad -\pi < x < \pi.$$

PE 5.13 Expand the function

$$f(x) = \frac{\pi - x}{2}$$

in a Fourier series on the interval $-\pi < x < \pi$.

PE 5.14 Using the result obtained in **PE 5.13**, show that

$$\sum_{n=1}^{\infty} \frac{1}{n^2} = \frac{\pi^2}{6}.$$

PE 5.15 Use the relations

$$\sin x = \frac{e^{ix} - e^{-ix}}{2i} \qquad \text{and} \qquad \cos x = \frac{e^{ix} + e^{-ix}}{2}$$

to show that the Fourier series for a function $f(x)$ can be written in the following form:

$$f(x) = \sum_{k=-\infty}^{\infty} c_k \, e^{ikx}, \qquad -\pi < x < \pi,$$

where k assumes only integer values. This is the so-called complex Fourier series.

PE 5.16 Find the complex Fourier series for $f(x) = x$ on the interval $-\pi < x < \pi$.

PE 5.17 Obtain the Fourier series for a periodic function given by

$$f(x) = \begin{cases} -\pi \ \text{if} \ -\pi < x < 0, \\ \ \ x \ \text{if} \ 0 < x < \pi. \end{cases}$$

PE 5.18 Expand the function $f(x) = \dfrac{x^2}{4}$ on the interval $-\pi < x < \pi$ to show that

$$1 + \frac{1}{4} + \frac{1}{9} + \frac{1}{16} + \ldots = \frac{\pi^2}{6}.$$

PE 5.19 Using the result obtained in **PE 5.18**, show that

$$1 - \frac{1}{4} + \frac{1}{9} - \frac{1}{16} + \ldots = \frac{\pi^2}{12}.$$

PE 5.20 Using the two previous results, show that

$$1 + \frac{1}{9} + \frac{1}{25} + \frac{1}{49} + \ldots = \frac{\pi^2}{8}.$$

PE 5.21 Obtain the Fourier series for the function $f(x) = \cos \alpha x$, with $\alpha \neq 0, \pm 1, \pm 2, \pm 3 \ldots$, on the interval $-\pi \leq x \leq \pi$, and show that

$$\pi \cotg \alpha \pi = \frac{1}{\alpha} + \frac{2\alpha}{\alpha^2 - 1} + \frac{2\alpha}{\alpha^2 - 4} + \frac{2\alpha}{\alpha^2 - 9} + \ldots$$

PE 5.22 Beginning with the series

$$x = 2 \left(\sin x - \frac{\sin 2x}{2} + \frac{\sin 3x}{3} - \cdots \right),$$

on the interval $-\pi < x < \pi$, show through term-by-term integration that

$$x^2 = \frac{\pi^2}{3} + 4 \sum_{k=1}^{\infty} (-1)^k \frac{\cos kx}{k^2}.$$

PE 5.23 Using the result obtained in **PE 5.22**, calculate the following sums:

$$\text{(a)} \ \sum_{k=1}^{\infty} \frac{1}{k^2} (-1)^{k+1} = \frac{\pi^2}{12};$$

$$\text{(b)} \ \sum_{k=1}^{\infty} \frac{1}{k^2} = \frac{\pi^2}{6}.$$

PE 5.24 Let $f(x)$ be an odd function on the interval $-\pi < x < \pi$. Obtain the corresponding Parseval identity. Discuss the particular case $f(x) = x$.

PE 5.25 Expand $f(x) = (\pi - x)/2$ in a Fourier series, on the interval $0 < x < 2\pi$.

PE 5.26 Using the result obtained in **PE 5.25**, calculate

$$S = \frac{3}{2\pi} \frac{S_1}{S_2},$$

where $S_1 = 1 + \frac{1}{4} + \frac{1}{9} + \ldots$ and $S_2 = 1 - \frac{1}{3} + \frac{1}{5} - \ldots$

PE 5.27 Expand the function $f(x) = |\sin x|$, on the interval $-\pi < x < \pi$, in a Fourier series.

PE 5.28 Using the result obtained in **PE 5.27**, show that, for $0 < x < \pi$, we have

$$\cos x = -\frac{8}{\pi} \sum_{k=1}^{\infty} \frac{k \sin 2kx}{1 - 4k^2}.$$

PE 5.29 Expand the function e^u in powers of u. Then, obtain Fourier expansions for the periodic functions in the θ variable,

$$e^{\frac{\lambda}{2} e^{i\theta}} \quad \text{and} \quad e^{-\frac{\lambda}{2} e^{-i\theta}},$$

on $0 \leq \theta \leq 2\pi$, with λ a positive real parameter.

PE 5.30 Let $n \in \mathbb{N}$. Show that

$$\frac{1}{\pi} \int_{-\pi}^{\pi} \frac{\sin(n + 1/2)s}{2\sin(s/2)}\, ds = \frac{1}{\pi} \int_{-\pi}^{\pi} \left[\frac{1}{2} + \sum_{k=1}^{n} \cos ks\right] ds.$$

PE 5.31 Let $n \in \mathbb{N}$. Show that

$$\int_{-\pi}^{\pi} \frac{\sin(n + 1/2)s}{2\sin(s/2)}\, ds = \pi.$$

PE 5.32 On the interval $-\pi \le x \le \pi$, expand the function

$$f(x) = \begin{cases} 0 & -\pi \le x \le 0 \\ x & 0 \le x \le \pi/2 \\ \pi - x & \pi/2 \le x \le \pi \end{cases}$$

in a Fourier series.

PE 5.33 Using the result obtained in **PE 5.32**, calculate the sum

$$S_0 = 1 - \frac{1}{9} - \frac{1}{25} + \frac{1}{49} + \frac{1}{81} - \cdots$$

PE 5.34 Knowing that $\omega T = 2\pi$, expand in a Fourier series the function

$$u(x) = \begin{cases} 0 & \text{for } -T/2 < x < 0, \\ E_0 \sin \omega x & \text{for } 0 < x < T/2. \end{cases}$$

PE 5.35 A triangular wave can be represented by

$$f(x) = \begin{cases} -x & \text{for } -\pi < x < 0, \\ x & \text{for } 0 < x < \pi, \end{cases}$$

with $f(x + 2\pi) = f(x)$; this is a periodic function with period 2π. Express $f(x)$ in a Fourier series.

PE 5.36 Using the result obtained in **PE 5.35**, calculate the sum

$$S = 1 + \frac{1}{9} + \frac{1}{25} + \cdots$$

PE 5.37 (RLC Electrical Circuit) Find, using an adequate Fourier series, the stationary current $I(t)$ in an RLC electrical circuit (**PE 1.26**) for which $R = 100$ ohms, $L = 10$ henries, $C = 10^{-2}$ farads and

$$E(t) = \begin{cases} 100(\pi t + t^2) \text{ if } -\pi < t < 0, \\ 100(\pi t - t^2) \text{ if } 0 < t < \pi, \end{cases}$$

with $E(t + 2\pi) = E(t)$.

PE 5.38 Given that, for $0 \le x \le \pi$,

$$\sum_{k=0}^{\infty} \frac{\cos[(2k+1)x]}{(2k+1)^2} = \frac{\pi^2 - 2\pi x}{8},$$

calculate the sum

$$\sum_{k=0}^{\infty} \frac{\sin[(2k+1)x]}{(2k+1)^3}.$$

PE 5.39 Consider the function $F(x)$ on the interval $-\pi < x < \pi$, satisfying the relation $F(x + 2\pi) = F(x)$ and given by

$$F(x) = \frac{x}{12}(\pi^2 - x^2).$$

Solve the ordinary differential equation

$$\frac{\mathrm{d}^2}{\mathrm{d}x^2} y(x) + \omega^2 y(x) = F(x),$$

where ω^2 is a constant and $|\omega| \ne 1, 2, 3, \ldots$

PE 5.40 Show that the coefficients a_k in the Fourier–Bessel expansion of the function $f(x) = 1$, for $0 < x < 1$, are given by

$$a_k = \frac{-2}{\lambda_k \mathcal{J}_0'(\lambda_k)},$$

where λ_k are the roots of $\mathcal{J}_0(x) = 0$ and the prime denotes the derivative.

PE 5.41 Using the result obtained in the preceding exercise, show that

$$\frac{\mathcal{J}_0(\lambda_1 x)}{\lambda_1 \mathcal{J}_1(\lambda_1)} + \frac{\mathcal{J}_0(\lambda_2 x)}{\lambda_2 \mathcal{J}_1(\lambda_2)} + \ldots = \frac{1}{2}.$$

PE 5.42 Express the function $f(x) = x^3$ on the interval $0 < x < 2$ in a Fourier–Bessel series involving the Bessel function $\mathcal{J}_3(x)$.

PE 5.43 Do as in **PE 5.42**, considering the interval $0 < x < R$ for the function

$$f(x) = \begin{cases} C \text{ if } 0 < x < a, \\ 0 \text{ if } a < x < R, \end{cases}$$

where C is a constant.

PE 5.44 Let $\lambda \in \mathbb{R}$. For $\mathcal{J}_m(\lambda)$ the Bessel function of order m, use the result obtained in **PE 5.29** to show that

$$e^{i\lambda \sin\theta} = \sum_{m=-\infty}^{\infty} \mathcal{J}_m(\lambda)\, e^{im\theta}.$$

PE 5.45 Using the expression obtained in the preceding exercise, calculate the integral

$$\int_0^{2\pi} e^{i(\lambda \sin\theta - n\theta)}\, d\theta.$$

PE 5.46 Show that

$$\frac{1-x^2}{4} = \sum_{k=1}^{\infty} \frac{\mathcal{J}_2(\lambda_k)}{\mathcal{J}_1^2(\lambda_k)} \frac{\mathcal{J}_0(\lambda_k x)}{\lambda_k^2}$$

on the interval $0 < x < 1$, where λ_k is the k-th positive root of the equation $\mathcal{J}_0(\lambda) = 0$.

PE 5.47 Show that the *Legendre polynomials* $\mathcal{P}_0(x) = 1$, $\mathcal{P}_1(x) = x$, $\mathcal{P}_2(x) = \dfrac{3x^2-1}{2}$ are orthogonal on the interval $[-1, 1]$.

PE 5.48 Show that

$$x^2 = \frac{1}{3}\mathcal{P}_0(x) + \frac{2}{3}\mathcal{P}_2(x),$$

where $\mathcal{P}_0(x)$ and $\mathcal{P}_2(x)$ are the Legendre polynomials of **PE 5.47**.

PE 5.49 Obtain the Fourier–Legendre series for the function

$$f(x) = \begin{cases} 0 \text{ if } -1 < x < 0, \\ 1 \text{ if } 0 < x < 1. \end{cases}$$

To this end, use the integral

$$\int_0^1 \mathcal{P}_l(x)\, dx = \begin{cases} 1 \text{ if } l = 0; \\ 0 \text{ if } l \text{ is even}; \\ \Omega \text{ if } l \text{ is odd}; \end{cases}$$

where $\Omega \equiv \dfrac{\Gamma(3/2)}{\Gamma\left(\frac{l+3}{2}\right)\Gamma\left(\frac{2-l}{2}\right)}$.

PE 5.50 Do as in the preceding exercise for the function

$$f(x) = \begin{cases} 0 \ \text{if} \ -1 < x < 0, \\ x \ \text{if} \ 0 < x < 1. \end{cases}$$

Use the following integral:

$$\int_0^1 x^{\mu} \mathcal{P}_\nu(x)\, \mathrm{d}x = \frac{\sqrt{\pi}\, 2^{-\mu-1} \Gamma(1+\mu)}{\Gamma\left(1+\frac{\mu-\nu}{2}\right)\Gamma\left(\frac{\mu+\nu+3}{2}\right)}$$

for $\mathrm{Re}(\mu) > -1$.

References

1. E. Capelas de Oliveira, W.A. Rodrigues Jr., *Funções Analíticas com Aplicações* (Analytic Functions with Applications), Editora (Livraria da Física, São Paulo, 2006)
2. T.W. Körner. *Fourier Analysis* (Cambridge University Press, Cambridge, 1990)
3. G.P. Tolstov, *Fourier Series* (Dover Publications, New York, 1976)

Chapter 6
Laplace and Fourier Transforms

But dull minds are never either intuitive or mathematical.

1623 – Blaise Pascal – 1662

The use of the so-called *integral transforms* to obtain solutions of differential equations consists of transforming a given differential equation into a simpler equation, solving this new equation and, finally, calculating the corresponding inverse transform to obtain the solution of the initial differential equation. This *method of transforms* is used to obtain particular solutions of differential equations, ordinary and partial. We discuss here only two such transforms, the Laplace transform and the Fourier transform [1–3].

6.1 Laplace Transform

The Laplace transform of a function $f(t)$, $t \geq 0$, which we will denote by $F(s)$ or by $\mathscr{L}[f(t)]$, is defined by the following improper integral:

$$F(s) \equiv \mathscr{L}[f(t)] = \int_0^\infty e^{-st} f(t)\, dt,$$

where s, the so-called *transformed variable*, must obey $\mathrm{Re}(s) > 0$, in order for this improper integral to converge.

As already mentioned, the integral transform methodology requires the use of the inverse transform to retrieve the solution of the original problem. Before discussing the inverse transform, we present some properties of the Laplace transform, stated as theorems.

Theorem 6.1 (Linearity) *If $\mathscr{L}[f(t)]$ and $\mathscr{L}[g(t)]$ are the Laplace transforms of functions $f(t)$ and $g(t)$, respectively, then*

$$\mathcal{L}[c_1 f(t) + c_2 g(t)] = c_1 \mathcal{L}[f(t)] + c_2 \mathcal{L}[g(t)],$$

for any constants c_1 and c_2.

Theorem 6.2 (Displacement) *Let $a \in \mathbb{R}$. If the Laplace transform of the function $f(t)$ is $F(s)$, then*

$$\mathcal{L}[e^{at} f(t)] = F(s - a),$$

with $\mathrm{Re}(s) > a$.

Theorem 6.3 (Scale) *If the Laplace transform of function $f(t)$ is $F(s)$, then*

$$\mathcal{L}[f(ct)] = \frac{1}{c} F\left(\frac{s}{c}\right),$$

for $c > 0$.

The next theorem involves the concept of a *function of exponential order*. We say that a function $f(t)$ is of exponential order γ as $t \to \infty$, or simply that $f(t)$ is of exponential order, if there exist real constants $M > 0$ and $\gamma > 0$ such that $|f(t)| < M \exp(\gamma t)$ as $t \to \infty$. This property is denoted by

$$f(t) = \mathcal{O}(e^{\gamma t}).$$

We can now formulate the next theorem, about the Laplace transform of the first order derivative $f'(t)$ of a function $f(t)$, which is important in discussing linear ordinary and partial differential equations.

Theorem 6.4 (Differentiation) *Let $f(t)$ be a continuous function and let its first-order derivative $f'(t)$ be continuous by parts on the interval $0 \le t \le T$ for $T > 0$. Suppose that $f(t)$ is of exponential order as $t \to \infty$. Then the Laplace transform of $f'(t)$ exists and is given by*

$$\mathcal{L}[f'(t)] = s\mathcal{L}[f(t)] - f(0).$$

If $f'(t)$ is also of exponential order, this theorem implies that the Laplace transform of the second order derivative, $f''(t)$, is given by

$$\mathcal{L}[f''(t)] = s^2\mathcal{L}[f(t)] - sf(0) - f'(0). \tag{6.1}$$

Theorem 6.5 (Integration) *If $F(s)$ is the Laplace transform of function $f(t)$, then*

$$\mathcal{L}\left[\int_0^t f(\tau)\,\mathrm{d}\tau\right] = \frac{1}{s}F(s),$$

provided that $\displaystyle\int_0^t f(\tau)\,\mathrm{d}\tau$ *is of exponential order.*

Finally, before we discuss the calculation of the inverse Laplace transform, we define the so-called *convolution product of two functions*, denoted by $\star$. The function

$$f(t) \star g(t) \equiv \int_0^t f(t - \tau) g(\tau)\, d\tau$$

is known as the *convolution* of functions $f(t)$ and $g(t)$.

Theorem 6.6 (Convolution) *Let $F(s)$ and $G(s)$ be the Laplace transforms of functions $f(t)$ and $g(t)$, respectively. Then,*

$$\mathscr{L}[f(t) \star g(t)] = F(s)G(s),$$

that is, the Laplace transform of the convolution $f(t) \star g(t)$ is equal to the product of the corresponding Laplace transforms, $F(s)G(s)$.

We can now introduce the inverse Laplace transform, given by the following integral on the complex plane:

$$\mathscr{L}^{-1}[F(s)] \equiv f(t) = \begin{cases} \dfrac{1}{2\pi i} \displaystyle\int_{\gamma - i\infty}^{\gamma + i\infty} e^{st} F(s)\, ds & \text{if } t > 0, \\[2ex] 0 & \text{if } t < 0, \end{cases} \tag{6.2}$$

where the integration is to be carried along the straight line $s = \gamma > 0$ on the complex plane, with $s = x + iy$. The complex number γ must be chosen in such a way that the straight line $s = \gamma$ be on the right-hand side of all singularities— poles, branching points, and/or essential singularities—of the integrand, being otherwise arbitrary. This contour is known as *Bromwich* [1875 – Thomas John Ianson Bromwich – 1929] *contour*, and is shown in Fig. 6.1, with the respective orientation.

Fig. 6.1 Bromwich contour, used to evaluate integral Eq. (6.2)

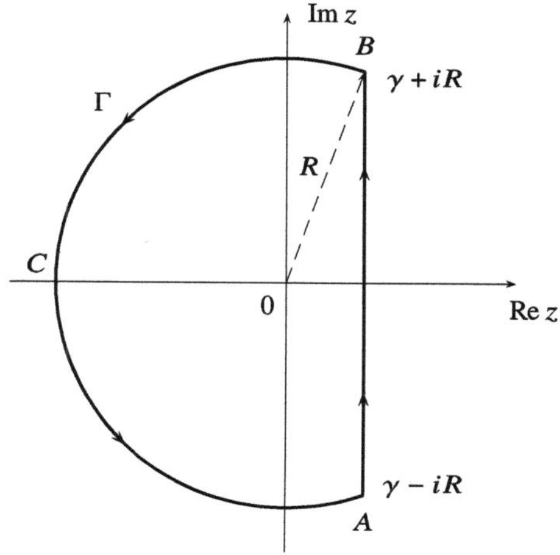

Suppose that all singularities of $F(s)$ are poles and are found on the left-hand side of the straight line $s = \gamma$, where γ is a constant. Using the residue theorem, discussed in Chap. 3, we then have

$$f(t) = \sum \text{Residues of } e^{st} F(s) \text{ at the poles of } F(s).$$

6.2   Fourier Transform

Just as in the case of the Laplace transform, we introduce the Fourier transform by means of a theorem.

Theorem 6.7 *Let $f(t)$ be a continuous function, smooth by parts and absolutely integrable. If we define*

$$F(\alpha) \equiv \mathscr{F}[f(t)] = \frac{1}{\sqrt{2\pi}} \int_{-\infty}^{\infty} f(t) \, e^{i\alpha t} \, dt,$$

we then have, for all t,

$$f(t) = \mathscr{F}^{-1}[F(\alpha)] = \frac{1}{\sqrt{2\pi}} \int_{-\infty}^{\infty} F(\alpha) \, e^{-i\alpha t} \, d\alpha \, .$$

The function $F(\alpha)$ is known as the *Fourier transform* of $f(t)$ and $f(t)$ is known as the *inverse Fourier transform* of $F(\alpha)$, where α is the so-called transformed variable.

Theorem 6.8 (Linearity)   *If $\mathscr{F}[f(t)]$ and $\mathscr{F}[g(t)]$ are the Fourier transforms of $f(t)$ and $g(t)$, respectively, then*

$$\mathscr{F}[c_1 f(t) + c_2 g(t)] = c_1 \mathscr{F}[f(t)] + c_2 \mathscr{F}[g(t)],$$

for all constants c_1 and c_2.

Theorem 6.9 (Displacement)   *If the Fourier transform of a function $f(t)$ is $\mathscr{F}[f(t)]$, then*

$$\mathscr{F}[f(t - c)] = e^{i\alpha c} \, \mathscr{F}[f(t)],$$

for $c > 0$.

Theorem 6.10 (Scale) *If the Fourier transform of a function $f(t)$ is $\mathscr{F}[f(t)]$, then*

$$\mathscr{F}[f(ct)] = \frac{1}{|c|}\mathscr{F}(\alpha/c),$$

for $c \neq 0$.

Theorem 6.11 (Differentiation) *Let $f(t)$ be a continuous and smooth by parts function on the interval $(-\infty, \infty)$, with $f(t)$ going to zero as $|t| \to \infty$. If the function $f(t)$ and its first derivative $f'(t)$ are absolutely integrable, then*

$$\mathscr{F}[f'(t)] = -i\alpha\mathscr{F}[f(t)].$$

If the first $(n-1)$ derivatives of $f(t)$ are continuous and its n-th derivative is continuous by parts, the aforementioned result can be extended, that is,

$$\mathscr{F}[f^{(n)}(t)] = (-i\alpha)^n\mathscr{F}[f(t)].$$

Theorem 6.12 (Convolution) *If the functions $F(\alpha)$ and $G(\alpha)$ are the Fourier transforms of $f(t)$ and $g(t)$, respectively, then the Fourier transform of their convolution $f(t) \star g(t)$ is equal to the product of the corresponding Fourier transforms,*

$$\mathscr{F}[f(t) \star g(t)] = F(\alpha)G(\alpha).$$

Analogously to the case of Fourier series in sine and cosine, presented in Chap. 5, there exist also Fourier transforms in sine and cosine, respectively for odd and even functions. These Fourier transforms are extensively used when a region is semi-infinite. Before we define such Fourier transforms, we present the so-called Fourier integral theorem:

Theorem 6.13 *If a function $f(t)$ is smooth by parts on every finite interval of the real straight line and absolutely integrable on $(-\infty, \infty)$, then*

$$\frac{1}{2}[f(t^+) + f(t^-)] = \frac{1}{\pi}\int_{\alpha=0}^{\infty}\left\{\int_{t'=-\infty}^{\infty} f(t')\cos[\alpha(t - t')]\,dt'\right\}\,d\alpha,$$

with α the transformed variable.

Odd Extension: (Fourier sine transform)

Consider a function $f(t)$ defined on $0 \leq t < \infty$. Let $f(t)$ be extended as an odd function on $(-\infty, \infty)$ satisfying the conditions of the Fourier integral theorem. If at the points of continuity of function f we have

$$F_s(\alpha) = \sqrt{\frac{2}{\pi}}\int_0^{\infty} f(t)\sin(\alpha t)\,dt,$$

then

$$f(t) = \sqrt{\frac{2}{\pi}} \int_0^\infty F_s(\alpha) \sin(\alpha t)\, d\alpha \ .$$

Even Extension: (Fourier cosine transform)

Consider a function $f(t)$ defined on the interval $0 \leq t < \infty$. Let $f(t)$ be extended as an even function on the interval $(-\infty, \infty)$ satisfying the conditions of the integral Fourier theorem. If at the points of continuity of the function f, we have

$$F_c(\alpha) = \sqrt{\frac{2}{\pi}} \int_0^\infty f(t) \cos(\alpha t)\, dt \ ,$$

then

$$f(t) = \sqrt{\frac{2}{\pi}} \int_0^\infty F_c(\alpha) \cos(\alpha t)\, d\alpha \ .$$

Theorem 6.14 (Differentiation) *If the function $f(t)$ and its first derivative are null as $t \to \infty$, and if $F_c(\alpha)$ is the Fourier cosine transform of $f(t)$ and $F_s(\alpha)$ is its Fourier sine transform, then:*

$$\mathscr{F}_c[f''(t)] = -\alpha^2 F_c(\alpha) - \sqrt{\frac{2}{\pi}} f'(0)$$

and

$$\mathscr{F}_s[f''(t)] = \sqrt{\frac{2}{\pi}} \alpha f(0) - \alpha^2 F_s(\alpha).$$

Finally, we also have finite Fourier transforms in sine and cosine, which are quite useful in the calculation of solutions of partial differential equations when we have finite intervals, as we will see in some applications in Chap. 9.

Definition 6.2.1 (Finite Fourier Sine Transform)

The *finite Fourier sine transform* of function $f(t)$, for $0 < t < l$, is defined as

$$F_s(n) = \int_0^l f(t) \sin \frac{n\pi t}{l}\, dt,$$

with $n = 1, 2, \ldots$ The function $f(t)$ is known as the inverse finite Fourier sine transform of $F_s(n)$ and is given by

$$f(t) = \frac{2}{l} \sum_{n=1}^\infty F_s(n) \sin \frac{n\pi t}{l} \ .$$

Definition 6.2.2 (Finite Fourier Cosine Transform)

The *finite Fourier cosine transform* of $f(t)$ (**PE 6.35**), for $0 < t < l$, is given by

$$F_c(n) = \int_0^l f(t) \cos \frac{n\pi t}{l} \, dt \, , \quad n = 0,\, 1,\, 2 \ldots$$

The inverse finite Fourier cosine transform of $F_c(n)$ is then

$$f(t) = \frac{1}{l} F_c(0) + \frac{2}{l} \sum_{n=1}^{\infty} F_c(n) \cos \frac{n\pi t}{l} \, .$$

6.3 Solved Exercises

SE 6.1 Let $a \in \mathbb{R}$. Evaluate the Laplace transform of function $f(t) = \cos at$.
Solution: We must evaluate the improper real integral

$$\mathscr{L}[\cos at] = \int_0^{\infty} e^{-st} \cos at \, dt \, .$$

To evaluate this integral, we use the Euler relation to write the cosine function in terms of exponential functions. Another way to evaluate this integral is by means of integration by parts. Thus,

$$\mathscr{L}[\cos t] = \frac{1}{2} \int_0^{\infty} e^{-st} [e^{iat} + e^{-iat}] \, dt \, ,$$

which can be rewritten in the form

$$\mathscr{L}[\cos t] = \frac{1}{2} \int_0^{\infty} e^{-t(s-ia)} \, dt + \frac{1}{2} \int_0^{\infty} e^{-t(s+ia)} \, dt \, ,$$

whose integration provides

$$\mathscr{L}[\cos t] = \frac{1}{2} \frac{1}{(s-ia)} + \frac{1}{2} \frac{1}{(s+ia)} \, .$$

Simplifying, we get the Laplace transform of $f(t) = \cos at$,

$$\mathscr{L}[\cos t] = \frac{s}{s^2 + a^2} \, .$$

Taking the inverse Laplace transform in this last expression we can write

$$\mathscr{L}^{-1}\left[\frac{s}{s^2 + a^2}\right] = \cos at \, .$$

SE 6.2 Using the Laplace transform, obtain a particular solution for the linear, second order, nonhomogeneous ordinary differential equation

$$\frac{d^2}{dt^2}x(t) + \omega^2 x(t) = t^2 \tag{6.3}$$

satisfying the initial conditions $x(0) = 0$ and $x'(0) = 0$, where ω^2 is a positive constant.

Solution: Let $X(s)$ be the Laplace transform of $x(t)$, given by

$$X(s) = \int_0^\infty e^{-st} x(t)\, dt \,.$$

Then, applying the Laplace transform to both members of Eq. (6.3) we have

$$\int_0^\infty e^{-st} x''(t)\, dt + \int_0^\infty \omega^2 e^{-st} x(t)\, dt = \int_0^\infty e^{-st} t^2\, dt \,. \tag{6.4}$$

For the first integral we have (cf. Eq. (6.1))

$$\int_0^\infty e^{-st} x''(t)\, dt = s^2 X(s) - sx(0) - x'(0),$$

and using the initial conditions we can write

$$\int_0^\infty e^{-st} x''(t)\, dt = s^2 X(s) \,.$$

The integral on the second member of Eq. (6.4) can be calculated, for example, using integration by parts; then, the resulting algebraic equation in variable $X(s)$ is

$$s^2 X(s) + \omega^2 X(s) = \frac{2}{s^3} \,.$$

Solving this algebraic equation for $X(s)$, we get

$$X(s) = \frac{2}{s^3(s^2 + \omega^2)} \,.$$

The expression on the right side can be written as a sum of partial fractions,

$$\frac{2}{s^3(s^2 + \omega^2)} = \frac{-2/\omega^4}{s} + \frac{2/\omega^2}{s^3} + \frac{(2/\omega^4)s}{s^2 + \omega^2} \,.$$

We must now evaluate the corresponding inverse Laplace transform, that is,

$$x(t) = \mathcal{L}^{-1}\left[\frac{2}{s^3(s^2+\omega^2)}\right]$$

$$= -\frac{2}{\omega^4}\mathcal{L}^{-1}\left[\frac{1}{s}\right] + \frac{2}{\omega^2}\mathcal{L}^{-1}\left[\frac{1}{s^3}\right] + \frac{2}{\omega^4}\mathcal{L}^{-1}\left[\frac{s}{s^2+\omega^2}\right].$$

Carrying the calculations and using the residue theorem, we finally get

$$x(t) = -\frac{2}{\omega^4} + \frac{t^2}{\omega^2} + \frac{2}{\omega^4}\cos\omega t,$$

which is the solution sought.

SE 6.3 (Branch Point) Let u_0 and k be two positive constants. Evaluate

$$u(x,t) = \frac{1}{2\pi i}\int_{\gamma-i\infty}^{\gamma+i\infty} e^{st}\, f(s)\, ds$$

for $f(s) = \dfrac{u_0}{s}\exp\left(-\sqrt{\dfrac{s}{k}}x\right)$.

Solution: Write $\dfrac{x}{\sqrt{k}} = \alpha$. Then

$$u(x,t) = \frac{1}{2\pi i}\int_{\gamma-i\infty}^{\gamma+i\infty} \frac{u_0}{s}\, e^{st-\alpha\sqrt{s}}\, ds, \tag{6.5}$$

where $s = 0$ is a branch point. Consider now the contour on the complex plane shown in Fig. 6.2, known as *modified Bromwich contour*. The integral in Eq. (6.5) is given by

$$\frac{u_0}{2\pi i}\oint_\gamma \frac{1}{s} e^{st-\alpha\sqrt{s}}\, ds = \frac{u_0}{2\pi i}\left\{\int_{ab} + \int_{bde} + \int_{eh} + \right.$$

$$\left. + \int_{HJK} + \int_{KL} + \int_{LNA}\right\} e^{st-\alpha\sqrt{s}}\,\frac{ds}{s} = 0.$$

The second equality is a consequence of the residue theorem (Chap. 3) and of the fact that the singularity (the branch point) is outside the contour Γ formed by the line segments AB ($s = \gamma$), EH and LK; the arcs of circumference of radius R centered at the origin BDE and LNA; and the circumference of radius ϵ, also centered at the origin, HJK.

As $R \to \infty$ the integrals along arcs BDE and LNA go to zero, so that

$$\frac{1}{2\pi i}\left\{\int_{AB} + \int_{EH} + \int_{KL} + \int_{HJK}\right\} e^{st-\alpha\sqrt{s}}\,\frac{ds}{s} = 0.$$

Fig. 6.2 Modified Bromwich
contour, used in the
calculation of the integral
Eq. (6.5)

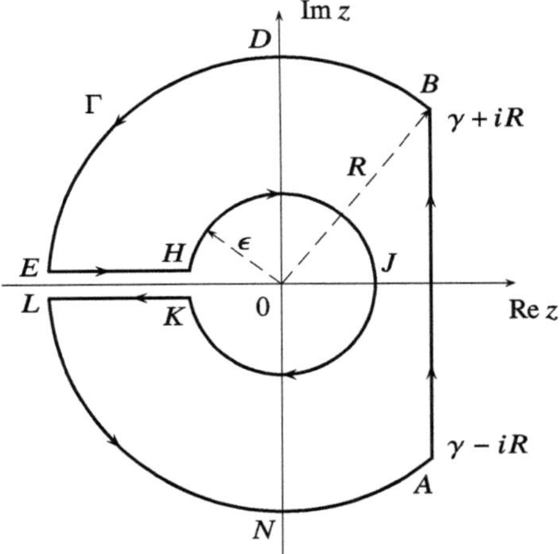

Since we want to know the value of the integral along AB, we take the limits
$R \to \infty$ and $\epsilon \to 0$, obtaining

$$\frac{1}{2\pi i}\int_{AB} = -\frac{1}{2\pi i}\int_{\gamma-i\infty}^{\gamma+i\infty}\frac{ds}{s}\,e^{st-\alpha\sqrt{s}} =$$

$$= \lim_{\substack{R\to\infty\\ \epsilon\to 0}}\frac{1}{2\pi i}\left\{\int_{EH}+\int_{HJK}+\int_{KL}\right\}\frac{ds}{s}\,e^{st-\alpha\sqrt{s}}. \tag{6.6}$$

Calculating separately each integral, we find:

(a) Along segment EH:

$$\int_{EH}\frac{ds}{s}\,e^{st-\alpha\sqrt{s}} = \int_{-R}^{-\epsilon}\frac{ds}{s}\,e^{st-\alpha\sqrt{s}} = \int_{R}^{\epsilon}\frac{dx}{x}\,e^{-xt-\alpha i\sqrt{x}},$$

where we have introduced $s = x\,e^{i\pi}$.

(b) Along segment KL:

$$\int_{KL}\frac{ds}{s}\,e^{st-\alpha\sqrt{s}} = \int_{-\epsilon}^{-R}\frac{ds}{s}\,e^{st-\alpha\sqrt{s}} = \int_{\epsilon}^{R}\frac{dx}{x}\,e^{-xt+\alpha i\sqrt{x}},$$

where $s = x\,e^{-i\pi}$.

(c) Along circumference HJK, where $s = \epsilon\,e^{i\theta}$:

$$\int_{HJK}\frac{ds}{s}\,e^{st-\alpha\sqrt{s}} = i\int_{\pi}^{-\pi}e^{\epsilon\,e^{i\theta}\,t-\alpha\sqrt{\epsilon}\,e^{i\theta/2}}\,d\theta.$$

Going back to Eq. (6.6) and taking adequate limits, the aforementioned results imply that:

$$-\frac{1}{2\pi i} \int_{\gamma-i\infty}^{\gamma+i\infty} \frac{ds}{s}\, e^{st-\alpha\sqrt{s}} = -1 + \frac{1}{\pi} \int_0^\infty \frac{dx}{x}\, e^{-xt} \sin\alpha\sqrt{x}\,.$$

Hence, the corresponding inverse Laplace transform of $f(s)$ is

$$u(x,t) = u_0 \left(1 - \frac{1}{\pi} \int_0^\infty \frac{dx}{x}\, e^{-xt} \sin\alpha\sqrt{x}\right)\,.$$

Introducing a new variable defined by $x = y^2$, we obtain the integral

$$I = \int_0^\infty \frac{dx}{x}\, e^{-xt} \sin\alpha\sqrt{x} = 2 \int_0^\infty \frac{dy}{y}\, e^{-ty^2} \sin\alpha y\,,$$

which finally leads us to the interesting result

$$u(x,t) = u_0\, \mathrm{erfc}\left(\frac{x}{2\sqrt{kt}}\right),$$

where erfc is know as *complementary error function*, which is given by the integral

$$\mathrm{erfc}(y) = 1 - \sqrt{\frac{2}{\pi}} \int_0^y e^{-u^2}\, du\,.$$

SE 6.4 Find the Fourier transform of

$$f(t) = \begin{cases} 1 & \text{for} \quad |t| \leq k, \\ 0 & \text{for} \quad |t| > k, \end{cases}$$

with $k > 0$, and show that

$$\int_{-\infty}^\infty \frac{\sin k\alpha \cos \alpha t}{\alpha}\, d\alpha = \begin{cases} \pi & \text{for } |t| < k, \\ \frac{\pi}{2} & \text{for } |t| = k, \\ 0 & \text{for } |t| > k. \end{cases}$$

Solution: The Fourier transform of function $f(t)$ is

$$F(\alpha) = \int_{-\infty}^\infty f(t)\, e^{-i\alpha t}\, dt = 2\frac{\sin\alpha k}{k}, \quad \text{for} \quad \alpha \neq 0\,.$$

For $\alpha = 0$, we have $F(0) = 2k$. Using the Fourier integral theorem, we get

$$F(\alpha) = \int_{-\infty}^{\infty} F(t)\,e^{-i\alpha t}\,dt \quad \text{and} \quad f(t) = \frac{1}{2\pi}\int_{-\infty}^{\infty} F(\alpha)\,e^{i\alpha t}\,d\alpha\,.$$

The aforementioned results allow us to write the following expression:

$$\frac{1}{2\pi}\int_{-\infty}^{\infty}\frac{2\sin k\alpha}{\alpha}\,e^{i\alpha t}\,d\alpha = \begin{cases} 1 & \text{for } |t| < k, \\ \frac{1}{2} & \text{for } |t| = k, \\ 0 & \text{for } |t| > k. \end{cases}$$

This integral can be written as

$$\frac{1}{\pi}\int_{-\infty}^{\infty}\frac{\sin k\alpha}{\alpha}\cos\alpha t\,d\alpha + \frac{i}{\pi}\int_{-\infty}^{\infty}\frac{\sin k\alpha}{\alpha}\sin\alpha t\,d\alpha\,.$$

Notice that the second integral aforementioned is null because the integrand is an odd function and the integration interval is symmetric. We thus have

$$\int_{-\infty}^{\infty}\frac{\sin k\alpha}{\alpha}\cos\alpha t\,d\alpha = \begin{cases} \pi & \text{for } |t| < k, \\ \frac{\pi}{2} & \text{for } |t| = k, \\ 0 & \text{for } |t| > k, \end{cases}$$

which is the desired result.

SE 6.5 Obtain the Fourier transform of the Dirac [1902 – Paul Adrien Maurice Dirac – 1984] delta function, denoted by $\delta(t)$.

Solution: Consider an impulse function $p(t)$ defined by the relation

$$p(t) = \begin{cases} h & \text{if } \quad a - \epsilon < t < a + \epsilon \\ 0 & \text{if } \quad |t| \geq a + \epsilon \end{cases}$$

where h is a large positive number, $a > 0$ and ϵ is a small positive constant. This type of function appears, for example, in problems in which we have a force with large magnitude acting during a short time interval. The Fourier transform of this function is

$$\mathscr{F}[p(t)] = \frac{1}{\sqrt{2\pi}}\int_{-\infty}^{\infty} p(t)\,e^{i\alpha t}\,dt$$

$$= \frac{1}{\sqrt{2\pi}}\int_{a-\epsilon}^{a+\epsilon} h\,e^{i\alpha t}\,dt = \frac{2h\epsilon}{\sqrt{2\pi}}\,e^{i\alpha a}\,\frac{\sin\alpha\epsilon}{\alpha\epsilon}\,.$$

Now, if we choose $h = 1/2\epsilon$, we have

$$I(\epsilon) = \int_{-\infty}^{\infty} p(t)\,dt = \int_{a-\epsilon}^{a+\epsilon}\frac{1}{2\epsilon}\,dt = 1,$$

which is a constant independent of ϵ. Taking the limit in which ϵ goes to zero, this particular function $p_\epsilon(t)$, with $2h\epsilon = 1$, satisfies the relations

$$\lim_{\epsilon \to 0} p_\epsilon(t) = 0 \quad \text{for} \quad t \neq a$$

and

$$\lim_{\epsilon \to 0} I(\epsilon) = 1.$$

We thus arrived at a function given by (the correct term for such a *function* is *distribution*)

$$\delta(t - a) = 0 \quad \text{for} \quad t \neq a,$$

and

$$\int_{-\infty}^{\infty} \delta(t - a)\, dt = 1,$$

which is the so-called *Dirac delta function*. We evaluate the Fourier transform of the Dirac delta function as the limit of the Fourier transform of the impulse function $p_\epsilon(t)$, that is, we take

$$\mathscr{F}[\delta(t - a)] = \lim_{\epsilon \to 0} \mathscr{F}[p_\epsilon(t)]$$

$$= \lim_{\epsilon \to 0} \frac{e^{i\alpha a}}{\sqrt{2\pi}} \frac{\sin \alpha\epsilon}{\alpha\epsilon}.$$

Since $\lim_{\epsilon \to 0} \dfrac{\sin \alpha\epsilon}{\alpha\epsilon} = 1$, we finally obtain

$$\mathscr{F}[\delta(t - a)] = \frac{e^{i\alpha a}}{\sqrt{2\pi}},$$

which, in the case $a = 0$, provides the desired result

$$\mathscr{F}[\delta(t)] = \frac{1}{\sqrt{2\pi}}.$$

SE 6.6 Use the integral representation of the Bessel function of order zero,

$$\mathscr{J}_0(x) = \frac{1}{2\pi} \int_0^{2\pi} \cos(x \cos \theta)\, d\theta,$$

to calculate the Laplace transform of $\mathcal{J}_0(x)$, that is, show that

$$\int_0^\infty e^{-sx}\, \mathcal{J}_0(x)\, dx = \frac{1}{\sqrt{1+s^2}}$$

for $\mathrm{Re}(s \pm i) > 0$.

Solution: Substituting the integral representation given above into the integral for the Laplace transform, we can write

$$\mathcal{L}[\mathcal{J}_0(t)] = \int_0^\infty \frac{1}{2\pi} \int_0^{2\pi} \cos(t\,\cos\theta)\, d\theta\; e^{-st}\, dt$$

or, changing the order of integration,

$$\mathcal{L}[\mathcal{J}_0(t)] = \int_0^{2\pi} \frac{1}{2\pi}\, d\theta \int_0^\infty \cos\beta t\; e^{-st}\, dt,$$

where we defined $\beta = \cos\theta$. Integrating in variable t we obtain

$$\mathcal{L}[\mathcal{J}_0(t)] = \frac{s}{2\pi} \int_0^{2\pi} \frac{d\theta}{s^2 + \cos^2\theta}\,.$$

In order to evaluate this integral, before using the residue theorem, we remember a formula involving the double arc,

$$\cos^2\theta = \frac{1 + \cos 2\theta}{2},$$

which allows us to write the relation

$$\mathcal{L}[\mathcal{J}_0(t)] = \frac{s}{\pi} \int_{-\pi}^{\pi} \frac{d\theta}{2s^2 + 1 + \cos\theta}\,.$$

This integral can be solved by means of a parametrization of the form

$$z = e^{i\theta},$$

which yields $\cos\theta = (z + z^{-1})/2$, i.e., a circumference C centered at the origin with unitary radius. With this parametrization we obtain the following integral on the complex plane:

$$\mathcal{L}[\mathcal{J}_0(t)] = \frac{2s}{\pi i} \oint_C \frac{dz}{z^2 + 2\alpha z + 1},$$

where $\alpha = 1 + 2s^2$. The denominator has two real and distinct roots, but only one of them is in the inner region of the contour, the circumference centered at the origin and radius 1. Then, using the residue theorem, we can write

$$\mathcal{L}[\mathcal{J}_0(t)] = \frac{2s}{\pi i} 2\pi i \lim_{z \to z_1} (z - z_1) \frac{1}{(z - z_1)(z - z_2)} = \frac{4s}{z_1 - z_2},$$

where z_1 and z_2 are the poles of the integrand. Substituting their respective values we finally obtain

$$\mathcal{L}[\mathcal{J}_0(t)] = \frac{1}{\sqrt{s^2 + 1}}.$$

SE 6.7 The Fourier transform can be generalized for more than one dimension. For example, in three-dimensional space, we have the following transform (the point denotes the scalar product and the integral is carried over the entire space):

$$\mathcal{F}[f(\mathbf{x})] \equiv \phi(\mathbf{k}) = \int d^3x \, f(\mathbf{x}) \, e^{-i\mathbf{k}\cdot\mathbf{x}},$$

whose corresponding inverse transform is given by

$$\mathcal{F}^{-1}[\phi(\mathbf{k})] \equiv f(\mathbf{x}) = \int \frac{d^3k}{(2\pi)^3} \phi(\mathbf{k}) \, e^{i\mathbf{k}\cdot\mathbf{x}} \, .$$

As an example, find the Fourier transform of the wave function associated with state $2p$ of an electron in the hydrogen atom,

$$\Psi(\mathbf{x}) = \frac{1}{\sqrt{32\pi a_0^5}} z \, e^{-r/2a_0},$$

where a_0 is the radius of the Bohr [1885 – Niels Henrik David Bohr – 1962] first orbit and z is a cartesian coordinate.

Solution: Introducing spherical coordinates (r, θ, ϕ), we can write

$$J = \int_0^{2\pi} \int_0^\pi \int_0^\infty d\phi \, d\theta \, dr \, r^2 \sin\theta \, \frac{1}{\sqrt{32\pi a_0^5}} r \cos\theta \, e^{-r/2a_0} \, e^{-ikr\cos\theta},$$

where we take $\mathbf{k} \cdot \mathbf{x} = kr\cos\theta$, i.e., we choose $\mathbf{k}$ in the direction of the z axis. Integrating in variable ϕ and introducing the change of variable $\cos\theta = x$ we obtain

$$J = \frac{2\pi}{\sqrt{32\pi a_0^5}} \int_0^\infty dr \, e^{-\alpha r} r^3 \int_{-1}^1 x \, e^{-\beta x} \, dx,$$

where we have introduced the parameters $\alpha = 1/2a_0$ and $\beta = ikr$. Finally, integrating in variables x and r we get

$$J = \frac{2\pi/k}{\sqrt{32\pi a_0^5}}\frac{16}{i}\frac{\alpha k^2}{(\alpha^2 + k^2)^2} = \frac{32k}{i}\frac{\sqrt{32\pi a_0^5}}{(1 + 4a_0^2 k^2)^3}.$$

SE 6.8 Consider a semi-infinite spring fixed at its extreme point $x = 0$ and which is initially at rest. Let f_0 be an external concentrated force acting at the point $x = vt$. We want to know the displacement $u(x, t)$ of the spring at the point with coordinate x at time t, for all x, $t \geq 0$. Using the Laplace transform methodology to solve this problem, we arrive at the following transformed function:

$$U(x, s) = \begin{cases} f_0 \dfrac{v^2}{s^2}\dfrac{e^{-sx/v} - e^{-sx/c}}{c^2 - v^2} & v \neq c, \\[2ex] -f_0 x \dfrac{e^{-sx/c}}{2cs} & v = c, \end{cases}$$

where c is a constant and s is the parameter of the Laplace transform. Calculate the corresponding inverse transform, i.e., calculate the following integral on the complex plane:

$$u(x, t) = \frac{1}{2\pi i}\int_C e^{st}\, U(x, s)\, ds,$$

where contour C leaves all singularities to the left of the straight line $s = \gamma$ with γ real, in order to show that

$$u(x, t) = f_0 \begin{cases} \dfrac{v^2}{c^2 - v^2}\left[\left(t - \dfrac{x}{v}\right)\theta\left(t - \dfrac{x}{v}\right) - \left(t - \dfrac{x}{c}\right)\theta\left(t - \dfrac{x}{c}\right)\right] & v \neq c, \\[2ex] -\dfrac{x}{2c}\theta\left(t - \dfrac{x}{c}\right) & v = c. \end{cases}$$

Solution: We must calculate, for $v \neq c$, the integral

$$u(x, t) = \frac{1}{2\pi i}\int_C e^{st}\, f_0 \frac{v^2}{s^2}\frac{e^{-sx/v} - e^{-sx/c}}{c^2 - v^2}\, ds$$

or

$$u(x, t) = \frac{f_0}{2\pi i}\frac{v^2}{c^2 - v^2}\int_C \frac{e^{s(t-x/v)} - e^{s(t-x/c)}}{s^2}\, ds,$$

which presents a pole of double order at $s = 0$. Thus, using the residue theorem we get

$$u(x,t) = \frac{f_0}{2\pi i}\frac{c^2}{c^2 - v^2}2\pi i \lim_{s\to 0}\frac{d}{ds}\left[e^{s(t-x/v)} - e^{s(t-x/c)}\right]$$

$$= f_0\frac{c^2}{c^2 - v^2}\left\{\left(t - \frac{x}{v}\right)\theta\left(t - \frac{x}{v}\right) - \left(t - \frac{x}{c}\right)\theta\left(t - \frac{x}{c}\right)\right\},$$

where θ is the so-called Heaviside [1850 – Oliver Heaviside – 1925] function, defined by,

$$\theta(x - a) = \begin{cases} 1, & x \geq a, \\ 0, & x < a. \end{cases}$$

On the other hand, we must also calculate the integral for the case $v = c$,

$$u(x,t) = \frac{1}{2\pi i}\int_C e^{st}(-f_0)\frac{x}{2cs}\,e^{-sx/c}\,ds,$$

or

$$u(x,t) = -\frac{f_0}{2\pi i}\frac{x}{2c}\int_C e^{s(t-x/c)}\frac{ds}{s},$$

which presents a simple pole at $s = 0$. From the residue theorem, we can write

$$u(x,t) = -\frac{f_0}{2\pi i}\frac{x}{2c}2\pi i \lim_{s\to 0} e^{s(t-x/c)}$$

$$= -\frac{f_0}{2c}x\theta(t - x/c).$$

Finally, combining these two results, we arrive at the desired solution, which is given by

$$u(x,t) = f_0\begin{cases} \dfrac{c^2}{c^2 - v^2}\left[\left(t - \dfrac{x}{v}\right)\theta\left(t - \dfrac{x}{v}\right) - \left(t - \dfrac{x}{c}\right)\theta\left(t - \dfrac{x}{c}\right)\right] & \text{for } v \neq c, \\[4mm] \dfrac{-x}{2c}\theta(t - x/c) & \text{for } v = c. \end{cases}$$

SE 6.9 Use the convolution theorem to evaluate the Laplace transform

$$\mathscr{L}^{-1}\left[\frac{s}{(s^2 + a^2)^2}\right],$$

where a is a real constant.

Solution: In order to use the convolution theorem to evaluate this inverse Laplace transform, we begin writing the integrand in the form of a product:

$$\frac{s}{(s^2 + a^2)^2} = \frac{s}{s^2 + a^2} \frac{1}{s^2 + a^2}.$$

Then, using directly inversion formulas, we get

$$\mathscr{L}^{-1}\left[\frac{s}{s^2 + a^2}\right] = \cos at \qquad \text{and} \qquad \mathscr{L}^{-1}\left[\frac{1}{s^2 + a^2}\right] = \frac{\sin at}{a}.$$

We now use the convolution theorem, that is,

$$\mathscr{L}^{-1}(f * g) = \int_0^t f(t - \xi)g(\xi)\, d\xi.$$

In this case, we find the following expression:

$$\mathscr{L}^{-1}\left[\frac{s}{(s^2 + a^2)^2}\right] = \int_0^t \cos a\xi\, \sin[a(t - \xi)]\frac{d\xi}{a} =$$

$$= \frac{\sin at}{a}\int_0^t \cos^2 a\xi\, d\xi - \frac{\cos at}{a}\int_0^t \sin a\xi\, \cos a\xi\, d\xi,$$

which, after evaluating the integrals, yields

$$\mathscr{L}^{-1}\left[\frac{s}{(s^2 + a^2)^2}\right] = \frac{t}{2a}\sin at.$$

SE 6.10 Consider the ordinary differential equation

$$\frac{d^2}{dt^2}x(t) - k^2 x(t) = f(t),$$

where $k^2 = $ constant and $0 \le t < \infty$, and $x(t)$ satisfies the conditions

$$\frac{d}{dt}x(t)|_{t=0} = b \qquad \text{and} \qquad x(\infty) < \infty,$$

with b a real constant. Use the Fourier cosine transform to show that

$$x(t) = -\frac{b}{k}e^{-kt} - \frac{1}{2k}\int_0^\infty f(\xi)\left(e^{-k|t-\xi|} + e^{-k|t+\xi|}\right) d\xi.$$

Solution: The Fourier cosine transform for $f(t)$ is given by

$$F_C(\alpha) = \sqrt{\frac{2}{\pi}}\int_0^\infty f(t)\cos \alpha t\, dt,$$

while for the second derivatives we have

$$\mathscr{F}[f''(x)] = -\alpha^2 F_C(\alpha) - \sqrt{\frac{2}{\pi}}\, f'(0),$$

or, in this case,

$$F_C(\alpha) = -\frac{f_0 + b\sqrt{2/\pi}}{\alpha^2 + k^2},$$

where we have defined

$$f_0 = \sqrt{\frac{\pi}{2}} \int_0^\infty f(t) \cos \alpha t \, dt.$$

The inverse Fourier cosine transform is given by

$$x(t) = \sqrt{\frac{2}{\pi}} \int_0^\infty F_C(\alpha) \cos \alpha t \, d\alpha$$

and substituting the expression for the Fourier cosine transform $F_C(\alpha)$ into the last expression we have

$$x(t) = -\frac{2b}{\pi} \int_0^\infty \frac{\cos \alpha t}{\alpha^2 + k^2} \, d\alpha - \frac{2}{\pi} \int_0^\infty f(\xi) \, d\xi \int_0^\infty \frac{\cos \alpha \xi \, \cos \alpha t}{\alpha^2 + k^2} \, d\alpha .$$

Using the residue theorem to solve the first integral, we get

$$x(t) = -\frac{b}{k} e^{-kt} - \frac{2}{\pi} \int_0^\infty f(\xi) \, d\xi \int_0^\infty \frac{\cos \alpha \xi \, \cos \alpha t}{\alpha^2 + k^2} \, d\alpha .$$

To solve the second integral, in variable α, we use the following trigonometric relation:

$$\cos \alpha \xi \, \cos \alpha t = \frac{1}{2} \left[\cos(\alpha \xi + \alpha t) + \cos(\alpha \xi - \alpha t) \right].$$

Using again the residue theorem, we finally obtain

$$x(t) = -\frac{b}{k} e^{-kt} - \frac{1}{2k} \int_0^\infty f(\xi) \left(e^{-k|\xi + t|} + e^{-k|\xi - t|} \right) d\xi,$$

which is the desired result.

6.4   Proposed Exercises

PE 6.1 Let $\mu \in \mathbb{R}_+$. Evaluate the Laplace transform $\mathscr{L}[t^\mu]$.

PE 6.2 Show that

$$\mathscr{L}^{-1}\left[\frac{1}{(s^2+1)^2}\right] = \sin t - t \cos t.$$

PE 6.3 Let $a > 0$ be a real constant. Let $x(t)$ and $f(t)$ be two functions that admit the Laplace transform. Solve the initial value problem

$$\begin{cases} \dfrac{\mathrm{d}}{\mathrm{d}t}x(t) + ax(t) = f(t)\,, & t > 0 \\ \qquad\qquad x(0) = x_0\,, \end{cases}$$

where x_0 is a positive constant.

PE 6.4 Let $x(t)$ be a function that admits the Laplace transform. Solve the initial value problem

$$\begin{cases} \dfrac{\mathrm{d}^2}{\mathrm{d}t^2}x(t) + x(t) = t\,, & t > 0, \\ \qquad\qquad x(0) = 1, \\ \qquad\qquad x'(0) = 0\,. \end{cases}$$

PE 6.5 Let $g(x) = x$. Use the Laplace transform to solve the integral equation

$$y(x) = 1 + \int_0^x g(x-\xi)y(\xi)\,\mathrm{d}\xi\,.$$

PE 6.6 Let $\sigma > 0$. Evaluate the Fourier transform $\mathscr{F}[e^{-\sigma x^2}]$.

PE 6.7 Let $f(x)$ be a function defined on the interval $0 \le x < \infty$ and which admits the Fourier transform. Denoting by $F_S(k)$ and $F_C(k)$ the Fourier sine and cosine transforms, respectively, given by

$$F_S(k) = \sqrt{\frac{2}{\pi}}\int_0^\infty f(x)\sin kx\,\mathrm{d}x \qquad \text{and} \qquad F_C(k) = \sqrt{\frac{2}{\pi}}\int_0^\infty f(x)\cos kx\,\mathrm{d}x,$$

show the following results:

$$\mathscr{F}_S[f'(x)] = -kF_C(k) \qquad \text{and} \qquad \mathscr{F}[f''(x)] = \sqrt{\frac{2}{\pi}}kf'(0) - k^2 F_S(k),$$

where the primes denote derivatives.

PE 6.8 Let $F(k)$ and $G(k)$ be the Fourier transforms of $f(x)$ and $g(x)$, respectively. The Parseval identity is given by

$$\int_{-\infty}^{\infty} f(x)g(x)\,dx = \int_{-\infty}^{\infty} F(k)\overline{G(k)}\,dk,$$

where $\overline{G(k)}$ denotes de conjugate of $G(k)$. Use the relation

$$\mathscr{F}\left[\frac{1}{1+x^2}\right] = \sqrt{\frac{\pi}{2}}\,\exp(-|k|)$$

to evaluate the integral

$$\Lambda = \int_0^{\infty} \frac{dx}{(x^2+1)^2}\,.$$

PE 6.9 Let $f(x)$ be a continuous function on the finite interval $0 < x < a$. We define the Fourier finite cosine transform and the corresponding inverse by

$$\mathscr{F}_C[f(x)] := F_C(n) = \int_0^a f(x)\cos\left(\frac{n\pi x}{a}\right)\,dx$$

and

$$\mathscr{F}_C^{-1}[F_C(n)] := f(x) = \frac{F_C(0)}{a} + \frac{2}{a}\sum_{n=1}^{\infty} F_C(n)\cos\left(\frac{n\pi x}{a}\right),$$

respectively. Show that, for $a = \pi$,

$$\mathscr{F}_C[f'(x)] = nF_S(n) + (-1)^n f(\pi) - f(0)$$

and

$$\mathscr{F}_C[f''(x)] = -n^2 F_C(n) + (-1)^n f'(\pi) - f'(0),$$

where $F_S(n)$ is the Fourier finite sine transform presented in **PE 6.10**.

PE 6.10 Let $f(x)$ be a continuous function on the finite interval $0 < x < a$. We define the Fourier finite sine transform and the corresponding inverse by

$$\mathscr{F}_S[f(x)] := F_S(n) = \int_0^a f(x)\sin\left(\frac{n\pi x}{a}\right)\,dx$$

and

$$\mathscr{F}_S^{-1}[F_S(n)] := f(x) = \frac{2}{a} \sum_{n=1}^{\infty} F_S(n) \sin\left(\frac{n\pi x}{a}\right),$$

respectively. Show that, for $a = \pi$,

$$\mathscr{F}_S[f'(x)] = -n F_C(n)$$

and

$$\mathscr{F}_C[f''(x)] = -n^2 F_S(n) + n[(-1)^{n+1} f(\pi) - f(0)],$$

where $F_C(n)$ is the Fourier finite cosine transform as in **PE 6.9**.

PE 6.11 Prove the first five theorems of this chapter, concerning the properties of the Laplace transform.

PE 6.12 Prove the following expression for the second derivative of the Laplace transform:

$$\mathscr{L}[f''(t)] = s^2 \mathscr{L}[f(t)] - sf(0) - f'(0).$$

PE 6.13 Calculate the Laplace transform for

$$x(t) = \frac{-2}{\omega^4} + \frac{t^2}{\omega^2} + \frac{2}{\omega^4} \cos \omega t$$

and compare the result with **SE 6.2**.

PE 6.14 Perform the calculations used to evaluate $X(s)$ in **SE 6.2**.

PE 6.15 Calculate the Laplace transform for $f(t) = t \cos at$, with $a > 0$.

PE 6.16 Perform the calculation of the inverse Laplace transforms in **SE 6.2**.

PE 6.17 Show that

$$\mathscr{L}^{-1}\left\{\frac{s^2}{(s^2 + a^2)^2}\right\} = \frac{1}{2}\left(t \cos at + \frac{1}{a} \sin at\right),$$

where $a > 0$ is a constant.

PE 6.18 Using the Laplace transform, solve the ordinary differential equation

$$xy'' + (1 - x)y' - y = 0.$$

Compare the result obtained with **PE 2.29**.

PE 6.19 Suppose that a function $f(t)$ is periodic with period T. Consider also that the function is continuous by parts on the interval $[0, T]$. Show that the Laplace transform of $f(t)$ is given by

$$\mathscr{L}[f(t)] = \frac{F_1(s)}{1 - e^{-Ts}},$$

where

$$F_1(s) = \int_0^T e^{-s\xi} f(\xi)\, d\xi$$

is the Laplace transform of function $f(t)$ on the first period.

PE 6.20 Using the result obtained in the preceding exercise, calculate the Laplace transform for the function

$$f(t) = \begin{cases} 1 & \text{if } 0 < t < 2, \\ -1 & \text{if } 2 < t < 4, \end{cases}$$

with $f(t + 4) = f(t)$.

PE 6.21 Calculate the Laplace transform $F(s)$ associated with the function

$$I(x) = \int_0^\infty \frac{\cos xt}{1 + t^2}\, dt,$$

with $x \geq 0$.

PE 6.22 Using the result obtained in the preceding exercise, calculate explicitly $I(x)$ and the value of the integral

$$\int_0^\infty \frac{\cos t}{1 + t^2}\, dt\,.$$

PE 6.23 Calculate the Laplace transform of the function

$$f(t) = \frac{\sin t}{t},$$

with $t > 0$.

PE 6.24 Use the Laplace transform to calculate the integral

$$I(x) = \int_0^\infty \frac{1 - \cos xt}{t^2}\, dt\,.$$

PE 6.25 Calculate the Laplace transform of the Bessel function of order zero.

PE 6.26 Using the result obtained in the preceding exercise, obtain an integral representation for $\mathcal{J}_0(x)$.

PE 6.27 Find *one* function satisfying the integro-differential equation

$$2\int_0^x f(t)f'(x-t)\,\mathrm{d}t = x\sin x,$$

where the prime denotes the derivative with respect to x.

PE 6.28 Solve the integro-differential equation

$$y(x) = \phi(x) + \int_0^x g(x-\xi)y(\xi)\,\mathrm{d}\xi$$

for $\phi(x) = 1$ and $g(x) = x$.

PE 6.29 Consider the first order linear partial differential equation

$$\frac{\partial}{\partial x}\omega(x,t) + x\frac{\partial}{\partial t}\omega(x,t) = 0,$$

with the conditions $\omega(x,0) = 0$ and $\omega(0,t) = t$. Solve this differential equation by means of the Laplace transform.

PE 6.30 Prove the Theorems 6.8 to 6.11 about the Fourier transform.

PE 6.31 Find the Fourier transform of $f(t) = \exp(-|t|)$.

PE 6.32 Obtain the Fourier sine and cosine transforms for the function $f(t) = \mathrm{e}^{-t}$.

PE 6.33 Show that

$$\mathscr{F}_s[f''(t)] = \sqrt{\frac{2}{\pi}}\alpha f(0) - \alpha^2 F_s(\alpha),$$

where $F_s(\alpha)$ is the Fourier sine transform of $f(t)$.

PE 6.34 Obtain expressions for the Fourier sine and cosine transforms associated with the first derivative $f'(t)$ of a function $f(t)$.

PE 6.35 Show that if the Fourier finite cosine transform of a function $f(t)$ for $-l < t < l$, with period equal to $2l$, is given by

$$F_c(n) = \int_0^l f(t)\cos\frac{n\pi t}{l}\,\mathrm{d}t\,, \qquad n = 0,\ 1,\ 2,\ldots$$

then

$$f(t) = \frac{1}{l}F_c(0) + \frac{2}{l}\sum_{n=1}^{\infty} F_c(n)\cos\frac{n\pi t}{l},$$

where $f(t)$ is the corresponding inverse Fourier transform $F_c(n)$.

PE 6.36 Find the Fourier finite sine and cosine transforms for $f(t) = t^2$ on the interval $0 \le t < l$.

PE 6.37 Find the Fourier transform of the function

$$f(t) = \begin{cases} 1 - t^2 & \text{for} \quad |t| < 1, \\ 0 & \text{for} \quad |t| > 1. \end{cases}$$

PE 6.38 In a way analogous to what happens with the Fourier series, we have the so-called Parseval identity associated with Fourier transforms. Let $F(\alpha)$ and $G(\alpha)$ be the Fourier transforms of functions $f(t)$ and $g(t)$, respectively. Show that

$$\int_{-\infty}^{\infty} F(\alpha)G(\alpha)\,\mathrm{d}\alpha = \int_{-\infty}^{\infty} f(-y)g(y)\,\mathrm{d}y.$$

PE 6.39 Consider the function

$$f(t) = \begin{cases} 1, & 0 \le t < 1, \\ 0, & t \ge 1. \end{cases}$$

Find the Fourier sine and cosine transforms of $f(t)$ in order to show that

$$\int_0^{\infty} \left(\frac{1 - \cos t}{t}\right)^2 \mathrm{d}t = \int_0^{\infty} \frac{\sin^4 t}{t^2}\,\mathrm{d}t = \frac{\pi}{2}.$$

PE 6.40 Using the relation

$$\int_{-\infty}^{\infty} \frac{e^{isx}}{(1+s^2)^{3/2}}\,\mathrm{d}s = 2x\,\mathscr{K}_1(x),$$

where $\mathscr{K}_1(x)$ is a modified Bessel function, show that

$$\int_0^{\infty} x^2 [\mathscr{K}_1(x)]^2\,\mathrm{d}x = \frac{3\pi^2}{32}.$$

PE 6.41 A free particle is described in Quantum Mechanics by a *plane wave*

$$\psi_k(x,t) = \exp\left[i\left(kx - \frac{\hbar}{2m}k^2 t\right)\right],$$

where $\hbar$ and m are positive constants. Combining waves of the same adjacent momentum with a weight factor of amplitude given by $\varphi(k)$, we can construct a *wave packet*:

$$\Psi(x, t) = \int_{-\infty}^{\infty} \varphi(k) \exp\left[i \left(kx - \frac{\hbar}{2m} k^2 t \right) \right] dk .$$

Knowing that $\Psi(x, 0) = \exp(-x^2/2a^2)$, where a is a constant, obtain an expression for function $\varphi(k)$.

PE 6.42 Using the result of the preceding exercise, obtain the explicit form of the function $\Psi(x, t)$.

PE 6.43 The function that describes the response of a physical system governed by an ordinary differential equation and subject to certain boundary conditions is known as *Green's function*, and is denoted by $G(x|\xi)$. Using the Fourier transform method, we obtain for the Green's function associated with a wave equation the expression

$$G(x|\xi) = \frac{1}{2\pi} \int_{-\infty}^{\infty} \frac{e^{ik\xi} e^{-ikx}}{k_0^2 - k^2} \, dk,$$

where $k_0^2 > 0$, and ξ is a parameter. Obtain explicitly $G(x|\xi)$ considering that $x > \xi$ and using a contour that avoids the pole $k = -k_0$, that is, for which the contribution to the integral is due to the pole $k = k_0$ only.

PE 6.44 Considering the result obtained in the precedent exercise, what must be changed if we consider the case $x < \xi$?

PE 6.45 Consider the ordinary differential equation

$$\frac{d^2}{dt^2} f(t) + k \frac{d}{dt} f(t) + \omega_0^2 f(t) = \varphi(t),$$

where k and ω_0^2 are positive constants and $t > 0$. Suppose that $k^2 < 4\omega_0^2$. Taking the Fourier transform on both sides of this equation and denoting the Fourier transforms of $f(t)$ and $\varphi(t)$, respectively, by $F(\omega)$ and $\Phi(\omega)$, solve the equation for $F(\omega)$.

PE 6.46 Using the result obtained in the preceding exercise, obtain explicitly $G(t)$ when

$$f(t) = \int_{-\infty}^{\infty} \varphi(\tau) G(t - \tau) \, d\tau.$$

PE 6.47 Consider the ordinary differential equation

$$\frac{d^2}{dx^2} G(x|\xi) - k^2 G(x|\xi) = \delta(x - \xi),$$

for $-\infty < x < \infty$, where ξ is a parameter and $k^2 > 0$.

Use the Fourier transform to express the function $G(x|\xi)$ in integral form.

PE 6.48 Calculate explicitly the integral appearing in **PE 6.47**.

PE 6.49 Let $x, y \in \mathbb{R}$. We define the Fourier transform of a function with two independent variables $f(x, y)$ as

$$g(x, y) = \frac{1}{2\pi} \int_{-\infty}^{\infty} \int_{-\infty}^{\infty} f(\xi, \eta)\, e^{i(\xi x + \eta y)}\, d\xi\, d\eta.$$

Use plane polar coordinates to show that if $f(\xi, \eta)$ is a function that depends only on variable ρ, that is, if function $f(\xi, \eta) = u(\rho)$, then $g(x, y)$ is also a function that depends only on variable r, $g(x, y) = v(r)$. Take $\rho^2 = \xi^2 + \eta^2$ and $r^2 = x^2 + y^2$.

PE 6.50 Use the result obtained in **PE 6.49** to calculate $v(r)$, knowing that $u(\rho) = \dfrac{1}{\rho}$.

References

1. L. Debnath, D. Battha, *Integral Transforms and Their Applications*, 2nd edn. (Chapman & Hall/CRC, Boca Raton, 2017)
2. I.N. Sneddon, *Fourier Transform* (McGraw-Hill, New York, 1951)
3. I.N. Sneddon, *The Use of Integral Transforms* (McGraw-Hill, New York, 1972)

Chapter 7
Sturm–Liouville Systems

All differential equations discussed so far are particular cases of an extensive class of linear, second-order ordinary differential equations whose coefficients depend on the independent variable. In this chapter, we will discuss the so-called *Sturm–Liouville systems*, each of which is composed of exactly one linear, second-order ordinary differential equation, together with prescribed boundary conditions given at the extremes of an interval of the real line, inside which we must look for the solution of that ordinary differential equation. These systems, also called *Sturm–Liouville problems*, are of vital importance to the discussion of Green's functions.

7.1 Sturm–Liouville Systems

Let $a_1(x)$, $a_2(x)$ and $a_3(x)$ be real functions with $a_1(x) \neq 0$ for all $x \in \mathbb{R}$. Consider the following homogeneous, linear, second-order ordinary differential equation:

$$a_1(x)\frac{\mathrm{d}^2 u}{\mathrm{d}x^2} + a_2(x)\frac{\mathrm{d}u}{\mathrm{d}x} + [a_3(x) + \lambda]u = 0, \tag{7.1}$$

where λ is a parameter that does not depend on x and $u = u(x)$.

Introducing the functions $p(x)$, $q(x)$ and $s(x)$ defined by

$$p(x) = \exp\left(\int^x \frac{a_2(x')}{a_1(x')}\,\mathrm{d}x'\right), \quad q(x) = \frac{a_3(x)}{a_1(x)}p(x) \quad \text{and} \quad s(x) = \frac{p(x)}{a_1(x)}$$

into the preceding equation we get

© The Author(s), under exclusive license to Springer Nature Switzerland AG 2025
E. Capelas de Oliveira, J. E. Maiorino, *Analytical Methods in Applied Mathematics*, Problem Books in Mathematics, https://doi.org/10.1007/978-3-031-74794-6_7

$$\frac{\mathrm{d}}{\mathrm{d}x}\left[p(x)\frac{\mathrm{d}u}{\mathrm{d}x}\right] + [q(x) + \lambda s(x)]u = 0,$$

which is known as *Sturm–Liouville equation*. This ordinary differential equation can be written using the so-called self adjoint operator

$$\mathcal{L} \equiv \frac{\mathrm{d}}{\mathrm{d}x}\left[p(x)\frac{\mathrm{d}}{\mathrm{d}x}\right] + q(x). \tag{7.2}$$

The original differential equation, written in terms of this operator, is then given by

$$\mathcal{L}[u] + \lambda s(x)u = 0,$$

where we have assumed that $p(x)$, $q(x)$ and $s(x)$ are real-valued functions and also that functions $q(x)$ and $s(x)$ are continuous and function $p(x)$ is continuously differentiable on a finite closed interval $[a, b]$.

The Sturm–Liouville equation is called *regular* on the closed interval $[a, b]$ if the functions $p(x)$ and $s(x)$ are strictly positive on this interval. On the other hand, in the case in which the interval is semi-infinite or infinite, or when one (or two) of the functions $p(x)$ and $s(x)$ are null at one or both extremes of the finite interval, we say that the Sturm–Liouville equation is singular.

Example 7.1 Let $y = y(x)$ on $(0, \infty)$. Consider the zero order Bessel equation

$$x^2 y'' + xy' + x^2 y = 0.$$

Write this linear ordinary differential equation in the form of a Sturm–Liouville equation and classify it.

Comparing this Bessel equation with the general form Eq. (7.1), we have: $a_1(x) = x^2$, $a_2(x) = x$ and $a_3(x) = x^2$. Then, evaluating $p(x)$, $q(x)$ and $s(x)$ we have,

$$p(x) = \exp\left(\int^x \frac{\xi}{\xi^2}\,\mathrm{d}\xi\right) = \exp\left(\int^x \frac{\mathrm{d}\xi}{\xi}\right) = \exp(\ln x) = x\,,$$

$$q(x) = \frac{x^2}{x^2}p(x) = x \qquad \text{and} \qquad s(x) = \frac{x}{x^2} = \frac{1}{x}$$

respectively. We can thus write the differential equation in the form

$$\frac{\mathrm{d}}{\mathrm{d}x}\left[x\frac{\mathrm{d}}{\mathrm{d}x}y(x)\right] + xy(x) = 0,$$

which is a singular Sturm–Liouville equation. $\qquad\qquad\square$

Let α_1, α_2, β_1 and β_2 be real numbers and $u = u(x)$ the solution of the Sturm–Liouville equation. Thus, the Sturm–Liouville equation, together with the so-called *separate extremes conditions*

$$\alpha_1 u(a) + \alpha_2 u'(a) = 0,$$

$$\beta_1 u(b) + \beta_2 u'(b) = 0,$$

constitute the so-called *Sturm–Liouville system* or *Sturm–Liouville problem*.

In analogy to systems of algebraic equations, the values of the parameter λ for which the Sturm–Liouville system has a nontrivial solution are called the *eigenvalues* associated with the system and the corresponding solutions are their *eigenfunctions*. The set of all eigenvalues associated with a regular Sturm–Liouville system is called the *spectrum* of the system.

Example 7.2 Let $x \in \mathbb{R}$ and consider the inverval $[0, \pi]$. Obtain the eigenvalues and the eigenfunctions associated with the Sturm–Liouville system

$$\begin{cases} y''(x) + \lambda^2 y(x) = 0, \\ y(0) = 0 = y(\pi). \end{cases} \tag{7.3}$$

The general solution of the Sturm–Liouville equation is

$$y(x) = A \sin \lambda x + B \cos \lambda x,$$

where A and B are arbitrary constants. Imposing the first condition, $y(0) = 0$, we get $B = 0$. For the second condition we have

$$y(\pi) = A \sin \lambda \pi = 0.$$

Thus, for $A = 0$ we have only the trivial solution while for $A \neq 0$ we get a trigonometric equation,

$$\sin \lambda \pi = \sin k \pi,$$

with $k = 1, 2, \ldots$ We conclude that the eigenvalues are $\lambda_k = k$ with $k = 1, 2, \ldots$ and the eigenfunctions are $y_k(x) = \sin kx$ with $k = 1, 2, \ldots$ $\qquad\square$

After these specific examples, we mention some properties of the eigenvalues and eigenfunctions of Sturm–Liouville systems. For the proofs, see reference [1].

Theorem 7.1 *Assume that the coefficients $p(x)$, $q(x)$ and $s(x)$ are continuous functions on the interval $[a, b]$. Let $u_j(x)$ and $u_k(x)$ be continuously differentiable eigenfunctions corresponding to distinct eigenvalues λ_j and λ_k. Then, $u_j(x)$ and $u_k(x)$ are orthogonal with respect to the so-called weight function $s(x)$ on the interval $[a, b]$.*

Theorem 7.2 *All eigenvalues associated with a regular Sturm–Liouville system with $s(x) > 0$ on the interval $[a, b]$ are real.*

Theorem 7.3 *All regular Sturm–Liouville systems possess an infinite sequence of real eigenvalues $\lambda_1 < \lambda_2 < \lambda_3 < \ldots$ with $\lim_{n \to \infty} \lambda_n = \infty$. The corresponding eigenfunctions $u_n(x)$ are univocally determined up to a constant and have exactly n zeros on the open interval (a, b). Moreover, these functions form an orthogonal and complete system with respect to the weight function $s(x)$.*

Theorem 7.4 *Any function $f(x)$, smooth by parts on the interval $[a, b]$, that satisfies the separate extremes conditions of the regular Sturm–Liouville system can be expressed in terms of an absolutely and uniformly convergent series*

$$f(x) = \sum_{n=1}^{\infty} c_n u_n(x),$$

where the coefficients c_n are given by

$$c_n = \frac{\displaystyle\int_a^b s(x) f(x) u_n(x)\, dx}{\displaystyle\int_a^b s(x) u_n^2(x)\, dx},$$

with $n = 1, 2, 3\ldots$

Example 7.3 Let $f(x)$ be a smooth function on the interval $[0, \pi]$ satisfying the conditions Eq. (7.3). We can write $f(x)$ in terms of the eigenfunction of Example 7.2, i.e.,

$$f(x) = \sum_{k=1}^{\infty} c_k \sin kx$$

where the coefficients are given by

$$c_k = \frac{\displaystyle\int_0^{\pi} \lambda^2 f(x) \sin kx\, dx}{\displaystyle\int_0^{\pi} \lambda^2 \sin^2 kx\, dx}.$$

Evaluating the integral and simplifying we have

$$c_k = \frac{2}{\pi} \int_0^{\pi} f(x) \sin kx\, dx,$$

which is the desired result. $\square$

7.2   Green's Function

In this section, we present a brief discussion of the so-called generalized functions, known as distributions, and introduce the concept of a Green's function associated with a linear differential operator [3, 4].

7.2.1   Green's Function × Generalized Function

Let $0 \leq x \leq \ell$. Suppose that we want to solve a nonhomogeneous ordinary differential equation, which can be put in the form

$$\mathfrak{L}u(x) = f(x), \tag{7.4}$$

where $\mathfrak{L}$ is a linear operator and $f(x)$ is called the source term, a known function. Suppose that $f(x)$ can be approximated by a collection of functions $f(\xi_1)$, $f(\xi_2)$, $\ldots$, $f(\xi_n)$ corresponding to the sources at $x = \xi_1, x = \xi_2, \ldots, x = \xi_n$, respectively, and all of them in $0 \leq x \leq \ell$. We define $\mathscr{G}(x|\xi_k)$ as the solution of Eq. (7.4) when there exists a unity point source at ξ_k. Also, the solution of Eq. (7.4) for $f(\xi_k)$ is $\mathscr{G}(x|\xi_k)f(\xi_k)$. If we have n point sources, the solution takes the form

$$u(x) = \sum_{k=1}^{n} \mathscr{G}(x|\xi_k)f(\xi_k).$$

On the other hand, considering a great number of point sources, i.e., taking the limit $n \to \infty$ and $\xi_{k+1} - \xi_k \to 0$ for all k, we can replace the sum by an integral, so that

$$u(x) = \int_0^{\ell} \mathscr{G}(x|\xi)f(\xi)\,\mathrm{d}\xi,$$

where $\mathscr{G}(x|\xi)$ is the Green's function associated to the problem.

The classical problem is: how does one determine the Green's function? This problem is formulated with the help of the so-called generalized function, a distribution. Before we introduce formally the Green's function, we present a particular example.

Example 7.4 (Laplacian in Spherical Coordinates × Coulombian Potential)
Let r, θ, ϕ be spherical coordinates. The Laplace operator (Laplacian), denoted by Δ, written in spherical coordinates, if we consider a function that depends only on the radial coordinate, is given by

$$\Delta = \frac{1}{r^2}\frac{\mathrm{d}}{\mathrm{d}r}\left(r^2\frac{\mathrm{d}}{\mathrm{d}r}\right).$$

Consider a Coulombian potential $\Psi(r) = 1/r$ satisfying $\Delta\Psi(r) = 0$ for all points with $r \neq 0$. At $r = 0$, $\Delta\Psi(r)$ does not exist but we can show, by means of the Gauss theorem (divergence theorem), that

$$\int_V \Delta\left(\frac{1}{r}\right) dV = -4\pi,$$

where V is the volume of a region limited by a spherical surface S containing the origin $r = 0$ [1]. We can thus conclude that $\Delta\left(\frac{1}{r}\right)$ is a *function* that satisfies the following properties:

(a) it is not defined at $r = 0$;
(b) it is null if $r \neq 0$;
(c) its integral on each sphere containingthe origin $r = 0$ is equal to -4π.

A *function* having these three properties cannot be considered an ordinary function, so it is called a distribution or a generalized function. Thus, we can write

$$\int_V \Delta\left(\frac{1}{r}\right) dV = \begin{cases} -4\pi, & \text{if } V \text{ contains } r = 0; \\ 0, & \text{if } V \text{ not contains } r = 0. \end{cases}$$

This can be expressed as

$$\Delta\left(\frac{1}{r}\right) = -4\pi\,\delta(r),$$

where $\delta(r)$ is the so-called Dirac delta function, which is a generalized function [1]. $\square$

Green's functions must satisfy a set of boundary conditions usually imposed at $|x| \to \infty$ or at some finite boundary points. For instance, the causal boundary condition in one dimension is $\mathscr{G}_{\mathrm{ret}}(x|\xi) = 0$ for $x < \xi$, which specifies the retarded Green's function. We can also introduce $\mathscr{G}_{\mathrm{adv}}(x|\xi)$, called advanced Green's function, depending on the boundary conditions associated with a particular problem.

It is important to note that the nonhomogeneous ordinary differential equation

$$\mathscr{L}\mathscr{G}(x|\xi) = \delta(x - \xi)$$

defines a Green's function only up to a solution of the corresponding homogeneous ordinary differential equation; the boundary conditions are necessary to fix the Green's function uniquely. Thus, the general solution of Eq. (7.4) can be written as

$$u(x) = u_h(x) + \int^x \mathscr{G}(x|\xi) f(\xi) \, d\xi,$$

where $u_h(x)$ is a general solution of the homogeneous equation, $\mathfrak{L} u_h(x) = 0$.

As an interpretation, we can say that the Green's function $\mathscr{G}(x|\xi)$ represents the effect on x due to the action of a delta function on ξ.

Before we discuss the effective calculation of the Green's function, we will give an example, using Fourier transform, of a problem that may arise if we do not treat a function as a generalized function.

Example 7.5 Consider formally the problem

$$\mathfrak{L}\mathscr{G}(x|\xi) = \delta(x - \xi),$$

where the linear operator is given by $\mathfrak{L} = \frac{d^2}{dx^2} + \lambda^2$, with λ^2 a real positive parameter. Taking the Fourier transform on both sides, we find the algebraic equation

$$(\lambda^2 - k^2) g(k) = 1,$$

where k is the parameter associated with the Fourier transform and where we use the notation

$$\mathscr{F}[\mathscr{G}(x|\xi)] = g(k).$$

We also used the result for the Fourier transform of a delta function,

$$\mathscr{F}[\delta(x)] = 1.$$

To solve the ordinary differential equation, we must evaluate the corresponding inverse Fourier transform, i.e., the integral

$$\mathscr{G}(x|\xi) = \frac{1}{2\pi} \int_{-\infty}^{\infty} e^{ik(x-\xi)} \frac{dk}{\lambda^2 - k^2}.$$

This is a meaningless divergent integral. The solution consists of treating $g(k)$ as a distribution, i.e., to evaluate the integral in the space of distributions, in which the integral involves the so-called principal value. This theory is beyond the scope of this text [1, 2]. $\qquad\square$

We now return to the problem of finding the Green's function associated with a linear differential operator; more specifically, the one assocciated with the Sturm–Liouville system.

7.2.2 Green's Function: One Dimensional Case

As we have mentioned before, a Sturm–Liouville system can be regular or singular. The case of regular systems will be discussed in the solved exercises. The fundamental characteristic of a singular system is that in this case the conditions do not appear in an explicit way. It is usual to describe the conditions by imposing that the function $u(x)$ be limited at a (singular) extreme point. We will see in Chap. 9 an example of this case, when we discuss the transversal vibrations of a thin membrane. We will now study the solution of the boundary value problems associated with ordinary differential equations by the so-called Green's function methodology, as mentioned in **PE 1.29**.

We consider a linear nonhomogeneous ordinary differential equation written in the self-adjoint form, i.e., as a Sturm–Liouville differential equation,

$$\mathfrak{L}[u(x)] = -f(x),$$

on the interval $[a, b]$, where

$$\mathfrak{L} = \frac{d}{dx}\left[p(x)\frac{d}{dx}\right] + q(x), \tag{7.5}$$

with the homogeneous boundary conditions (separate conditions)

$$a_1 u(a) + a_2 u'(a) = 0,$$
$$b_1 u(b) + b_1 u'(b) = 0,$$

where constants a_1 and a_2, as well as b_1 and b_2, are not all null. We will suppose that $f(x)$ and $q(x)$ are two continuous functions and that $p(x)$ is a continuously differentiable function, and also that none of them assume a zero value on the interval $[a, b]$.

We introduce the Green's function, denoted by $\mathscr{G}(x|\xi)$, for the linear differential expression Eq. (7.5) $\mathfrak{L}[u(x)]$ with the boundary conditions aforementioned, as a function that satisfies the following three conditions:

(a) $\mathscr{G}(x|\xi)$ is continuous along the diagonal of the square $a \leq x, \xi \leq b$, i.e., at $x = \xi$.

(b) At the point $x = \xi$ the first derivative of $\mathscr{G}(x|\xi)$ has a jump discontinuity given by

$$\frac{d}{dx}\mathscr{G}(x|\xi)\bigg|_{x=\xi_-}^{x=\xi_+} = -\frac{1}{p(\xi)}.$$

(c) For fixed ξ, $\mathscr{G}(x|\xi)$ satisfies the boundary conditions. Besides, $\mathscr{G}(x|\xi)$ is a solution of the corresponding homogeneous differential equation

$$\mathfrak{L}[u(x)] = 0,$$

in each of the intervals $a \leq x \leq \xi$ and $\xi \leq x \leq b$.

Theorem 7.5 *Let $x \in \mathbb{R}$ and let $[a, b]$ be a closed interval. If the function $f(x)$ is continuous on this interval, then the function*

$$u(x) = \int_a^b \mathscr{G}(x|\xi) f(\xi) \, d\xi$$

is a solution of the following boundary value problem

$$\mathfrak{L}[u(x)] = -f(x)$$

$$a_1 u(a) + a_2 u'(a) = 0$$
$$b_1 u(b) + b_2 u'(b) = 0,$$

where constants a_1, a_2, b_1 and b_2 are not all null.

From the preceding theorem, it is clear that solving a nonhomogeneous ordinary differential equation is equivalent to finding its corresponding Green's function. We will now consider the construction of the Green's function associated with such an equation.

First, we suppose that the homogeneous equation satisfying the separate extremes conditions has *only* the trivial solution. We then construct the solution $u_1(x)$ of the homogeneous differential equation

$$\mathfrak{L}[u(x)] = 0 \tag{7.6}$$

satisfying the condition $a_1 u(a) + a_2 u'(a) = 0$. The general solution is $c_1 u_1(x)$, where c_1 is an arbitrary constant.

By an analogous reasoning, we can say that $c_2 u_2(x)$, with c_2 an arbitrary constant, is the general solution of Eq. (7.6) satisfying the condition $b_1 u(b) + b_2 u'(b) = 0$. Thus, $u_1(x)$ and $u_2(x)$ exist on the interval (a, b) and are linearly independent.

Indeed, if these functions were linearly dependent, then $u_1(x) = c u_2(x)$ for some constant c, showing that $u_1(x)$ satisfies both conditions at $x = a$ and $x = b$, which contradicts the hypothesis about the trivial solution.

Consequently, the Green's function can be sought in the following form

$$\mathscr{G}(x|\xi) = \begin{cases} c_1(\xi) u_1(x) \text{ for } x < \xi \\ c_2(\xi) u_2(x) \text{ for } x > \xi. \end{cases}$$

Then, from the continuity of the Green's function at $x = \xi$ we have

$$c_2(\xi) u_2(\xi) = c_1(\xi) u_1(\xi).$$

On the other hand, from the discontinuity of the first derivatives of the Green's function at $x = \xi$ we get

$$\frac{\mathrm{d}}{\mathrm{d}x} \mathcal{G}(x|\xi) \Big|_{x=\xi_-}^{x=\xi_+} = c_2(\xi)u_2'(\xi) - c_1(\xi)u_1'(\xi) = -\frac{1}{p(\xi)}.$$

Solving these last two equations for $c_1(\xi)$ and $c_2(\xi)$ we obtain

$$c_1(\xi) = \frac{-u_2(\xi)}{p(\xi)W(u_1, u_2; \xi)},$$

$$c_2(\xi) = \frac{-u_1(\xi)}{p(\xi)W(u_1, u_2; \xi)},$$

where $W(u_1, u_2; \xi)$ is the Wronskian, introduced in Chap. 1.

Since $u_1(x)$ and $u_2(x)$ are linearly independent functions, their Wronskian is different from zero. Since $p(\xi)W(u_1, u_2; \xi)$ is constant, as can be seen in **PE 7.12**, we obtain for the Green's function

$$\mathcal{G}(x|\xi) = -\frac{1}{C} \begin{cases} u_1(x)u_2(\xi) \text{ for } x \leq \xi, \\ u_2(x)u_1(\xi) \text{ for } x \geq \xi, \end{cases}$$

where C is an arbitrary constant.

Example 7.6 Let $f(x)$ be a continuous function on the closed interval $[0, \pi]$. Obtain the Green's function associated with the following Sturm–Liouville system:

$$\begin{cases} \dfrac{\mathrm{d}^2}{\mathrm{d}x^2} y(x) = -f(x), \\[2mm] y(0) = 0 = y(\pi). \end{cases}$$

The general solution of the corresponding homogeneous ordinary differential equation is $y(x) = Ax + B$, where A and B are two arbitrary constants. It is easy to show that the unique solution satisfying the boundary conditions is the trivial solution.

Thus, we consider two linearly independent solution $y_1(x) = Ax$, which satisfies the condition $y_1(0) = 0$, and $y_2 = B(\pi - x)$, satisfying the condition $y_2(\pi) = 0$. We can then write for the Green's function

$$\mathcal{G}(x|\xi) = \frac{1}{W(y_1, y_2; \xi)} \begin{cases} Ax \cdot B(\pi - \xi) \text{ for } 0 \leq x \leq \xi, \\ B(\pi - x) \cdot A\xi \text{ for } \xi \leq x \leq \pi, \end{cases}$$

where $W(y_1, y_2; \xi)$ is the Wronskian. Evaluating the Wronskian we find

$$W(y_1, y_2; \xi) = \begin{vmatrix} Ax & B(\pi - x) \\ A & -B \end{vmatrix} = -ABx - AB(\pi - x) = -AB\pi,$$

and after simplification, we finally get

$$\mathscr{G}(x|\xi) = -\frac{1}{\pi} \begin{cases} x \cdot (\pi - \xi) \text{ for } 0 \leq x \leq \xi, \\ \xi \cdot (\pi - x) \text{ for } \xi \leq x \leq \pi, \end{cases}$$

which is the desired Green's function. $\qquad\square$

Example 7.7 Using the result of Example 7.6, solve the nonhomogeneous linear ordinary differential equation

$$\frac{d^2}{dx^2} y(x) = -x \,.$$

Since we know the Green's function, we just need to evaluate the integral

$$y(x) = \int_0^\pi \mathscr{G}(x|\xi) f(\xi) \, d\xi \,,$$

where $\mathscr{G}(x|\xi)$ is the Green's function and $f(x) = x$. Substituting the Green's function and $f(x)$ and rearranging we have

$$y(x) = \frac{1}{\pi} \int_0^x \xi(\pi - x)\xi \, d\xi + \frac{1}{\pi} \int_x^\pi x(\pi - \xi)\xi \, d\xi \,,$$

whose solution yields

$$y(x) = -\frac{x}{6}(x^2 - \pi^2),$$

which is the solution of the boundary value problem. $\qquad\square$

Note that this boundary value problem is very simple and can be solved by direct integration because the solution of the ordinary differential equation is immediate.

Before discussing the case in which we do not have only the trivial solution, we state two theorems related to Green's function.

Theorem 7.6 *If the homogeneous boundary value problem has only the trivial solution, then the Green's function exists and is unique.*

Theorem 7.7 *The Green's function associated with a boundary value problem is symmetric, i.e., $G(x|x') = G(x'|x)$.*

7.3   Generalized Green's Function

If the homogeneous problem associated with a certain Sturm–Liouville problem has a nontrivial solution, then any solution of the differential equation $\mathfrak{L}[u(x)] = -f(x)$ that satisfies one of the boundary conditions will also satisfy the other condition and the corresponding Green's function does not exist.

The theorem stated further provides a way to find a different type of Green's function, called generalized Green's function, for nonhomogeneous problems associated with homogeneous problems, which have nontrivial solutions.

Theorem 7.8 *Let $x \in \mathbb{R}$ and $[a, b]$ a closed interval. We will suppose that $f(x)$ and $q(x)$ are two continuous functions and that $p(x)$ is a continuously differentiable function, and also that none of them assumes a zero value on the interval $[a, b]$. The nonhomogeneous boundary value problem*

$$\frac{\mathrm{d}}{\mathrm{d}x}\left[p(x)\frac{\mathrm{d}}{\mathrm{d}x}u(x) \right] + q(x)u(x) = -f(x),$$

$$\begin{aligned} a_1 u(a) + a_2 u'(a) &= 0, \\ b_1 u(b) + b_2 u'(b) &= 0, \end{aligned} \tag{7.7}$$

where the constants a_1, a_2, b_1 and b_2 are not all null, has as one of its solutions the function

$$u(x) = \int_a^b \mathscr{G}(x|\xi) f(\xi)\,\mathrm{d}\xi + A u_1(x)$$

if, and only if,

$$\int_a^b f(\xi) u_1(\xi)\,\mathrm{d}\xi = 0,$$

where A is an arbitrary constant and $u_1(x)$ is the nontrivial solution associated with the corresponding homogeneous problem. The Green's function $\mathscr{G}(x|\xi)$, called generalized Green's function, is a solution of the differential equation

$$\mathfrak{L}[\mathscr{G}(x|\xi)] = C u_1(x) u_1(\xi),$$

except at the point $x = \xi$; C is a constant, and $\mathscr{G}(x|\xi)$ satisfies the homogeneous boundary conditions Eq. (7.7). $\mathscr{G}(x|\xi)$ is continuous at $x = \xi$ and its derivative $\mathscr{G}'(x|\xi)$ is continuous everywhere except at the point $x = \xi$, where it presents a jump of magnitude $-1/p(\xi)$. Moreover, $\mathscr{G}(x|\xi)$ satisfies the condition

$$\int_a^b \mathscr{G}(x|\xi) u_1(x)\,\mathrm{d}x = 0.$$

We will talk again about Green's function in Chap. 10, where we will associate the Green's function with eigenvalue problems and partial differential equations discussed in Chap. 9.

7.4 Solved Exercises

SE 7.1 Let $x \neq k\pi/2$ for $k = 1, 2, 3, \ldots$, $y_1(x) = e^x$ and $y_2(x) = \sin x \, e^x$. Are these two functions linearly independent?
Solution: To answer the question, we must evaluate the Wronskian

$$W(y_1, y_2; x) = \begin{vmatrix} e^x & \sin x \, e^x \\ e^x & \cos x \, e^x + \sin x \, e^x \end{vmatrix} = \cos x \, e^{2x}.$$

As the Wronskian is different from zero, these two functions are linearly independent.

SE 7.2 Let $m, n \in \mathbb{N}$ with $m \neq n$, $y_1(x) = \sin mx$ and $y_2(x) = \sin nx$. Evaluate the Wronskian.
Solution: We must evaluate the determinant

$$W(y_1, y_2; x) = \begin{vmatrix} \sin mx & \sin nx \\ m \cos mx & n \cos nx \end{vmatrix} = n \sin mx \cos nx - m \sin nx \cos mx.$$

For $m \neq 0$ and/or $n \neq 0$, the Wronskian is always different from zero, and these two functions are linearly independent. On the other hand, if $m = 0$ and/or $n = 0$ the two functions are linearly dependent.

SE 7.3 Put the ordinary differential equation

$$\frac{d^2}{dx^2}y(x) + 2x\frac{d}{dx}y(x) + (1 + \lambda)y(x) = 0, \tag{7.8}$$

with λ a parameter independent of variable x, in the form of a Sturm–Liouville system.
Solution: Identifying this differential equation with (7.1), we have

$$a_1(x) = 1, \quad a_2(x) = 2x \text{ and } a_3(x) = 1.$$

Next, we must calculate the function $p(x)$, i.e.,

$$p(x) = \exp\left(\int^x \frac{2x'}{1} \, dx'\right) = \exp(x^2). \tag{7.9}$$

Then, multiplying the ordinary differential equation by $p(x)$, we can write

$$e^{x^2} \frac{d^2}{dx^2} y(x) + 2x\, e^{x^2} \frac{d}{dx} y(x) + e^{x^2} y(x) + \lambda\, e^{x^2} y(x) = 0, \qquad (7.10)$$

which can be written as

$$\frac{d}{dx}\left[e^{x^2} \frac{d}{dx} y(x) \right] + e^{x^2} y(x) + \lambda\, e^{x^2} y(x) = 0, \qquad (7.11)$$

an ordinary differential equation in the Sturm-Liouville form, with parameters given by $p(x) = q(x) = s(x) = \exp(x^2)$.

SE 7.4 Determine the eigenvalues and eigenfunctions associated with the Sturm–Liouville equation

$$\frac{d^2}{dx^2} u(x) + \lambda^2 u(x) = 0,$$

with $0 \leq x \leq 1$ and the boundary conditions

$$u(0) = 0 \quad \text{and} \quad u(1) = 0,$$

where λ is a positive parameter.

Solution: The general solution of the ordinary differential equation is given by

$$u(x) = A \sin \lambda x + B \cos \lambda x,$$

where A and B are arbitrary constants.

Using the first boundary condition, we get

$$u(0) = B = 0,$$

which allows us to write the solution $u(x) = A \sin \lambda x$.

$$u(1) = A \sin \lambda = 0,$$

from where we conclude that if $A = 0$, the only solution is the trivial solution. Thus, for $A \neq 0$, we must solve the following trigonometric equation

$$\sin \lambda = 0,$$

whose solution is $\lambda_k = k\pi$, with $k = 1, 2, \ldots$

Finally, the eigenvalues are given by

$$\lambda_k = k\pi$$

and the corresponding eigenfunctions are

$$u_k(x) = \sin k\pi x,$$

with $k = 1, 2, \ldots$

SE 7.5 Consider the ordinary differential equation, known as Euler equation, discussed in Chap. 1,

$$x^2 u'' + xu' + \lambda^2 u = 0, \qquad (7.12)$$

on the interval $1 \leq x \leq$ e, with the following boundary conditions at the extremes of the interval:

$$u(1) = 0 \quad \text{and} \quad u(\text{e}) = 0.$$

(a) Write the ordinary differential equation in the Sturm–Liouville form. (b) Solve the eigenvalue problem.

Solution: (a) Identifying Eq. (7.12) with Eq. (7.1) we have $p(x) = x$, $q(x) = 0$ and $s(x) = 1/x$. Thus, the Sturm–Liouville form for Eq. (7.12) is

$$\frac{\mathrm{d}}{\mathrm{d}x}\left[x\frac{\mathrm{d}}{\mathrm{d}x}u(x)\right] + \frac{\lambda^2}{x}u(x) = 0.$$

(b) As the differential equation is an Euler equation, we have the corresponding auxiliary equation (an algebraic equation)

$$m^2 + \lambda^2 = 0,$$

whose general solution is given by

$$u(x) = c_1 x^{i\lambda} + c_2 x^{-i\lambda},$$

where c_1 and c_2 are arbitrary constants. Using the trigonometric identity

$$x^{ia} = \mathrm{e}^{ia\ln x} = \cos(a\ln x) + i\sin(a\ln x),$$

we can write this expression in terms of trigonometric functions and we get

$$u(x) = A\cos(\lambda\ln x) + B\sin(\lambda\ln x),$$

where A and B are arbitrary constants.

The boundary condition at the extreme $x = 1$ furnishes

$$u(1) = A = 0$$

and the condition at $x =$ e implies that

$$B \sin \lambda = 0.$$

If $B = 0$, we have only the trivial solution. Thus, in order to find nontrivial solutions ($B \neq 0$), we must have

$$\sin \lambda = 0.$$

Therefore, the eigenvalues are,

$$\lambda_n = n\pi, \quad n = 1, 2, 3 \ldots$$

and the corresponding eigenfunctions are

$$u_n(x) = \sin(n\pi \ln x), \quad n = 1, 2, 3 \ldots$$

SE 7.6 (a) Find the Green's function associated with the following problem:

$$\begin{cases} \dfrac{d^2 u}{dx^2} = -6x, \\ u(0) = 0, \\ u(1) = 0. \end{cases}$$

(b) Obtain the solution $u(x)$ using the answer of the preceding item.
Solution: (a) For a fixed value of ξ, the Green's function satisfies the homogeneous ordinary differential equation

$$\frac{d^2}{dx^2}\mathscr{G}(x|\xi) = 0,$$

on $0 < x < \xi$ and $\xi < x < 1$, and also the boundary conditions

$$\mathscr{G}(0|\xi) = 0 \quad \text{and} \quad \mathscr{G}(1|\xi) = 0.$$

The discontinuity of the first derivative of $\mathscr{G}(x|x')$ allows us to write

$$\frac{d}{dx}\mathscr{G}(x|\xi)\Bigg|_{x=\xi_-}^{x=\xi_+} = -\frac{1}{p(\xi)}.$$

The general solution of the homogeneous differential equation is

$$u(x) = Ax + B,$$

where A and B are arbitrary constants. Thus, the Green's function is given by

$$\mathscr{G}(x|\xi) = \begin{cases} x(1-\xi) & \text{for } 0 \le x \le \xi \\ (1-x)\xi & \text{for } \xi \le x \le 1. \end{cases}$$

(b) To find the solution, it is enough to evaluate the corresponding integrals, that is, by means of Theorem 7.5:

$$\begin{aligned} u(x) &= \int_0^x \mathscr{G}(x|\xi) f(\xi)\, d\xi + \int_x^1 \mathscr{G}(x|\xi) f(\xi)\, d\xi \\ &= \int_0^x (1-x)\xi\, 6\xi\, d\xi + \int_x^1 (1-\xi)x\, 6\xi\, d\xi \\ &= x(1-x^2). \end{aligned}$$

It is easy to verify, by direct substitution, that this function is the solution of the problem.

SE 7.7 Obtain the eigenvalues associated with the following Sturm–Liouville system:

$$\begin{cases} \dfrac{d^2}{dt^2}x(t) + \dfrac{2}{t}\dfrac{d}{dt}x(t) + \lambda^2 x(t) = 0, \\ x(1) = 0, \\ x(\pi) = 0. \end{cases}$$

Solution: We consider a solution of the type

$$x(t) = \frac{1}{t}v(t),$$

from which we get the ordinary differential equation

$$\frac{d^2}{dt^2}v(t) + \lambda^2 v(t) = 0$$

and new boundary conditions $v(1) = 0$ and $v(\pi) = 0$. We must then analyze the three possible cases for the value of parameter λ.

First, for $\lambda^2 = 0$ we get

$$\frac{d^2}{dt^2}v(t) = 0,$$

whose general solution is $v(t) = At + B$. Using the boundary conditions, we then conclude that $A = B = 0$, i.e., we have only the trivial solution.

Now, for $\lambda^2 < 0$, putting $\lambda^2 = -\mu^2$ with $\mu^2 > 0$ we obtain

$$\frac{d^2}{dt^2}v(t) - \mu^2 v(t) = 0,$$

whose general solution is

$$v(t) = A \sinh\mu t + B \cosh\mu t .$$

Again, using the boundary conditions we conclude that $A = B = 0$, the trivial solution only.

Finally, for the case in which we have $\lambda^2 > 0$ the general solution is given by

$$v(t) = A \sin \lambda t + B \cos \lambda t.$$

Applying the boundary conditions we can write:

$$v(1) = A \sin\lambda + B \cos\lambda = 0 \quad \Longrightarrow \operatorname{tg}\lambda = -B/A,$$
$$v(\pi) = A \sin\pi\lambda + B \cos\pi\lambda = 0 \Longrightarrow \operatorname{tg}\pi\lambda = -B/A.$$

From these results we obtain the following transcendental equation

$$\operatorname{tg}\pi\lambda = \operatorname{tg}\lambda,$$

whose roots are the eigenvalues associated with the Sturm–Liouville system.

SE 7.8 Use the method of Green's function to reduce the ordinary differential equation

$$(1 + x^2)\frac{d^2}{dx^2}y(x) + 2x\frac{d}{dx}y(x) = \lambda y(x),$$

satisfying the boundary conditions $y(0) = y'(1) = 0$, to an integral equation. *Solution*: Write the differential equation given above in the form

$$\frac{d}{dx}\left[(1 + x^2)\frac{d}{dx}y(x)\right] = \lambda y(x).$$

In order to calculate the Green's function, we must find two linearly independent solutions of the corresponding homogeneous differential equation. We have

$$\frac{d}{dx}y(x) = 0 \Longrightarrow y_1(x) = C = \text{constant}$$

and

$$(1 + x^2)\frac{d}{dx}y(x) = A \Longrightarrow y_2(x) = A \operatorname{arctg}x.$$

We see that $y_1(x)$ satisfies the condition $y'(1) = 0$ while $y_2(x)$ satisfies the other condition, $y(0) = 0$.

The Wronskian of these solutions is given by

$$W = \begin{vmatrix} A\,\text{arctg}x & C \\ \dfrac{A}{1+x^2} & 0 \end{vmatrix} = -\frac{AC}{1+x^2},$$

from which we can write

$$p(x)W = (1+x^2)\left(-\frac{AC}{1+x^2}\right) = -AC = \text{constant}.$$

Thus, we obtain for the Green's function

$$\mathscr{G}(x|\xi) = \begin{cases} \text{arctg}\,x & \text{for } 0 < x < \xi, \\ \text{arctg}\,\xi & \text{for } \xi < x < 1. \end{cases}$$

Therefore,

$$y(x) = \lambda \int_0^1 \mathscr{G}(x|\xi)\,y(\xi)\,\mathrm{d}\xi,$$

which is the desired integral equation.

7.5   Proposed Exercises

PE 7.1 Let $x \in \mathbb{R}$ and $a, b, c \in \mathbb{R}$. Put the hypergeometric equation

$$x(1 - x)y'' + [c - (a + b + 1)x]y' - aby = 0,$$

where $y = y(x)$ and the prime denotes differentiation with respect to x, into the Sturm-Liouville form.

PE 7.2 Let $x \in \mathbb{R}$ and $a, c \in \mathbb{R}$. Put the confluent hypergeometric equation

$$xy'' + (c - x)y' - ay = 0,$$

where $y = y(x)$ and the prime denotes differentiation with respect to x, into the Sturm-Liouville form.

PE 7.3 Let $x \in \mathbb{R}$ and $\mu \in \mathbb{C}$ such that $\pm\mu \neq 0, 1, 2, \ldots$ Let $J_\mu(x)$ and $J_{-\mu}(x)$ be two Bessel functions of the first kind of order μ and $-\mu$, respectively. Evaluate their Wronskian and show that

$$W\left[J_\mu(x),\, J_{-\mu}(x)\right] = -\frac{2}{\pi x}\,\sin(\mu\pi)\,.$$

PE 7.4 Let $x \in \mathbb{R}$. Obtain the eigenvalues and eigenfunctions for the regular Sturm–Liouville system

$$\begin{cases} y'' + \lambda y = 0, \\ y(0) = 0 = y'(1), \end{cases}$$

where $y = y(x)$ and λ is a parameter.

PE 7.5 Let $x \in \mathbb{R}$. Obtain the eigenvalues and eigenfunctions for the regular Sturm–Liouville system

$$\begin{cases} y'' + \lambda y = 0, \\ y'(0) = 0 = y'(1), \end{cases}$$

where $y = y(x)$ and λ is a parameter.

PE 7.6 Use the result of **PE 7.5** to obtain the corresponding normalized eigenfuntions.

PE 7.7 Let $x \in \mathbb{R}^*$ and λ a positive parameter. Put the second-order, homogeneous, linear ordinary differential equation

$$x^2 y''(x) - x y'(x) + (\lambda + 1) y(x) = 0$$

into the Sturm-Liouville form.

PE 7.8 For the equation in **PE 7.7**, impose the boundary conditions $y(1) = 0 = y(e)$ and solve the resulting Sturm-Liouville system, i.e., obtain the corresponding eigenvalues and eigenfunctions.

PE 7.9 Let $x \in \mathbb{R}$, with $0 \leq x \leq 1$, and $f(x)$ a continuous function. Obtain the Green's function for the following regular Sturm–Liouville system:

$$\begin{cases} y'' = -f(x), \\ y(0) = 0 = y'(1)\,. \end{cases}$$

PE 7.10 Use the methodology of Green's function to solve **PE 7.9**, considering $f(x) = 1$.

PE 7.11 Prove Theorems 7.1, 7.2 and 7.4.

PE 7.12 Consider the general Sturm–Liouville problem presented in Eq. (7.1). Show that $p(x)W(u_1, u_2; x) = \text{constant}$, where W is the Wronskian and u_1 and u_2 are two linearly independent solutions of the corresponding homogeneous ordinary differential equation.

PE 7.13 Prove Theorems 7.5, 7.6 and 7.7.

PE 7.14 Find the eigenvalues and eigenfunctions associated with the following regular Sturm–Liouville problem:

$$u'' + \lambda^2 u = 0,$$

$$u(0) = u(\pi) = 0.$$

PE 7.15 Find the eigenvalues and the eigenfunctions associated with the Sturm–Liouville system

$$u'' + \lambda^2 u = 0,$$

with boundary conditions

$$u(1) = 0 \quad \text{and} \quad u(0) + u'(0) = 0.$$

PE 7.16 Find the eigenvalues and eigenfunctions associated with the Sturm–Liouville system composed of the ordinary differential equation

$$u'' + \lambda^2 u = 0$$

and the boundary conditions

$$u(-1) = u(1) \quad \text{and} \quad u'(-1) = u'(1).$$

This system is known as *periodic Sturm–Liouville system.*

PE 7.17 Find the eigenvalues and the eigenfunctions associated with the Sturm–Liouville system

$$u'' + u' + (1 + \lambda^2)u = 0,$$

$$u(0) = u(1) = 0.$$

PE 7.18 Obtain the eigenvalues and eigenfunctions for the following system:

$$u'' - 3u' + 3(1 + \lambda)u = 0,$$

with $u(0) = u(\pi) = 0.$

PE 7.19 Let $u = u(x)$. Do as in the preceding exercise for the system

$$\frac{d}{dx}\left[(2+x)^2\frac{du}{dx}\right] = -\lambda u,$$

with $u(-1) = u(1) = 0$.

PE 7.20 Find the eigenvalues and eigenfunctions associated with the singular Sturm–Liouville problem on the interval $0 \leq x \leq 1$, composed of the ordinary differential equation

$$x^2 u'' + xu' + \lambda^2 u = 0$$

and the boundary conditions

$$u(1) = 0 \quad \text{and} \quad \lim_{x\to 0^+} |u(x)| < \infty.$$

PE 7.21 Find the eigenvalues and eigenfunctions associated with the singular Sturm–Liouville system on the interval $0 \leq x < \infty$

$$u'' + \lambda^2 u = 0,$$

$$u(0) = 0 \quad \text{and} \quad \lim_{x\to\infty} |u(x)| < \infty.$$

PE 7.22 Express the function $f(x) = \cos x$ on the interval $0 \leq x \leq \pi$ in terms of the eigenfunctions of the Sturm–Liouville problem composed of the ordinary differential equation

$$u'' + \lambda^2 u = 0$$

and the homogeneous boundary conditions

$$u(0) = 0 = u(\pi).$$

PE 7.23 If possible, expand the function $f(x) = \cos x$ on the interval $0 \leq x \leq \pi/2$ in terms of the eigenfunctions associated with the following Sturm–Liouville problem:

$$u'' + \lambda^2 u = 0,$$

with $u(0) = 0 = u(\pi/2)$.

PE 7.24 Construct the Green's function for the following problem:

$$u'' = 0,$$

$$u(0) = u(1) = 0.$$

PE 7.25 Construct the Green's function for the ordinary differential equation

$$u'' = -f(x),$$

satisfying the boundary conditions $u(0) = u'(1) = 0$.

PE 7.26 Calculate the Green's function associated with the Sturm-Liouville problem

$$\left(\frac{d^2}{dx^2} - \lambda\right) y(x) = 0,$$

with $y(0) = y(1)$ and $y'(0) = y'(1)$.

PE 7.27 Find the Green's function associated with the ordinary differential equation

$$(1 + x^2)y'' - 2xy' = 0,$$

imposing convenient boundary conditions.

PE 7.28 Construct the Green's function for the system

$$xu'' + u' = 0,$$

$$u(1) = 0 \quad \text{and} \quad \lim_{x \to 0} |u(x)| < \infty.$$

PE 7.29 Find the Green's function associated with the Sturm–Liouville problem

$$xu'' + u' = -f(x),$$

$$u(1) = 0 \quad \text{and} \quad \lim_{x \to 0^+} |u(x)| < \infty,$$

on the interval $0 < x < 1$.

PE 7.30 Consider the ordinary differential equation

$$\mathcal{L}[y(x)] + \lambda r(x)y(x) = 0,$$

where

$$\mathcal{L} \equiv \frac{d}{dx}\left[p(x)\frac{d}{dx}\right] + q(x),$$

together with the boundary conditions

$$hy(a) + h'y'(a) = 0 \quad \text{and} \quad ky(b) + k'y'(b) = 0,$$

where h, k, h' and k' are constants, not all simultaneously null. Show that, in this case, the ordinary differential equation with the boundary conditions is equivalent to an integral equation of the type

$$y(\xi) + \lambda \int_a^b \mathscr{G}(x|\xi) r(x) y(x) dx = 0,$$

which is called *homogeneous Fredholm* [1866 – Erik Ivar Fredholm – 1927] *equation*, where $\mathscr{G}(x|\xi)$ is the corresponding Green's function.

PE 7.31 Show that the ordinary differential equation

$$xy'' + \lambda y = 0,$$

whose solution satisfies the conditions $y(0) = y(1) = 0$, is analogous to an integral equation,

$$y(x) = -\lambda \int_0^1 \mathscr{G}(x|\xi) \frac{1}{\xi} y(\xi) d\xi.$$

PE 7.32 Calculate explicitly the function $\mathscr{G}(x|\xi)$ mentioned in the preceding exercise.

PE 7.33 Change the ordinary differential equation

$$y'' + \{g(x) - \eta^2\} y = 0,$$

where η is a positive known constant and $g(x)$ is a known function, into a Fredholm integral equation with the conditions $y(0) = 0$ and $[y'/y] = -\eta$ at $x = x_0$.

PE 7.34 Obtain explicitly the Green's function $\mathscr{G}(x|\xi)$ associated with the preceding exercise.

PE 7.35 Using the result of the preceding exercise, take

$$g(x) = V_0 \frac{e^{-x}}{x},$$

with V_0 constant, to obtain the following integral equation:

$$y(x) = \frac{V_0}{\eta} \int_0^{x_0} \mathscr{G}(x|\xi) \frac{e^{-\xi}}{\xi} y(\xi) \, d\xi .$$

PE 7.36 Write explicitly the Green's function $\mathscr{G}(x|\xi)$ associated with the preceding exercise.

PE 7.37 Change the ordinary differential equation

$$y'' + \left\{ V_0 \frac{e^{-x}}{x} - \eta^2 \right\} y = 0,$$

with $y(0) = 0 = y(\infty)$, into a Fredholm integral equation, writing explicitly the corresponding Green's function.

PE 7.38 Show that the Fredholm integral equation corresponding to the ordinary differential equation

$$y'' + \lambda x^2 y = 0,$$

with $y(0) = y(1) = 0$, is given by

$$y(x) = \lambda \left[\int_0^x \xi^3 (1 - x) y(\xi)\, d\xi + \int_x^1 x \xi^2 (1 - \xi) y(\xi)\, d\xi \right].$$

PE 7.39 Let $y = y(x)$. Transform the ordinary differential equation

$$y'' + \frac{y'}{x} - \lambda y = 0,$$

with $y'(0) = 0 = y(1)$, into an integral equation.

PE 7.40 Let $y = y(x)$. Do as in the preceding exercise for

$$y'' + \frac{y'}{2x} + \lambda \sqrt{x}\, y = 0,$$

with $y(0) = 0 = y(1)$.

PE 7.41 Let $y = y(x)$. The solution of the ordinary differential equation

$$y'' + \omega^2 y = g(x),$$

where $0 \le x \le 2\pi$ and ω is a positive constant, subject to the boundary conditions $y(0) = y(2\pi)$ and $y'(0) = y'(2\pi)$, can be written in the following form:

$$y(x) = \int_0^{2\pi} \mathscr{G}(x, \xi, \omega) g(\xi)\, d\xi.$$

Find the corresponding Green's function $\mathscr{G}(x, \xi, \omega)$ in a closed form.

PE 7.42 Reduce the following ordinary differential equation and the boundary conditions to an equivalent integral equation on the interval $0 \le x \le 1$:

$$u'' + \lambda^2 u = 0,$$

$$u(0) = u(1) = 0.$$

PE 7.43 Let $u = u(x)$. Do as in the preceding problem for

$$\frac{d}{dx}(xu') + \left(-\frac{1}{x} + \lambda^2 x\right) u = 0,$$

$$u(0) = u(1) = 0,$$

with λ a positive constant.

PE 7.44 Let $u = u(x)$ and $f(x)$ an integrable function. Solve the following regular Sturm–Liouville problem on the interval $0 < x < 1$:

$$u'' + u = f(x),$$

with $u(0) = u(1) = 0$.

PE 7.45 Solve the ordinary differential equation

$$\left(\frac{d^2}{dx^2} + \omega_0^2\right) u(x) = 1,$$

with $u(0) = u(\pi) = 0$ and ω_0 a positive constant, using the Green's function methodology.

PE 7.46 Let $u = u(x)$. Solve the following nonhomogeneous boundary value problem on the interval $0 < x < 1$:

$$u'' + u = 1,$$

$$u(0) = u(1) = 0.$$

PE 7.47 Let $u = u(x)$. Solve this nonhomogeneous boundary value problem on the interval $0 < x < 1$:

$$u'' = -x,$$

$$u(0) = 0 \quad \text{and} \quad u(1) + 2u'(1) = 0.$$

PE 7.48 Let $u = u(x)$ and $f(x)$ an integrable function. Find the Green's function associated with the following problem:

$$u'' = -f(x),$$

$$u(0) = 0 \quad \text{and} \quad u(1) = u'(1),$$

where $0 < x < 1$.

PE 7.49 Let $y = y(x)$. Show that the Green's function corresponding to the Legendre differential equation

$$\frac{\mathrm{d}}{\mathrm{d}x}\left[(1 - x^2)\frac{\mathrm{d}y}{\mathrm{d}x}\right] - \lambda y = 0,$$

where $y(-1)$ and $y(1)$ are both finite and λ is a positive constant, is given by

$$\mathscr{G}(x|\xi) = \begin{cases} \ln 2 - \dfrac{1}{2}\ln(1 - x)(1 + \xi) - \dfrac{1}{2}, & -1 \leq x \leq \xi, \\[2mm] \ln 2 - \dfrac{1}{2}\ln(1 + x)(1 - \xi) - \dfrac{1}{2}, & \xi \leq x \leq 1. \end{cases}$$

PE 7.50 Let $u = u(x)$ and $f(x)$ an integrable function. Using the Green's function obtained in the preceding exercise, solve the system

$$(1 - x^2)u'' - 2xu' = -f(x),$$

$$\lim_{x \to \pm 1} |u(x)| < \infty,$$

on the interval $-1 < x < 1$.

References

1. G. Arfken, *Mathematical Methods for Physicists* (Academic Press, New York, 1973)
2. E. Capelas de Oliveira, W.A. Rodrigues Jr., *Funções Analíticas com Aplicações* (Analytic Functions with Applications), Editora (Livraria da Física, São Paulo, 2006)
3. S. Hassani, *Foundations of Mathematical Physics* (Prentice-Hall, New York, 1991)
4. H. Sagan, *Boundary and Eigenvalue Problems in Mathematical Physics* (Dover Publications, New York, 1989)

Chapter 8
Partial Differential Equations

Mathematics consists of proving the most obvious thing in the least obvious way.

1887 – George Polyá – 1985

We have so far discussed only linear ordinary differential equations of first and second orders. In practice, there are several problems in which the equation to be solved involves more than one independent variable, for example, the differential equation describing the motion of a vibrating spring [1–3].

In this chapter, we discuss first- and second-order linear partial differential equations, with one dependent variable and two independent variables. The case in which we have more than two independent variables is treated in the next chapter, in which we present the method of separation of variables.

8.1 First-Order Partial Differential Equation

The most general form of a first-order partial differential equation with only two independent variables, x and y, whose dependent variable we denote by $u = u(x, y)$, is

$$F\left(x, y, u, \frac{\partial u}{\partial x}, \frac{\partial u}{\partial y}\right) = 0.$$

A solution of this partial differential equation on a domain Ω of $\mathbb{R}^2$ is a function $u = f(x, y)$ defined on Ω and satisfying the conditions: (a) For every $(x, y) \in \Omega$, there is a point of $\mathbb{R}^5$ $\left(x, y, u, \dfrac{\partial u}{\partial x} \equiv \alpha, \dfrac{\partial u}{\partial y} \equiv \beta\right)$ in the domain of function F, and (b) when substituting $u = f(x, y)$ in the partial differential equation, the resulting equation is an identity in x and y for every $(x, y) \in \Omega$.

© The Author(s), under exclusive license to Springer Nature Switzerland AG 2025
E. Capelas de Oliveira, J. E. Maiorino, *Analytical Methods in Applied Mathematics*,
Problem Books in Mathematics, https://doi.org/10.1007/978-3-031-74794-6_8

The partial differential equation is classified according to the form of F. As already mentioned, we are interested in linear differential equations, so the most general form is

$$P(x, y)\frac{\partial u}{\partial x} + Q(x, y)\frac{\partial u}{\partial y} + R(x, y)u = S(x, y),$$

with $u = u(x, y)$, that is, F is linear on $u(x, y)$ and on the first derivatives, $\frac{\partial u}{\partial x}$ and $\frac{\partial u}{\partial y}$, and all the coefficients depend only on the independent variables.

Unlike the case with ordinary differential equations, here we can have equations classified as quasilinear and semilinear, in addition to nonlinear ones.

8.2 Method of Characteristics

The so-called method of characteristics can be applied to both linear and nonlinear equations. It consists in adequately changing the original coordinates (x, y) into the characteristic coordinates (or only characteristics) (ξ, η), so that the partial differential equation is transformed into an ordinary differential equation (Chap. 1). After solving the differential equation in terms of coordinates (ξ, η), we turn to coordinates (x, y) to get the solution of the partial differential equation.

8.2.1 Quasi-linear First-Order Equations

We will illustrate this technique for the case of quasi-linear first-order partial differential equations, that is, for differential equations of the type

$$P(x, y)\frac{\partial u}{\partial x} + Q(x, y)\frac{\partial u}{\partial y} = R(x, y, u).$$

Consider families of curves satisfying the equation

$$\frac{dx}{P(x, y)} = \frac{dy}{Q(x, y)} = d\rho.$$

The curves are then defined by

$$\frac{dy}{dx} = \frac{Q(x, y)}{P(x, y)}.$$

Consider an infinitesimal element of arc length ds of the curve; then we can write (Pythagorean theorem)

$$(ds)^2 = (dx)^2 + (dy)^2 .$$

Combining the last two equations, we have

$$(ds)^2 = \frac{P^2}{Q^2}(dy)^2 + (dy)^2 = (dx)^2 + \frac{Q^2}{P^2}(dx)^2,$$

with $P = P(x, y)$ and $Q = Q(x, y)$. Thus

$$\frac{dx}{P} = \frac{dy}{Q} = \frac{ds}{\sqrt{P^2 + Q^2}} = d\rho . \tag{8.1}$$

Multiplying the partial differential equation by ds, using the previous equation and rearranging, we have

$$dx \frac{\partial u}{\partial x} + dy \frac{\partial u}{\partial y} = \frac{R}{\sqrt{P^2 + Q^2}} ds.$$

The left-hand side of this equation is the total differential of $u(x, y)$, so that using Eq. (8.1) we get

$$du = \frac{R}{\sqrt{P^2 + Q^2}} ds$$

or $du = R\, d\rho$. Going back to Eq. (8.1), we have

$$\frac{dx}{P} = \frac{dy}{Q} = \frac{du}{R} = d\rho$$

which, by integration along the curve, provides $u = u(x, y)$. These curves are known as **characteristic curves**.

Example 8.1 Find the solution of the problem composed of the first-order partial differential equation

$$x \frac{\partial u}{\partial x} + y \frac{\partial u}{\partial y},$$

with $u = u(x, y)$ and the condition $u(1, x) = 1$.

Comparing this equation with the partial differential equation in its most general form, we see that $P(x, y) = x$, $Q(x, y) = y$ and $R(x, y) = 1$, then

$$\frac{dx}{x} = \frac{dy}{y} = \frac{du}{1} .$$

It thus follows that the characteristics satisfy the first-order ordinary differential equation

$$\frac{dy}{dx} = \frac{y}{x},$$

whose solution provides the family of curves

$$y = Cx.$$

In this case, the curves are straight lines passing through the origin and forming an angle θ with the x axis such that $\tan \theta = C$.

For all such characteristic curves, we have

$$\frac{du}{1} = \frac{dy}{y},$$

whose integration yields $u = \ln y + D$, where D is another arbitrary constant which is normally different for each characteristic.

Since for each value of constant C there is a value for D, we have $D = f(C)$, where f relates C to D and must be determined. Getting back to the solution, we can write

$$\begin{aligned}u(x, y) &= \ln y + f(C)\\ &= \ln y + f(y/x),\end{aligned}$$

which is the general solution of the partial differential equation.

Note the difference between a first-order ordinary differential equation, in which an arbitrary constant emerges and a partial differential equation, in which an arbitrary function emerges. In order to determine this function, we must use the condition $u(1, x) = 1$; then $u(1, x) = \ln x + f(x) = 1$, hence $f(x) = 1 - \ln x$. Since we want $f(y/x)$, we finally get

$$f(y/x) = 1 - \ln y + \ln x,$$

which provides for the solution of the boundary value problem

$$u(x, y) = 1 + \ln x.$$

$\square$

This method is also useful for solving a second-order partial differential equation if it is possible to factorize the second-order partial differential equation in two first-order partial differential equations.

8.3 Second-Order Partial Differential Equation

In this section, we discuss linear second order partial differential equations with one dependent variable and two independent variables. The case in which there are more than two independent variables will be discussed in the next chapter, in which we present the method of separation of variables [3–5].

8.3.1 Classification

We call second-order linear partial differential equation all equations of the form

$$A\frac{\partial^2 u}{\partial x^2} + B\frac{\partial^2 u}{\partial x \partial y} + C\frac{\partial^2 u}{\partial y^2} + D\frac{\partial u}{\partial x} + E\frac{\partial u}{\partial y} + Fu = G, \qquad (8.2)$$

where the coefficients $A, \ldots, G$ are functions of the independent variables x and y and $u = u(x, y)$ is the dependent variable. We assume from the beginning that $u(x, y)$ and the coefficients appearing in Eq. (8.2) are continuously differentiable and that the coefficients A, B, and C are not simultaneously null.

The classification of a linear partial differential equation is based on the possibility of reducing it to the so-called canonical form, at any point of its domain, by means of a transformation of coordinates. This classification is analogous to the classification of quadratic equations. A linear partial differential equation in the form of Eq. (8.2) is called a *hyperbolic*, *parabolic*, or *elliptic* differential equation at a point $P(x_0, y_0)$ of its domain if the *discriminant*

$$\Delta \equiv B^2(x_0, y_0) - 4A(x_0, y_0)C(x_0, y_0)$$

is respectively *positive*, *zero*, or *negative* at that point. If this fact is true for all points of the domain, we say that the equation is hyperbolic, parabolic, or elliptic on the domain considered. When the discriminant depends on x and y, we say that the linear partial differential equation is of mixed type, that is, the discriminant assumes different signs at distinct points of the plane.

In the particular case of two independent variables, it is always possible to find a transformation of coordinates that leaves the differential equation invariant, i.e., that preserves the *form* of the differential equation, provided the Jacobian associated with the transformation is different from zero.

Let us then consider, for two independent variables, a general coordinate transformation given by

$$\begin{aligned} \xi &= \xi(x, y), \\ \eta &= \eta(x, y), \end{aligned} \qquad (8.3)$$

where we must suppose that ξ and η are twice continuously differentiable and the Jacobian

$$J \equiv \det \begin{bmatrix} \dfrac{\partial \xi}{\partial x} & \dfrac{\partial \xi}{\partial y} \\[2ex] \dfrac{\partial \eta}{\partial x} & \dfrac{\partial \eta}{\partial y} \end{bmatrix}$$

is different from zero on the domain considered, so that x and y can be obtained univocally from the equations for ξ and η. Then, introducing the transformations in Eq. (8.3) into the original linear partial differential equation, we obtain the new differential equation

$$\bar{A}\frac{\partial^2 u}{\partial \xi^2} + \bar{B}\frac{\partial^2 u}{\partial \xi \partial \eta} + \bar{C}\frac{\partial^2 u}{\partial \eta^2} + \bar{D}\frac{\partial u}{\partial \xi} + \bar{E}\frac{\partial u}{\partial \eta} + \bar{F}u = \bar{G}, \tag{8.4}$$

where the new coefficients are related to the old ones by

$$\bar{A} = A\left(\frac{\partial \xi}{\partial x}\right)^2 + B\frac{\partial \xi}{\partial x}\frac{\partial \xi}{\partial y} + C\left(\frac{\partial \xi}{\partial y}\right)^2; \tag{8.5}$$

$$\bar{B} = 2A\frac{\partial \xi}{\partial x}\frac{\partial \eta}{\partial x} + B\left(\frac{\partial \xi}{\partial x}\frac{\partial \eta}{\partial y} + \frac{\partial \xi}{\partial y}\frac{\partial \eta}{\partial x}\right) + 2C\frac{\partial \xi}{\partial y}\frac{\partial \eta}{\partial y}; \tag{8.6}$$

$$\bar{C} = A\left(\frac{\partial \eta}{\partial x}\right)^2 + B\frac{\partial \eta}{\partial x}\frac{\partial \eta}{\partial y} + C\left(\frac{\partial \eta}{\partial y}\right)^2; \tag{8.7}$$

$$\bar{D} = A\frac{\partial^2 \xi}{\partial x^2} + B\frac{\partial^2 \xi}{\partial x \partial y} + C\frac{\partial^2 \xi}{\partial y^2} + D\frac{\partial \xi}{\partial x} + E\frac{\partial \xi}{\partial y}; \tag{8.8}$$

$$\bar{E} = A\frac{\partial^2 \eta}{\partial x^2} + B\frac{\partial^2 \eta}{\partial x \partial y} + C\frac{\partial^2 \eta}{\partial y^2} + D\frac{\partial \eta}{\partial x} + E\frac{\partial \eta}{\partial y}; \tag{8.9}$$

$$\bar{F} = F \quad \text{and} \quad \bar{G} = G. \tag{8.10}$$

Notice that the new differential equation, Eq. (8.4), has the same form as the original one, i.e., it is a second-order, linear partial differential equation, Eq. (8.2). The classification of this linear, second order partial differential equation depends only on the coefficients A, B, and C at the point (x, y); for this reason we rewrite Eq. (8.2) and Eq. (8.4), respectively, as

$$A(x, y)\frac{\partial^2 u}{\partial x^2} + B(x, y)\frac{\partial^2 u}{\partial x \partial y} + C(x, y)\frac{\partial^2 u}{\partial y^2} = H$$

and

$$\bar{A}(\xi, \eta)\frac{\partial^2 u}{\partial \xi^2} + \bar{B}(\xi, \eta)\frac{\partial^2 u}{\partial \xi \partial \eta} + \bar{C}(\xi, \eta)\frac{\partial^2 u}{\partial \eta^2} = \bar{H},$$

where H and $\bar{H}$ are functions of x, y, u, $\partial u/\partial x$, $\partial u/\partial y$ and ξ, η, u, $\partial u/\partial \xi$, $\partial u/\partial \eta$, respectively.

8.3.2   *The Canonical Form*

Suppose that the functions A, B, and C are not simultaneously null. We can choose the new variables ξ and η so that the coefficients $\bar{A}$ and $\bar{C}$ are null. For this we must have

$$\bar{A} = A\left(\frac{\partial \xi}{\partial x}\right)^2 + B\frac{\partial \xi}{\partial x}\frac{\partial \xi}{\partial y} + C\left(\frac{\partial \xi}{\partial y}\right)^2 \equiv 0;$$

and

$$\bar{C} = A\left(\frac{\partial \eta}{\partial x}\right)^2 + B\frac{\partial \eta}{\partial x}\frac{\partial \eta}{\partial y} + C\left(\frac{\partial \eta}{\partial y}\right)^2 \equiv 0.$$

We note that these two differential equations have the same form. For this reason, we discuss only the differential equation

$$A\left(\frac{\partial \tau}{\partial x}\right)^2 + B\frac{\partial \tau}{\partial x}\frac{\partial \tau}{\partial y} + C\left(\frac{\partial \tau}{\partial y}\right)^2 = 0,$$

where the variable τ represents ξ or η. The last differential equation can be rewritten in the following form:

$$A\left(\frac{\partial \tau/\partial x}{\partial \tau/\partial y}\right)^2 + B\left(\frac{\partial \tau/\partial x}{\partial \tau/\partial y}\right) + C = 0.$$

Along a curve $\tau =$ constant on the (x, y) plane we have

$$d\tau = \frac{\partial \tau}{\partial x}dx + \frac{\partial \tau}{\partial y}dy = 0,$$

whence we obtain

$$\frac{\partial \tau/\partial x}{\partial \tau/\partial y} = -\frac{dy}{dx}.$$

This result allows us to write an algebraic equation in variable $\dfrac{dy}{dx}$,

$$A\left(\frac{dy}{dx}\right)^2 - B\left(\frac{dy}{dx}\right) + C = 0.$$

The roots of this second degree algebraic equation are

$$\frac{dy}{dx} = \frac{1}{2A}(B + \sqrt{\Delta}) \tag{8.11}$$

and

$$\frac{dy}{dx} = \frac{1}{2A}(B - \sqrt{\Delta}), \tag{8.12}$$

where $\Delta = B^2 - 4AC$.

We have thus obtained two first order differential equations, Eqs. (8.11) and (8.12), called *characteristic equations*, whose respective integrals are called *characteristic curves*. Since these equations are of first order, each of them admits an integration constant.

We must note that if coefficients A, B, and C are constant, the characteristic equations lead us to two families of straight lines, and the equation is of the same type at all points of its domain, since Δ will also be constant.

Example 8.2 Obtain two first-order differential equations associated with the linear second-order partial differential equation

$$\frac{\partial^2 u}{\partial x^2} + 3\frac{\partial^2 u}{\partial x \partial y} + 2\frac{\partial^2 u}{\partial y^2} + 3\frac{\partial u}{\partial x} + 2\frac{\partial u}{\partial y} + u = 0.$$

Comparing this equation with Eq. (8.2) and identifying their coefficients, we find $A = 1$, $B = 3$, and $C = 2$. Then, substituting into Eq. (8.11) and Eq. (8.12), we get

$$\frac{dy}{dx} = 2 \qquad \text{and} \qquad \frac{dy}{dx} = 1$$

respectively.   □

In what follows, we discuss the three types of second-order linear partial differential equations: hyperbolic, parabolic, and elliptic. The mixed-type partial differential equation will be discussed in the solved exercises.

8.3.3   *Equation of Hyperbolic Type*

If $\Delta > 0$, we have two distinct families of characteristic curves and the original partial differential equation reduces to the form

$$\frac{\partial^2 u}{\partial \xi \partial \eta} = H_1\left(\xi, \eta, u, \frac{\partial u}{\partial \xi}, \frac{\partial u}{\partial \eta}\right),$$

where $H_1 = \bar{H}/\bar{B}$, with $\bar{B} \neq 0$. This is the so-called first canonical form of the hyperbolic equation. Introducing a second pair of independent variables,

$$\begin{aligned} \alpha &= \xi + \eta, \\ \beta &= \xi - \eta, \end{aligned} \tag{8.13}$$

we obtain the *second canonical form*

$$\frac{\partial^2 u}{\partial \alpha^2} - \frac{\partial^2 u}{\partial \beta^2} = H_2\left(\alpha, \beta, u, \frac{\partial u}{\partial \alpha}, \frac{\partial u}{\partial \beta}\right).$$

A classical example of this type of partial differential equation is the differential equation associated with wave propagation, the so-called wave equation, which will be discussed in the next chapter.

Example 8.3 Consider A, B, and C nonsimultaneously null real constants, satisfying $B^2 > 4AC$. The second-order partial differential equation

$$A\frac{\partial^2 u}{\partial x^2} + B\frac{\partial^2 u}{\partial x \partial y} + C\frac{\partial^2 u}{\partial y^2} = H\left(u, x, y, \frac{\partial u}{\partial x}, \frac{\partial u}{\partial y}\right)$$

can be put in the first canonical form

$$\frac{\partial^2 u}{\partial \xi \partial \eta} = H_1\left(\xi, \eta, u, \frac{\partial u}{\partial \xi}, \frac{\partial u}{\partial \eta}\right),$$

where $H_1 = \bar{H}/\bar{B}$, with $\bar{B} \neq 0$. Using the transformation Eq. (8.13), it can be put in the second canonical form

$$\frac{\partial^2 u}{\partial \alpha^2} - \frac{\partial^2 u}{\partial \beta^2} = H_2\left(\alpha, \beta, u, \frac{\partial u}{\partial \alpha}, \frac{\partial u}{\partial \beta}\right),$$

with $\alpha = \xi + \eta$ and $\beta = \xi - \eta$. $\qquad\square$

8.3.4   *Equation of Parabolic Type*

If the discriminant $\Delta = 0$, the characteristic equations (8.11) and (8.12) are identical. In this case, there exists only one family of characteristic curves, and we obtain only one integral curve $\xi = $ constant (or $\eta = $ constant). Thus, the canonical form of a differential equation of parabolic type is given by

$$\frac{\partial^2 u}{\partial \eta^2} = H_3\left(\xi, \eta, u, \frac{\partial u}{\partial \xi}, \frac{\partial u}{\partial \eta}\right) \quad \text{for} \quad \bar{C} \neq 0$$

or

$$\frac{\partial^2 u}{\partial \xi^2} = \bar{H}_3\left(\xi, \eta, u, \frac{\partial u}{\partial \xi}, \frac{\partial u}{\partial \eta}\right) \quad \text{for} \quad \bar{A} \neq 0,$$

depending on whether we choose $\xi = $ constant or $\eta = $ constant, respectively.

The phenomena governed by the heat equation are the most representative of parabolic differential equations and will be discussed in the next chapter.

Example 8.4 Consider A, B, and C nonsimultaneously null real constants, satisfying $B^2 = 4AC$. The second-order partial differential equation

$$A\frac{\partial^2 u}{\partial x^2} + B\frac{\partial^2 u}{\partial x \partial y} + C\frac{\partial^2 u}{\partial y^2} = H\left(u, x, y, \frac{\partial u}{\partial x}, \frac{\partial u}{\partial y}\right)$$

can be put in the canonical forms

$$\frac{\partial^2 u}{\partial \eta^2} = H_3\left(\xi, \eta, u, \frac{\partial u}{\partial \xi}, \frac{\partial u}{\partial \eta}\right),$$

or

$$\frac{\partial^2 u}{\partial \xi^2} = \overline{H}_3\left(\xi, \eta, u, \frac{\partial u}{\partial \xi}, \frac{\partial u}{\partial \eta}\right),$$

respectively for $C \neq 0$ and $B \neq 0$. $\square$

8.3.5   *Equation of Elliptic Type*

In this case $\Delta < 0$ and the characteristic curves are not real. However, if the coefficients A, B, and C are analytic functions, we can consider the equation

$$A\left(\frac{dy}{dx}\right)^2 - B\left(\frac{dy}{dx}\right) + C = 0$$

for complex variables x and y. Since ξ and η are complex conjugate, we can introduce the real variables

$$\alpha = \frac{1}{2}(\xi + \eta) \quad \text{and} \quad \beta = \frac{1}{2i}(\xi - \eta),$$

obtaining, after all transformations, the differential equation

$$\frac{\partial^2 u}{\partial \alpha^2} + \frac{\partial^2 u}{\partial \beta^2} = H_0\left(\alpha, \beta, u, \frac{\partial u}{\partial \alpha}, \frac{\partial u}{\partial \beta}\right),$$

called canonical form of the elliptic equation.

The Laplace differential equation (or just Laplace equation) constitutes an example of elliptic differential equation and will also be discussed in the next chapter.

Example 8.5 Consider A, B, and C nonsimultaneously null real constants, satisfying $B^2 < 4AC$. The second-order partial differential equation

$$A\frac{\partial^2 u}{\partial x^2} + B\frac{\partial^2 u}{\partial x \partial y} + C\frac{\partial^2 u}{\partial y^2} = H\left(u, x, y, \frac{\partial u}{\partial x}, \frac{\partial u}{\partial y}\right)$$

can be put in the following canonical form:

$$\frac{\partial^2 u}{\partial \alpha^2} + \frac{\partial^2 u}{\partial \beta^2} = H_4\left(\alpha, \beta, u, \frac{\partial u}{\partial \alpha}, \frac{\partial u}{\partial \beta}\right).$$

with $2\alpha = \xi + \eta$ and $2i\beta = \xi - \eta$. $\square$

8.4 Solved Exercises

SE 8.1 The well-known two-dimensional Laplace equation, written in Cartesian coordinates, is given by

$$\frac{\partial^2}{\partial x^2}u(x, y) + \frac{\partial^2}{\partial y^2}u(x, y) \equiv \nabla^2 u(x, y) \equiv \Delta u(x, y) = 0. \tag{8.14}$$

Introduce plane polar coordinates, defined by the relations

$$x = r\cos\theta \quad \text{and} \quad y = r\sin\theta, \tag{8.15}$$

with $r > 0$ and $0 \le \theta \le 2\pi$, to write the Laplace equation in coordinates r and θ. Classify the resulting partial differential equation.

Solution: The Jacobian associated with the transformation is given by the expression

$$J = \begin{vmatrix} \partial x/\partial r & \partial x/\partial \theta \\ \partial y/\partial r & \partial y/\partial \theta \end{vmatrix} = \begin{vmatrix} \cos\theta & -r\sin\theta \\ \sin\theta & r\cos\theta \end{vmatrix} = r, \tag{8.16}$$

which is always different from zero. Inverting the transformation, we can write

$$r = (x^2 + y^2)^{1/2} \quad \text{and} \quad \theta = \arctan\frac{y}{x}. \tag{8.17}$$

Calculating the first derivatives, we obtain

$$\frac{\partial}{\partial x} = \frac{\partial r}{\partial x}\frac{\partial}{\partial r} + \frac{\partial \theta}{\partial x}\frac{\partial}{\partial \theta} = \cos\theta\frac{\partial}{\partial r} - \frac{1}{r}\sin\theta\frac{\partial}{\partial \theta} \tag{8.18}$$

and

$$\frac{\partial}{\partial y} = \frac{\partial r}{\partial y}\frac{\partial}{\partial r} + \frac{\partial \theta}{\partial y}\frac{\partial}{\partial \theta} = \sin\theta\frac{\partial}{\partial r} + \frac{1}{r}\cos\theta\frac{\partial}{\partial \theta}. \tag{8.19}$$

Next, calculating the second derivatives we get

$$\frac{\partial^2}{\partial x^2} = \cos^2\theta\frac{\partial^2}{\partial r^2} + \frac{1}{r}\sin^2\theta\frac{\partial}{\partial r} - \frac{2}{r}\sin\theta\cos\theta\frac{\partial^2}{\partial r\partial\theta} +$$
$$+ \frac{1}{r^2}\sin^2\theta\frac{\partial^2}{\partial\theta^2} + \frac{2}{r^2}\sin\theta\cos\theta\frac{\partial}{\partial\theta}$$

and

$$\frac{\partial^2}{\partial y^2} = \sin^2\theta\frac{\partial^2}{\partial r^2} + \frac{1}{r}\cos^2\theta\frac{\partial}{\partial r} + \frac{2}{r}\sin\theta\cos\theta\frac{\partial^2}{\partial r\partial\theta} +$$
$$+ \frac{1}{r^2}\cos^2\theta\frac{\partial^2}{\partial\theta^2} - \frac{2}{r^2}\sin\theta\cos\theta\frac{\partial}{\partial\theta}.$$

Adding the last two expressions, we get

$$\frac{\partial^2}{\partial r^2}u(r,\theta) + \frac{1}{r}\frac{\partial}{\partial r}u(r,\theta) + \frac{1}{r^2}\frac{\partial^2}{\partial\theta^2}u(r,\theta) = 0 \tag{8.20}$$

which is the Laplace equation written in terms of plane polar coordinates (**PE 9.19**). It is already in the canonical form.

To classify the last equation according to its type, we calculate its discriminant, that is,

$$\Delta = 0^2 - 1\cdot\frac{4}{r^2} = -\frac{4}{r^2} < 0, \tag{8.21}$$

which is always negative. We thus conclude that the equation is of elliptic type.   □

SE 8.2 Let $B, C \in \mathbb{R}^*$ with $B^2 = C$. Solve the partial differential equation

$$\frac{\partial^2 u}{\partial x^2} + 2B\frac{\partial^2 u}{\partial x \partial y} + C\frac{\partial^2 u}{\partial y^2} + B\frac{\partial u}{\partial x} + B^2\frac{\partial u}{\partial y} = 0,$$

with $u = u(x, y)$.

We first show that this partial differential equation is of parabolic type:

$$\Delta = 4B^2 - 4C = 4(B^2 - C) = 0.$$

The characteristic coordinates ξ, η can be written in terms of coordinates x, y as

$$\begin{cases} \xi = y, \\ \eta = y - Bx. \end{cases}$$

Using the chain rule we obtain for the first derivatives

$$\frac{\partial}{\partial x} = -B\frac{\partial}{\partial \eta} \quad \text{and} \quad \frac{\partial}{\partial y} = \frac{\partial}{\partial \xi} + \frac{\partial}{\partial \eta}.$$

For the second derivatives, the relations are

$$\frac{\partial^2}{\partial x^2} = B^2\frac{\partial^2}{\partial \eta^2}, \quad \frac{\partial^2}{\partial y^2} = \frac{\partial^2}{\partial \xi^2} + 2\frac{\partial^2}{\partial \xi \partial \eta} + \frac{\partial^2}{\partial \eta^2}, \quad \frac{\partial^2}{\partial x \partial y} = -B\frac{\partial^2}{\partial \xi \partial \eta} - B\frac{\partial^2}{\partial \eta^2}.$$

Substituting these expressions into the partial differential equation and simplifying, we get

$$\frac{\partial^2}{\partial \xi^2}u + \frac{\partial}{\partial \xi}u = 0,$$

where $u = u(\xi, \eta)$. This partial differential equation is in the canonical form. To obtain its solution, we first introduce a change of dependent variable, $\frac{\partial u}{\partial \xi} = v = v(\xi, \eta)$, so that

$$\frac{\mathrm{d}}{\mathrm{d}\xi}v + v = 0$$

which is a first-order partial differential equation whose solution is given by

$$v(\xi, \eta) = f(\eta)\, e^{-\xi},$$

where $f(\eta)$ is an arbitrary function depending on η only. Turning back to dependent variable $u(\xi, \eta)$, we perform another integration to obtain

$$u(\xi, \eta) = -f(\eta)\, e^{-\xi} + g(\eta)\,,$$

where $g(\eta)$ is another arbitrary function depending on η only. Using coordinates x, y, we finally have

$$u(x, y) = -f(y - Bx)\, e^{-y} + g(y - Bx)\,,$$

which is the desired result.   □

SE 8.3 Consider the partial differential equation

$$\frac{\partial^2 u}{\partial x^2} + 4x^2 \frac{\partial^2 u}{\partial y^2} = 0\,, \tag{8.22}$$

with $x \neq 0$. (a) Classify this partial differential equation. (b) Obtain the corresponding characteristic equation. (c) Determine the canonical form.

Solution: (a) Identifying the terms of this equation with those used in Eq. (8.2), we have $A = 1$, $B = 0$, and $C = 4x^2$; hence

$$\Delta = B^2 - 4AC = -4.1.4x^2 = -16x^2 < 0.$$

So, the equation is of elliptic type.

(b) To obtain the characteristic equation, we use (cf. Eq. (8.11) and Eq. (8.12))

$$\frac{dy}{dx} = \frac{B \pm \sqrt{\Delta}}{2A}\,,$$

whence we get

$$\frac{dy}{dx} = 2ix \quad \text{and} \quad \frac{dy}{dx} = -2ix\,.$$

(c) Integrating these ordinary differential equations, we find

$$y - ix^2 = c_1 \quad \text{and} \quad y + ix^2 = c_2,$$

where c_1 and c_2 are constants. Introducing (i.e., defining) the new variables ξ and η given by the relations

$$y - ix^2 = \xi$$
$$y + ix^2 = \eta$$

we obtain for the variables α and β:

$$\alpha = \frac{1}{2}(\xi + \eta) = y,$$

$$\beta = \frac{1}{2i}(\xi - \eta) = -x^2.$$

We then calculate explicitly the operators associated with the partial derivatives with respect to these new variables. For the first derivatives, we get

$$\frac{\partial}{\partial x} = \frac{\partial \alpha}{\partial x}\frac{\partial}{\partial \alpha} + \frac{\partial \beta}{\partial x}\frac{\partial}{\partial \beta} = -2x\frac{\partial}{\partial \beta} = -2\sqrt{-\beta}\frac{\partial}{\partial \beta},$$

$$\frac{\partial}{\partial y} = \frac{\partial \alpha}{\partial y}\frac{\partial}{\partial \alpha} + \frac{\partial \beta}{\partial y}\frac{\partial}{\partial \beta} = \frac{\partial}{\partial \alpha},$$

and for the second derivatives, we find

$$\frac{\partial^2}{\partial x^2} = 4\sqrt{-\beta}\frac{\partial}{\partial \beta}\left(\sqrt{-\beta}\frac{\partial}{\partial \beta}\right) = -4\beta\frac{\partial^2}{\partial \beta^2} - 2\frac{\partial}{\partial \beta},$$

$$\frac{\partial^2}{\partial y^2} = \frac{\partial^2}{\partial \alpha^2}.$$

Substituting these expressions into Eq. (8.22), we obtain

$$-4\beta\frac{\partial^2 u}{\partial \beta^2} - 2\frac{\partial u}{\partial \beta} - 4\beta\frac{\partial^2 u}{\partial \alpha^2} = 0.$$

Thus,

$$\frac{\partial^2 u}{\partial \alpha^2} + \frac{\partial^2 u}{\partial \beta^2} = -\frac{1}{2\beta}\frac{\partial u}{\partial \beta},$$

which is the desired canonical form. Notice that $\beta \neq 0$, as $x \neq 0$. $\qquad\square$

SE 8.4 Consider the following partial differential equation:

$$3\frac{\partial^2}{\partial x^2}u(x, y) + 5\frac{\partial^2}{\partial x\partial y}u(x, y) + 2\frac{\partial^2}{\partial y^2}u(x, y) + \frac{\partial}{\partial x}u(x, y) + \frac{\partial}{\partial y}u(x, y) = -\frac{1}{3}.$$

We ask the following: (a) Classify the differential equation according to its type; (b) reduce the differential equation to the corresponding canonical form, and (c) obtain the general solution.

Solution: (a) To classify this differential equation, we calculate the corresponding discriminant, i.e.,

$$\Delta = 5^2 - 4.3.2 = 25 - 24 = 1 > 0$$

and we conclude that the equation is of hyperbolic type.

(b) To reduce it to canonical form, we must first obtain the characteristic equations:

$$\frac{dy}{dx} = \frac{5 \pm \sqrt{1}}{2.3}.$$

Integrating it, we obtain the respective characteristic curves, given by

$$y = x + c_1 \qquad \text{and} \qquad y = \frac{2}{3}x + c_2 .$$

Introducing the characteristic coordinates defined by these relations, we get

$$\xi = y - x \qquad \text{and} \qquad \eta = y - \frac{2}{3}x.$$

Finally, calculating the derivatives, substituting them into the differential equation and rearranging, we get

$$\frac{\partial^2}{\partial \xi \partial \eta} u(\xi, \eta) - \frac{\partial}{\partial \eta} u(\xi, \eta) = 1$$

which is the first canonical form of an equation of hyperbolic type.

(c) To find the general solution, we introduce the change of dependent variable

$$\frac{\partial}{\partial \eta} u(\xi, \eta) = v(\xi, \eta)$$

and we obtain, omitting the independent variables, the following first-order partial differential equation:

$$\frac{\partial v}{\partial \xi} - v = 1.$$

Another change of dependent variable of the form $v = w + \alpha$, where α is a parameter to be adequately chosen, leads us to the first order differential equation

$$\frac{\partial w}{\partial \xi} - w - \alpha = 1 ,$$

where we have put $\alpha = -1$. Then, integrating the corresponding homogeneous differential equation

$$\frac{\partial w}{\partial \xi} - w = 0$$

we get

$$w = f(\eta)\, e^{\xi}\,,$$

where $f(\eta)$ is a function that depends only on variable η. Going back to variable v, we get

$$v = f(\eta)\, e^{\xi} - 1\,.$$

Finally, we can write for the dependent variable u,

$$\frac{\partial u}{\partial \eta} = f(\eta)\, e^{\xi} - 1$$

and integrating again, we obtain

$$u(\xi, \eta) = e^{\xi} \int^{\eta} f(\eta')d\eta' - \eta + G(\xi)$$

or

$$u(\xi, \eta) = e^{\xi}\, F(\eta) - \eta + G(\xi)\,.$$

In terms of the original variables, we can write

$$u(x, y) = e^{y-x}\, F\!\left(y - \frac{2}{3}x\right) - y + \frac{2}{3}x + G(y - x)\,,$$

where F and G are two arbitrary, twice continuously differentiable functions. $\square$

We conclude this chapter observing that it is not always easy (or useful) to obtain the general solution of a partial differential equation. Nevertheless, if the canonical form is simple, we can almost always obtain such a general solution (cf. **PE 8.45** to **PE 8.50**).

8.5   Proposed Exercises

PE 8.1 Let $x, y \in \mathbb{R}^*$ and $u = u(x, y)$. Classify the partial differential equation

$$(1 + xy)\frac{\partial u}{\partial x} + \frac{\partial u}{\partial y} + u^2 = 2022\,.$$

PE 8.2 Let $x, y \in \mathbb{R}^*$ and $u = u(x, y)$. Solve the boundary value problem

$$\begin{cases} x\dfrac{\partial u}{\partial x} - y\dfrac{\partial u}{\partial y} = 0, \\ \qquad\qquad u(1, 1) = 0. \end{cases}$$

PE 8.3 Let $x, y \in \mathbb{R}^*$ and $u = u(x, y)$. For the partial differential equation

$$x\frac{\partial u}{\partial x} - y\frac{\partial u}{\partial y} = 1,$$

obtain the characteristic curves.

PE 8.4 For the partial differential equation of the previous exercise, obtain its general solution.

PE 8.5 Let $x, y \in \mathbb{R}^*$ and $u = u(x, y)$. Solve the boundary value problem

$$\begin{cases} x\dfrac{\partial u}{\partial x} - y\dfrac{\partial u}{\partial y} = 1, \\ \qquad\qquad u(1, y) = 1. \end{cases}$$

PE 8.6 Let $x, y \in \mathbb{R}^*$ and $u = u(x, y)$. Solve the partial differential equation

$$x^2\frac{\partial u}{\partial x} + y^2\frac{\partial u}{\partial y} = 1.$$

PE 8.7 Let $x, y \in \mathbb{R}^*$ and $u = u(x, y)$. Solve the boundary value problem

$$\begin{cases} x^2\dfrac{\partial u}{\partial x} + y^2\dfrac{\partial u}{\partial y} = 1, \\ \qquad\qquad u(x, 2x) = 1. \end{cases}$$

PE 8.8 Let $x, y \in \mathbb{R}^*$ and $u = u(x, y)$. Classify the partial differential equation

$$\frac{\partial u}{\partial x} + \frac{\partial u}{\partial y} = u^2.$$

PE 8.9 Let $x, y \in \mathbb{R}^*$ and $u = u(x, y)$. Obtain the general solution of the partial differential equation

$$\frac{\partial u}{\partial x} + \frac{\partial u}{\partial y} = u^2.$$

PE 8.10 Let $x, y \in \mathbb{R}$ and $u = u(x, y)$. Classify the second-order partial differential equation

$$x^2 \frac{\partial^2 u}{\partial x^2} + 2xy \frac{\partial^2 u}{\partial x \partial y} + y^2 \frac{\partial^2 u}{\partial y^2} = xy.$$

PE 8.11 Obtain explicitly the expressions for the coefficients $\bar{A}, \bar{B}, \ldots, \bar{G}$ appearing in Eqs. (8.4) to (8.10).

PE 8.12 Prove the relation $\bar{B}^2 - 4\bar{A}\bar{C} = J^2(B^2 - 4AC)$, where J is the Jacobian determinant associated with the coordinate transformation, given by Eq. (8.13).

PE 8.13 Show that in the expression $H_1 = \bar{H}/\bar{B}$, for an equation of hyperbolic type, we always have $\bar{B} \neq 0$.

PE 8.14 Obtain explicitly, from the first canonical form of an equation of hyperbolic type, the corresponding second canonical form.

PE 8.15 Obtain explicitly the canonical form for an equation of parabolic type.

PE 8.16 Obtain the canonical form associated with the equation of elliptic type.

PE 8.17 Show that when the coefficients of a partial differential equation are constant, the transformation of coordinates that puts this partial equation into the corresponding canonical form generates two families of straight lines.

PE 8.18 Let $u = u(x, y)$. Classify, according to its type, the so-called Tricomi [1897 – Francesco Giacomo Tricomi – 1978] equation, also called equation of mixed type:

$$\frac{\partial^2 u}{\partial x^2} + x \frac{\partial^2 u}{\partial y^2} = 0.$$

PE 8.19 Considering $x > 0$, classify according to the type the equation

$$x \frac{\partial^2 u}{\partial x^2} + \frac{\partial^2 u}{\partial y^2} = x^2,$$

with $u = u(x, y)$.

PE 8.20 Classify according to the type the differential equation

$$\sin^2 x \frac{\partial^2 u}{\partial x^2} + \sin 2x \frac{\partial^2 u}{\partial x \partial y} + \cos^2 x \frac{\partial^2 u}{\partial y^2} = 0,$$

where $u = u(x, y)$.

PE 8.21 Show that the partial differential equation

$$\frac{\partial^2 u}{\partial x^2} + 4 \frac{\partial^2 u}{\partial x \partial y} + 3 \frac{\partial^2 u}{\partial y^2} + 3 \frac{\partial u}{\partial x} - \frac{\partial u}{\partial y} + 2u = 0,$$

where $u = u(x, y)$, is of hyperbolic type.

PE 8.22 Let $u = u(x, y)$. Show that the differential equation

$$\frac{\partial^2 u}{\partial x^2} + 2\frac{\partial^2 u}{\partial x \partial y} + \frac{\partial^2 u}{\partial y^2} + 5\frac{\partial u}{\partial x} + 3\frac{\partial u}{\partial y} + u = 0$$

is of parabolic type.

PE 8.23 Determine the regions of the xy plane where this Tricomi equation

$$\frac{\partial^2 u}{\partial x^2} + y\frac{\partial^2 u}{\partial y^2} = 0,$$

is of elliptic type, parabolic type, or hyperbolic type, with $u = u(x, y)$.

PE 8.24 Determine the regions on the xy plane where the equation

$$(xy - 1)\frac{\partial^2 u}{\partial x^2} + (x + 2y)\frac{\partial^2 u}{\partial x \partial y} + \frac{\partial^2 u}{\partial y^2} + xy^2 u = 0,$$

for $u = u(x, y)$ is hyperbolic, parabolic, or elliptic.

PE 8.25 For $u(x, y)$, classify the partial differential equation

$$y\frac{\partial^2 u}{\partial x^2} + 2x\frac{\partial^2 u}{\partial x \partial y} + y\frac{\partial^2 u}{\partial y^2} = 0,$$

sketching a graph representation of each region.

PE 8.26 Find the characteristic equations, the characteristic curves, and the characteristic coordinates associated with the partial differential equation

$$\frac{\partial^2 u}{\partial x^2} + \frac{\partial^2 u}{\partial y^2} - 8\frac{\partial u}{\partial x} + \frac{\partial u}{\partial y} = 2,$$

where $u = u(x, y)$.

PE 8.27 Do as in the preceding exercise for the differential equation

$$\frac{\partial^2 u}{\partial x^2} + y^2\frac{\partial^2 u}{\partial y^2} = y,$$

with $u = u(x, y)$ and $y > 0$.

PE 8.28 Reduce the partial differential equation presented in **PE 8.21** to its second canonical form.

PE 8.29 Reduce the partial differential equation given in **PE 8.22** to the corresponding canonical form.

PE 8.30 Show that the differential equation for $u = u(x, y)$

$$\frac{\partial^2 u}{\partial x^2} - 6\frac{\partial^2 u}{\partial x \partial y} + 12\frac{\partial^2 u}{\partial y^2} + 4\frac{\partial u}{\partial x} - u = 0,$$

is of elliptic type and reduce it to the corresponding canonical form.

PE 8.31 Verify that $u(x, y) = \ln \sqrt{x^2 + y^2}$ satisfies the so-called two-dimensional Laplace partial differential equation:

$$\frac{\partial^2 u}{\partial x^2} + \frac{\partial^2 u}{\partial y^2} = 0.$$

PE 8.32 Verify that the so-called Gaussian function

$$u(x, t) = \frac{1}{\sqrt{4\pi t}} \exp\left(\frac{-x^2}{4t}\right)$$

satisfies the heat equation (also called *diffusion equation*)

$$\frac{\partial u}{\partial t} = \frac{\partial^2 u}{\partial x^2}$$

for $t \neq 0$ and $u = u(x, t)$.

PE 8.33 Let $u = u(x, t)$. Show that the sinusoidal wave given by the expression $u = \sin(x + t)$ satisfies the wave equation

$$\frac{\partial^2 u}{\partial t^2} - \frac{\partial^2 u}{\partial x^2} = 0.$$

PE 8.34 Consider the elliptic equation with constant coefficients for $u = u(x, y)$,

$$\frac{\partial^2 u}{\partial x^2} + \frac{\partial^2 u}{\partial y^2} = c_1\frac{\partial u}{\partial x} + c_2\frac{\partial u}{\partial y} + c_3 u + f.$$

Introducing the change of dependent variable

$$v = u\, e^{-(ax + by)}, \tag{8.23}$$

find a and b such that $v = v(x, y)$ obeys the partial differential equation

$$\frac{\partial^2 v}{\partial x^2} + \frac{\partial^2 v}{\partial y^2} = hv + g,$$

where h and g are functions of the coefficients of the original differential equation.

PE 8.35 Proceed as in **PE 8.34** for the hyperbolic equation

$$\frac{\partial^2 u}{\partial x \partial y} = a_1 \frac{\partial u}{\partial x} + a_2 \frac{\partial u}{\partial y} + a_3 u + f ,$$

that is, transform this equation into an equation satisfied by the function $v(x, y)$:

$$\frac{\partial^2 v}{\partial x \partial y} = h_1 v + g_1,$$

where h_1 and g_1 are related to the coefficients of the original equation, i.e., they are functions of those coefficients.

PE 8.36 Using a transformation as the one shown in **PE 8.34**, reduce the partial differential equation

$$\frac{\partial^2 u}{\partial x^2} - \frac{\partial^2 u}{\partial y^2} + 3 \frac{\partial u}{\partial x} - 2 \frac{\partial u}{\partial y} + u = 0 ,$$

with $u = u(x, y)$, to another partial differential equation with the form

$$\frac{\partial^2 v}{\partial \xi \partial \eta} = cv,$$

where c is a constant and $v = v(\xi, \eta)$.

PE 8.37 Do as in the preceding exercise for the differential equation

$$\frac{\partial^2 u}{\partial x^2} - \frac{\partial^2 u}{\partial y^2} = a \frac{\partial u}{\partial y} + b \frac{\partial u}{\partial x} + cu,$$

for $u = u(x, y)$, in the case in which the relation $4c + b^2 = 0$ is valid.

PE 8.38 Introduce the function $u(\xi, \eta) = \exp(\alpha \xi + \beta \eta) v(\xi, \eta)$ into the equation

$$\frac{\partial^2 u}{\partial \xi^2} + \frac{\partial^2 u}{\partial \eta^2} + \frac{\partial u}{\partial \xi} + \frac{\partial u}{\partial \eta} - 2u = 0$$

and choose the parameters α and β in such a way as to eliminate the terms involving the first derivative appearing in the resulting equation.

PE 8.39 Show that the terms involving $\dfrac{\partial u}{\partial \eta}$ and u in equation

$$\frac{\partial^2 u}{\partial \eta^2} - 2 \frac{\partial u}{\partial \xi} + \frac{\partial u}{\partial \eta} + u = 0$$

can be eliminated by using a particular function of the type used in the preceding exercise.

PE 8.40 Proceeding as in **PE 8.38**, show that the terms involving the first derivatives in the result of **PE 8.30** can be eliminated.

PE 8.41 Reduce the linear partial differential equation

$$\frac{\partial^2 u}{\partial x^2} - \frac{\partial^2 u}{\partial y^2} + \frac{1}{y}\frac{\partial u}{\partial y} = 0$$

to the corresponding second canonical form.

PE 8.42 Show that the linear partial differential equation satisfied by $u(x, y)$

$$\frac{\partial^2 u}{\partial x^2} + \frac{\partial^2 u}{\partial y^2} + \frac{1}{2}\frac{\partial u}{\partial y} = 0,$$

can be written in the form

$$\frac{\partial^2 u}{\partial \xi \partial \eta} + \frac{1}{4}\left(\frac{\partial u}{\partial \xi} + \frac{\partial u}{\partial \eta}\right) = 0,$$

with $u = u(\xi, \eta)$.

PE 8.43 Show that $u(x, y) = f(x + 2\sqrt{-y}) + g(x - 2\sqrt{-y})$, where f and g are arbitrary, continuously differentiable functions, defined in the region where the equation is of hyperbolic type, is solution of the equation

$$\frac{\partial^2 u}{\partial x^2} + \frac{1}{y}\frac{\partial^2 u}{\partial y^2} + \frac{1}{2}\frac{\partial u}{\partial y} = 0.$$

PE 8.44 Let $u = u(x, t)$. Verify that the solution of the partial differential equation

$$\frac{\partial u}{\partial t} = c^2\frac{\partial^2 u}{\partial x^2},$$

satisfying the condition $u(x, 0) = \cos x$ is given by

$$u(x, t) = \cos x \exp(-c^2 t),$$

where c^2 is a positive constant.

PE 8.45 Consider the partial differential equation

$$\frac{\partial^2 u}{\partial x^2} - \frac{1}{c^2}\frac{\partial^2 u}{\partial t^2} = 0,$$

where $u = u(x, t)$ and c^2 is a positive constant. Obtain the general solution of this partial differential equation.

PE 8.46 Find the general solution of the partial differential equation

$$x^2 \frac{\partial^2 u}{\partial x^2} + 2xy \frac{\partial^2 u}{\partial x \partial y} + y^2 \frac{\partial^2 u}{\partial y^2} = 0,$$

with $u = u(x, y)$.

PE 8.47 Do as in the preceding exercise for the partial differential equation

$$\frac{\partial^2 u}{\partial x^2} + \frac{10}{3} \frac{\partial^2 u}{\partial x \partial y} + \frac{\partial^2 u}{\partial y^2} = 0,$$

where $u = u(x, y)$.

PE 8.48 Obtain the general solution for the partial differential equation

$$r \frac{\partial^2 u}{\partial t^2} - r \frac{\partial^2 u}{\partial r^2} - 2 \frac{\partial u}{\partial r} = 0,$$

with $u = u(r, t)$.

PE 8.49 Let $u = u(x, y)$. Show that if the partial differential equation

$$A \frac{\partial^2 u}{\partial x^2} + 2B \frac{\partial^2 u}{\partial x \partial y} + C \frac{\partial^2 u}{\partial y^2} = 0$$

is of parabolic type, the change of variables

$$\begin{cases} \xi = x, \\ \eta = rx + y, \end{cases}$$

where $r = -B/A$ reduces the partial differential equation to the corresponding canonical form, which is given by

$$\frac{\partial^2 u}{\partial \xi^2} = 0,$$

with $u = u(\xi, \eta)$. Solve this partial differential equation.

PE 8.50 Show that if the partial differential equation presented in the preceding exercise is of hyperbolic type, the change of variables

$$\begin{cases} \xi = r_1 x + y, \\ \eta = r_2 x + y, \end{cases}$$

where r_1 and r_2 are roots of the algebraic equation $Ar^2 + 2Br + C = 0$, reduces the equation to the corresponding canonical form, which is given by

$$\frac{\partial^2 u}{\partial \xi \partial \eta} = 0,$$

with $u = u(\xi, \eta)$. Solve this partial differential equation.

References

1. M. Braun, *Differential Equations and Their Applications* (Springer, New York, 1983)
2. S.J. Farlow, *Partial Differential Equations for Scientists and Engineers* (Wiley, New York, 1982)
3. P.R. Garabedian, *Partial Differential Equations* (AMS Chelsea Publishing, Rhode Island, 1964)
4. M.A. Pinsky, *Partial Differential Equations and Boundary Value Problems with Applications* (McGraw-Hill, New York, 1991)
5. E. Zauderer, *Partial Differential Equations of Applied Mathematics* (Wiley, New York, 1983)

Chapter 9
The Method of Separation of Variables

> *As for everything else, so for a mathematical theory: beauty can be perceived but not explained.*
>
> *1821 – Arthur Cayley – 1895*

In the previous chapter, we discussed the general form of a linear, second-order partial differential equation, its characteristic equation, characteristic curves, and the so-called canonical forms. In this chapter, we are interested in a formal method to solve such linear partial differential equations. Besides the solution itself, we must also pay attention to the so-called initial conditions and boundary conditions.

Then, once the general solution is known and the initial conditions and boundary conditions are satisfied, we have the complete solution of a given problem. In usual problems, the initial and/or boundary conditions are given, and we must obtain the formal solution that satisfies the corresponding conditions. These solutions, here, are obtained by means of the so-called method of separation of variables, a powerful tool to discuss linear differential equations, also known as the *Fourier method*.

9.1 Basic Concepts

In the preceding chapter, we discussed the formal aspects of a homogeneous, linear second-order partial differential equation. We begin this chapter presenting the definitions of some basic concepts associated with these homogeneous, linear second-order partial differential equations and the method of separation of variables.

An equation involving one or more partial derivatives of an unknown function $u(x, y, \ldots)$ of two or more independent variables is called a partial differential equation. By definition, the order of the differential equation is equal to the order of the highest derivative appearing in it.

As for linear ordinary differential equations, we say that a partial differential equation is linear if it is of the first degree on the dependent variable $u(x, y, \ldots)$

© The Author(s), under exclusive license to Springer Nature Switzerland AG 2025
E. Capelas de Oliveira, J. E. Maiorino, *Analytical Methods in Applied Mathematics*,
Problem Books in Mathematics, https://doi.org/10.1007/978-3-031-74794-6_9

and also on its derivatives. Note that we cannot have, for example, terms like $u\frac{\partial u}{\partial x}$, for this characterizes nonlinearity.

If each term of a partial differential equation contains *only* the dependent variable or one of its derivatives, multiplied by any function of the independent variables, it is called *homogeneous*; otherwise, it is called *nonhomogeneous*. This means that in a homogeneous, linear partial differential equation, there do not appear terms formed only by functions of the independent variables.

A solution of a linear partial differential equation on some region $\mathcal{R}$ of the space of independent variables is a function that possesses all the partial derivatives appearing in the equation in some domain containing $\mathcal{R}$ and which satisfies the equation throughout the domain $\mathcal{R}$. In general, the full set of solutions of a partial differential equation is very large. We will see later that the unique solution of the partial differential equation corresponding to a given problem will be obtained by using the additional information appearing in a particular physical situation.

As in the study of linear ordinary differential equations, we present the following fundamental theorem called *superposition principle*.

Theorem 9.1 (Superposition Principle) *If u_1 and u_2 are any solutions of a homogeneous, linear partial differential equation on a region $\mathcal{R}$, we have that*

$$u = c_1 u_1 + c_2 u_2,$$

where c_1 and c_2 are arbitrary constants, is also a solution of the equation on that region.

Example 9.1 Let $x, y \in \mathbb{R}$ and $u = u(x, y)$. Consider the following homogeneous, linear second-order partial differential equation:

$$x^2 \frac{\partial^2 u}{\partial x^2} - y^2 \frac{\partial^2 u}{\partial y^2} = 0.$$

Knowing that $u_1(x, y) = \sin(xy)$ and $u_2(x, y) = \cos(xy)$ are solutions of this linear partial differential equation, show that

$$u(x, y) = A\,xy + B\sin(xy) + C\cos(xy),$$

with $A, B, C \in \mathbb{R}$, is also a solution.

Evaluating the first derivatives, we have

$$u_x = Ay + By\cos(xy) - Cy\sin(xy) \quad \text{and} \quad u_y = Ax + Bx\cos(xy) - Cx\sin(xy).$$

For the second derivatives, we get

$$u_{xx} = -By^2\sin(xy) - Cy^2\cos(xy) \quad \text{and} \quad u_{yy} = -Bx^2\sin(xy) - Cx^2\cos(xy).$$

Substituting these expressions into the partial differential equation and simplifying, we obtain the desired result. $\qquad\square$

9.2 The Method of Separation of Variables

The so-called method of separation of variables in product form consists in substituting a homogeneous, linear partial differential equation by a set of linear ordinary differential equations. We restrict our study to the case in which the linear partial differential equation has only two independent variables, which will serve as a model to discuss other problem that may contain more than two independent variables [1, 3–6]. Then, to solve a specific problem (equation + conditions), we proceed as follows:

(a) In applying the method of separation of variables, also called *product method*, we reduce the partial differential equation to two ordinary differential equations.
(b) We determine the solutions of these linear ordinary differential equations that satisfy the corresponding boundary conditions.
(c) These solutions will be combined in such a way that the result satisfies the linear partial differential equation and also the given conditions.

The sequence of three steps presented above constitutes to the so-called *Fourier method* for obtaining the solution of a homogeneous and linear partial differential equation.

In order to apply the method of separation of variables, we must first reduce the partial differential equation to the corresponding canonical form. Thus, consider the second-order, homogeneous, and linear partial differential equation

$$A\frac{\partial^2 u}{\partial \xi^2} + B\frac{\partial^2 u}{\partial \xi \partial \eta} + C\frac{\partial^2 u}{\partial \eta^2} + D\frac{\partial u}{\partial \xi} + E\frac{\partial u}{\partial \eta} + Fu = 0,$$

in which the coefficients $A, B, \ldots, F$ are functions of the independent variables ξ and η only, as is the dependent variable $u = u(\xi, \eta)$.

As we have already seen, it is always possible to find a coordinate transformation of the type $x = x(\xi, \eta)$ and $y = y(\xi, \eta)$, with a Jacobian different from zero, that reduces this partial differential equation to the corresponding canonical form, i.e., the form with no term involving a mixed derivative,

$$a(x, y)\frac{\partial^2 u}{\partial x^2} + c(x, y)\frac{\partial^2 u}{\partial y^2} + d(x, y)\frac{\partial u}{\partial x} + e(x, y)\frac{\partial u}{\partial y} + f(x, y)u = 0, \qquad (9.1)$$

with $u = u(x, y)$. So, $a = -c$ for a hyperbolic equation; $a = 0$ (or $c = 0$) for parabolic equations and $a = c$ in the case of an equation of elliptic type.

We then suppose that the solution $u(x, y)$ of Eq. (9.1) can be written in the form of a product

$$u(x, y) = R(x)T(y),$$

where the function $R(x)$ depends only on variable x and $T(y)$ depends only on variable y. Introducing $u(x, y)$ written in that form into the canonical form of the linear partial differential equation, we obtain an expression involving the functions R and T

$$aT\frac{\mathrm{d}^2 R}{\mathrm{d}x^2} + cR\frac{\mathrm{d}^2 T}{\mathrm{d}y^2} + dT\frac{\mathrm{d}R}{\mathrm{d}x} + eR\frac{\mathrm{d}T}{\mathrm{d}y} + fRT = 0, \tag{9.2}$$

where we omitted the functional dependence on x and y in the expressions for the coefficients and the functions R and T.

We now suppose that it is possible to find a function $p(x, y)$ such that, when we divide Eq. (9.2) by $p(x, y)$, we obtain an expression of the form

$$Ta_1(x)\frac{\mathrm{d}^2 R}{\mathrm{d}x^2} + Rb_1(y)\frac{\mathrm{d}^2 T}{\mathrm{d}y^2} + Ta_2(x)\frac{\mathrm{d}R}{\mathrm{d}x} + Rb_2(y)\frac{\mathrm{d}T}{\mathrm{d}y} + [a_3(x) + b_3(y)]RT = 0.$$

Dividing this expression by the product RT and rearranging, we then get

$$\frac{a_1}{R}\frac{\mathrm{d}^2 R}{\mathrm{d}x^2} + \frac{a_2}{R}\frac{\mathrm{d}R}{\mathrm{d}x} + a_3 = -\left(\frac{b_1}{T}\frac{\mathrm{d}^2 T}{\mathrm{d}y^2} + \frac{b_2}{T}\frac{\mathrm{d}T}{\mathrm{d}y} + b_3\right). \tag{9.3}$$

It is important to note that in this equality the left-hand side contains only functions of variable x, while the right-hand side involves only variable y. Differentiating both members of Eq. (9.3) with respect to x, we obtain

$$\frac{\mathrm{d}}{\mathrm{d}x}\left(\frac{a_1}{R}\frac{\mathrm{d}^2 R}{\mathrm{d}x^2} + \frac{a_2}{R}\frac{\mathrm{d}R}{\mathrm{d}x} + a_3\right) = 0.$$

It is important to note that the procedure and the final result are the same if we first differentiate with respect to variable y.

Integrating this differential equation, we find that

$$\frac{a_1}{R}\frac{\mathrm{d}^2 R}{\mathrm{d}x^2} + \frac{a_2}{R}\frac{\mathrm{d}R}{\mathrm{d}x} + a_3 = \lambda,$$

where the constant λ is known as *separation constant*.

With this result, we can then write

$$a_1\frac{\mathrm{d}^2 R}{\mathrm{d}x^2} + a_2\frac{\mathrm{d}R}{\mathrm{d}x} + (a_3 - \lambda)R = 0 \tag{9.4}$$

and

$$b_1\frac{\mathrm{d}^2 T}{\mathrm{d}y^2} + b_2\frac{\mathrm{d}T}{\mathrm{d}y} + (b_3 + \lambda)T = 0, \tag{9.5}$$

which constitutes a system of two linear ordinary differential equations. It should be clear that if we had begun with a partial differential equation with n independent variables, we would have obtained $(n - 1)$ separation constants and n ordinary differential equations.

Thus, $u(x, y)$ is a solution of the linear partial differential equation (9.1) if $R(x)$ and $T(y)$ are, respectively, solutions of the linear ordinary differential equations (9.4) and (9.5). This procedure constitutes the first step to solve a homogeneous, second-order, linear partial differential equation by means of the Fourier method.

Example 9.2 Let $x, y \in \mathbb{R}$ and $u = u(x, y)$. Use the method of separation of variables in the homogeneous, linear, second-order partial differential equation

$$(1 - x^2)\frac{\partial^2 u}{\partial x^2} + (1 - y^2)\frac{\partial^2 u}{\partial y^2} = 0$$

to obtain two linear ordinary differential equations linked through a separation constant.

Suppose that $u(x, y) = R(x)T(y) \equiv RT$. Introducing this function into the partial differential equation, we get

$$(1 - x^2)T R'' + (1 - y^2)RT'' = 0,$$

with the prime denoting ordinary derivatives. Dividing this expression by the product RT and rearranging, we can write

$$(1 - x^2)\frac{R''}{R} = -(1 - y^2)\frac{T''}{T}.$$

As the first member depends on variable x and the second member depends on variable y, we must have both members equal to a constant, that is,

$$(1 - x^2)R'' - \lambda R = 0 \qquad \text{and} \qquad (1 - y^2)T'' + \lambda T = 0$$

where λ is the separation constant. $\qquad\qquad\square$

The linear ordinary differential equations above carry in their respective general solutions two arbitrary integration constants. In order to find them, we must proceed to the second step, that is, we have to apply the boundary conditions.

9.3   Boundary Conditions

As we have seen above, we must determine the solutions of the two linear ordinary differential equations obtained by means of the method of separation of variables and then impose that they satisfy the boundary conditions of the original problem. Here, we discuss only three types of boundary conditions, namely:

(a) *Dirichlet conditions*, when we specify the value of the function for a certain $x = x_0$, that is, when it is given

$$u(x, y)|_{x=x_0} = \alpha,$$

where α is known.

(b) *Neumann conditions*, when we know the value of the first derivative of the function for a certain value $x = x_0$, i.e.,

$$\frac{\partial}{\partial x} u(x, y)\bigg|_{x=x_0} = \beta,$$

with β given.

(c) *Cauchy conditions*, in which we have the value, for $h \neq 0$, of

$$\frac{\partial}{\partial x} u(x, y) + h u(x, y)\bigg|_{x=x_0} = \gamma,$$

where γ is known.

Another type of conditions are the so-called *Robin* [1855 – Victor Gustave Robin – 1897] conditions, when one has one of the above three types of conditions for a part of the contour and another type for the other part of the contour.

Note that the correct choice of the coordinate system is very important to simplify the problem, i.e., to have separate conditions on the boundaries. Besides, the boundary conditions given at $x = x_0$ must contain derivatives of the dependent variable $u(x, y)$ with respect to x only, and its coefficients must depend only on x.

Example 9.3 The general solution of the linear partial differential equation

$$\frac{\partial^2 u}{\partial x^2} - \frac{\partial^2 u}{\partial t^2} = 0,$$

with $u = u(x, t)$ is given by

$$u(x, t) = f(x + t) + g(x - t)$$

where $f(\cdot)$ and $g(\cdot)$ are two arbitrary, twice differentiable functions. If we impose an initial condition $u(x, 0) = F(x)$, with $F(x)$ a known function (Dirichlet condition), we can write $F(x) = f(x) + g(x)$. On the other hand, if we impose another initial condition, $\frac{\partial}{\partial t} u(x, t)|_{t=0} = G(x)$, with $G(x)$ given (Neumann condition), we get $G(x) = f'(x) - g'(x)$, where the prime denotes ordinary derivatives.

Finally, the third step of the Fourier method consists in imposing that the solutions found must satisfy the corresponding linear ordinary differential equations and also the initial conditions associated with the specific problem. As an example of the application of the Fourier method, we discuss in the sequence the classical problem of a vibrating string fixed at its extremities (as a guitar string). □

9.4 Solved Exercises

SE 9.1 We will consider a classical problem involving the method of separation of variables, the one-dimensional wave equation. This problem consists in the study of the vibrational modes of a homogeneous string of length l, stretched with a constant tension along the axis x from 0 to l, and kept fixed at its two extremities. The homogeneous linear partial differential equation that describes the system is

$$\frac{\partial^2}{\partial t^2}u(x,t) - c^2\frac{\partial^2}{\partial x^2}u(x,t) = 0,$$

for $0 < x < l$ and $t > 0$, where c is the velocity of wave propagation on the string, considered constant, and $u(x,t)$, the dependent variable, represents the displacement of a point on the string from its equilibrium position $u = 0$. Note that, if the string has a constant linear mass density μ and is subject to a tension τ, the velocity c is given by $c^2 = \tau/\mu$.

As the extremities are fixed, the boundary conditions (Dirichlet conditions) are given by

$$u(0,t) = 0 \quad \text{and} \quad u(l,t) = 0.$$

Finally, the initial conditions are

$$u(x,0) = f(x), \quad 0 \le x \le l,$$

and

$$\left.\frac{\partial}{\partial t}u(x,t)\right|_{t=0} = g(x), \quad 0 \le x \le l;$$

where $f(x)$ and $g(x)$ correspond, respectively, to the initial displacement and the initial velocity (at $t = 0$) of the point of the string with coordinate x.

Solution: In order to solve this problem by means of the Fourier method, we proceed with the three steps described above.

First, we note that the homogeneous linear partial differential equations is already written in the canonical form. We can then search for a solution in the form of a function written as a product of two functions,

$$u(x,t) = R(x)T(t),$$

where $R(x)$ depends only on x and $T(t)$ depends only on t. Substituting $u(x,t)$, given above, into the homogeneous linear partial differential equation, we obtain two homogeneous linear ordinary differential equations, one of them in variable x,

$$\frac{d^2}{dx^2}R(x) - \lambda R(x) = 0,$$

and the other in variable t,

$$\frac{d^2}{dt^2}T(t) - \lambda c^2 T(t) = 0,$$

where λ is the separation constant.

Second, we must impose the boundary conditions. As these conditions require that $u(0,t) = u(l,t) = 0$, we must solve the homogeneous, linear ordinary differential equation

$$\frac{d^2}{dx^2}R(x) - \lambda R(x) = 0$$

with the conditions

$$R(0) = R(l) = 0,$$

which is just a *Sturm-Liouville problem* as in Chap. 7. Thus, we must look for values of λ that give rise to nontrivial solutions. The result is an infinite set of functions

$$R_n(x) = A_n \sin\left(\frac{n\pi}{l}x\right), \quad n = 1, 2, 3 \dots,$$

where A_n are constants independent of x.

On the other hand, for each $\lambda = \lambda_n = \left(\frac{n\pi}{l}\right)^2$ with $n = 1, 2, 3 \dots$, the general solution of the homogeneous, linear ordinary differential equation for $T(t)$ is given by

$$T_n(t) = B_n \cos\left(\frac{n\pi c}{l}t\right) + C_n \sin\left(\frac{n\pi c}{l}t\right),$$

where B_n and C_n are arbitrary constants independent of t.

Thus, the function

$$u_n(x,t) = R_n(x)T_n(t)$$
$$= \left\{a_n \cos\left(\frac{n\pi c}{l}t\right) + b_n \sin\left(\frac{n\pi c}{l}t\right)\right\} \sin\left(\frac{n\pi}{l}x\right),$$

where $a_n = A_n B_n$ and $b_n = A_n C_n$ are arbitrary constants, satisfies the one-dimensional wave equation and the condition of fixed extremes. Since the equation is linear and homogeneous, it follows from the superposition principle that the infinite series

$$u(x, t) = \sum_{n=1}^{\infty} u_n(x, t)$$

$$= \sum_{n=1}^{\infty} \left\{ a_n \cos \left(\frac{n\pi c}{l} t \right) + b_n \sin \left(\frac{n\pi c}{l} t \right) \right\} \sin \left(\frac{n\pi}{l} x \right)$$

is also a solution of the given problem.

Finally, in order to determine the constants appearing in the last expression, we will use the initial conditions. Imposing the first condition, $u(x, 0) = f(x)$, we get

$$u(x, 0) = \sum_{n=1}^{\infty} a_n \sin \left(\frac{n\pi}{l} x \right) = f(x). \tag{9.6}$$

From the second condition, $\left. \dfrac{\partial}{\partial t} u(x, t) \right|_{t=0} = g(x)$, we obtain

$$\left. \frac{\partial}{\partial t} u(x, t) \right|_{t=0} = \sum_{n=1}^{\infty} b_n' \sin \left(\frac{n\pi}{l} x \right) = g(x), \tag{9.7}$$

where we have defined $b_n' = b_n n\pi c/l$.

Note that the above expressions have the same form and are particular cases of the expansion of a function in a Fourier series, discussed in Chap. 5.

As an example we calculate explicitly the coefficients a_n and b_n' using the orthogonality property associated with the sine and cosine functions. In fact, we are assuming the working hypothesis that the functions $f(x)$ and $g(x)$ can be expressed in a Fourier series, which permits us to calculate the coefficients directly. Multiply both sides of Eq. (9.6) by $\sin \left(\dfrac{m\pi}{l} x \right)$ and integrate in variable x on the interval $[0, l]$:

$$\int_0^l \sum_{n=1}^{\infty} a_n \sin \left(\frac{n\pi}{l} x \right) \sin \left(\frac{m\pi}{l} x \right) \, dx = \int_0^l f(x) \sin \left(\frac{m\pi}{l} x \right) \, dx.$$

Supposing that the series is absolutely convergent, we interchange, in the last expression, the integration and the sum, obtaining

$$\sum_{n=1}^{\infty} a_n \int_0^l \sin \left(\frac{n\pi}{l} x \right) \sin \left(\frac{m\pi}{l} x \right) \, dx = \int_0^l f(x) \sin \left(\frac{m\pi}{l} x \right) \, dx. \tag{9.8}$$

The integral in the first member of Eq. (9.8) is equal to a Kronecker [**1823 – Leopold Kronecker – 1891**] delta, that is, it is different from zero only when $m = n$. Evaluating the integral in the first member, we obtain

$$a_n = \frac{2}{l} \int_0^l f(x) \sin\left(\frac{n\pi}{l}x\right) \, dx \, . \tag{9.9}$$

Applying the same procedure to Eq. (9.7), we find that the coefficients b_n are given by

$$b_n = \frac{2}{n\pi c} \int_0^l g(x) \sin\left(\frac{n\pi}{l}x\right) \, dx \, . \tag{9.10}$$

Finally, the solution of the problem of the vibrating string subject to the boundary and initial conditions presented is given by the expression

$$u(x, t) = \sum_{n=1}^{\infty} \left\{ a_n \cos\left(\frac{n\pi c}{l}t\right) + b_n \sin\left(\frac{n\pi c}{l}t\right) \right\} \sin\left(\frac{n\pi}{l}x\right) ,$$

where the coefficients are given by Eqs. (9.9) and (9.10) with $n = 1, 2, 3, \ldots$ It is possible to prove that this solution exists and is unique.

SE 9.2 Using the Fourier transform, find the distribution of temperatures on a semi-infinite rod, given that at the extremity $x = 0$ the rate of heat flow is equal to $g(t)$. The initial temperature is zero at all points of the rod.
Solution:

The mathematical problem is given by the homogeneous linear second-order partial differential equation

$$\frac{\partial}{\partial t}u(x, t) = \frac{\partial^2}{\partial x^2}u(x, t) , \qquad x > 0 \text{ and } t > 0 ,$$

with initial and boundary conditions $u(x, 0) = 0$ and $\frac{\partial}{\partial x}u(x, t)|_{x=0} = g(t)$, respectively.

First, to use the Fourier transform, we assume that $u(x, t)$ and $\frac{\partial u(x,t)}{\partial x}$ go to zero as x goes to infinite. Then, let $U(\alpha, t)$ be the cosine Fourier transform of $u(x, t)$, that is,

$$U(\alpha, t) = \sqrt{\frac{2}{\pi}} \int_0^\infty u(x, t) \cos \alpha x \, dx \, .$$

Transforming the homogeneous linear partial differential equation above and using the boundary conditions, we obtain a nonhomogeneous linear ordinary differential equation

$$\frac{\partial}{\partial t} U(\alpha, t) + \alpha^2 U(\alpha, t) = -\sqrt{\frac{2}{\pi}}\, g(t),$$

whose solution is given by

$$U(\alpha, t) = e^{-\alpha^2 t} \left\{ -\int_0^t \sqrt{\frac{2}{\pi}}\, g(\tau)\, e^{\alpha^2 \tau}\, d\tau + C \right\},$$

where C is a constant.

Using the initial condition, $u(x, 0) = 0$, we find that $C = 0$ and thus

$$U(\alpha, t) = -\sqrt{\frac{2}{\pi}} \int_0^t g(\tau),\, e^{-\alpha^2(t-\tau)}\, d\tau$$

whose inverse Fourier transform is given by

$$u(x, t) = -\frac{2}{\pi} \int_0^\infty \left\{ \int_0^t g(\tau)\, e^{-\alpha^2(t-\tau)}\, d\tau \right\} \cos\alpha x\, d\alpha.$$

Finally, integrating in variable α, using the result

$$\int_0^\infty e^{-Ax^2} dx = \frac{\sqrt{\pi}}{2\sqrt{A}}$$

with $A > 0$ and rearranging we can write

$$u(x, t) = -\frac{1}{\sqrt{\pi}} \int_0^t \frac{g(\tau)}{(t-\tau)^{1/2}}\, e^{-x^2/4(t-\tau)}\, d\tau.$$

SE 9.3 Let $u = u(x, y)$. Solve the Laplace equation on a rectangular region:

$$\frac{\partial^2 u}{\partial x^2} + \frac{\partial^2 u}{\partial y^2} = 0, \qquad 0 < x < a, \quad 0 < y < b,$$

with $a > 0$ and $b > 0$, satisfying the boundary conditions

$$u(x, 0) = 1, \quad u(x, b) = 0, \quad u_x(0, y) = 0 \quad \text{and} \quad u_x(a, y) = 0.$$

Solution: Introducing the separation of variables $u(x, y) = R(x)T(y)$ into the Laplace equation and rearranging, we can write

$$\frac{R''}{R} = -\frac{T''}{T} = \lambda,$$

where λ is a separation constant. As the boundary conditions in variable x are homogeneous, we obtain the following Sturm-Liouville problem:

$$\begin{cases} R'' - \lambda R = 0, & 0 < x < a, \\ R'(0) = R'(a) = 0. \end{cases}$$

Its solution is given by the set of eigenvalues $\lambda_n = -(n\pi/a)^2$ and the corresponding eigenfunctions

$$R_n(x) = A_n \cos\left(\frac{n\pi}{a}x\right), \tag{9.11}$$

with $n = 0, 1, 2, \ldots$ Using the eigenvalues, the ordinary differential equation in variable y can be written as

$$T'' - \left(\frac{n\pi}{a}\right)^2 T = 0.$$

Its solution, for $n = 0$, is $T(y) = \alpha y + \beta$, with α and β arbitrary constants. For $n = 1, 2, \ldots$, the solution is given by

$$T_n(y) = B_n \cosh\left(\frac{n\pi}{a}y\right) + C_n \sinh\left(\frac{n\pi}{a}y\right),$$

where B_n and C_n are arbitrary constants. This expression can be rewritten in a more adequate form by introducing $-\phi_n = \dfrac{a}{n\pi} \arctan\left(\dfrac{B_n}{C_n}\right)$, that is,

$$T_n(y) = D_n \sinh\left[\frac{n\pi}{a}(y - \phi_n)\right],$$

where $D_n = B_n / \cosh(n\pi\phi_n/a)$ is another arbitrary constant.

From the homogeneous condition $u(x, b) = R(x)T(y) = 0$, we have that $T(b) = 0$. We then get, for $n = 0$,

$$T_0(y) = \alpha(y - b). \tag{9.12}$$

For $n = 1, 2, \ldots$ the condition yields

$$T_n(b) = 0 = D_n \sinh\left[\frac{n\pi}{a}(y - \phi_n)\right] \qquad \Longrightarrow \qquad \phi_n = b.$$

The solution can then be written as

$$T_n(y) = D_n \sinh\left[\frac{n\pi}{a}(y - b)\right], \tag{9.13}$$

for $n = 1, 2, \ldots$ Thus, using Eqs. (9.11), (9.12), and (9.13), we obtain

$$u(x, y) = a_0(y - b) + \sum_{n=1}^{\infty} \sinh\left[\frac{n\pi}{a}(y - b)\right] \cos\left(\frac{n\pi}{a}x\right),$$

where we have introduced the notation a_n for the constants.

Finally, using the nonhomogeneous boundary condition $u(x, 0) = 1$, we have

$$1 = -ba_0 + \sum_{n=1}^{\infty} a_n \sinh\left[\frac{n\pi}{a}(-b)\right] \cos\left(\frac{n\pi}{a}x\right).$$

This is easily recognized as a Fourier series whose coefficients are given by

$$-2ba_0 = \frac{2}{a}\int_0^a 1 \cdot dx \qquad \Longrightarrow \qquad a_0 = -\frac{1}{b}$$

and

$$-a_n \sinh\left(\frac{n\pi}{a}b\right) = \frac{2}{a}\int_0^a 1 \cdot \cos\left(\frac{n\pi}{a}x\right) dx \qquad \Longrightarrow \qquad a_n = 0.$$

This allows us to write

$$u(x, y) = \frac{1}{b}(b - y),$$

which is the desired result.

SE 9.4 A uniform bar of length ℓ is fixed at an extreme $(x = 0)$. Suppose that a force

$$f(t) = \begin{cases} f_0 & t > 0 \\ 0 & t < 0 \end{cases}$$

where f_0 is constant, is suddenly applied at the extreme $x = \ell$. If the bar is initially at rest, use the Laplace transform to find its longitudinal displacement $u(x, t)$ for $t > 0$.

The movement of the bar is governed by the homogeneous, linear partial differential equation

$$\frac{\partial^2}{\partial t^2}u(x, t) = a^2 \frac{\partial^2}{\partial x^2}u(x, t),$$

where $0 < x < \ell$ and a^2 is a positive constant; moreover, $u(x, t)$ satisfies the conditions

$$u(x, 0) = u(0, t) = 0, \quad \frac{\partial}{\partial t}u(x, t)|_{t=0} = 0, \quad \text{and} \quad \frac{\partial}{\partial x}u(x, t)|_{x=\ell} = \frac{1}{E}f_0,$$

where E is a positive constant.

Solution: Let $U(x, s)$ be the Laplace transform of $u(x, t)$ in the time variable, that is,

$$U(x, s) = \int_0^\infty u(x, t)\, e^{-st}\, dt.$$

We transform the homogeneous linear partial differential equation into the homogeneous linear ordinary differential equation

$$\frac{d^2}{dx^2}U(x, s) - \frac{s^2}{a^2}U(x, s) = 0,$$

satisfying the transformed conditions (boundary conditions)

$$U(0, s) = 0 \quad \text{and} \quad \frac{\partial}{\partial x}U(\ell, s) = \frac{1}{sE}f_0.$$

The general solution of this linear ordinary differential equation is given by

$$U(x, s) = A\, e^{xs/a} + B\, e^{-xs/a},$$

with A and B constants. Applying the boundary conditions, we obtain a system of two algebraic equations for A and B,

$$A + B = 0$$

and

$$A\left(\frac{s}{a}e^{\ell s/a}\right) + B\left(-\frac{s}{a}e^{-\ell s/a}\right) = \frac{1}{sE}f_0.$$

Solving the system and simplifying, we can write for the solution

$$U(x, s) = \frac{af_0}{Es^2}\frac{e^{xs/a} - e^{-xs/a}}{e^{\ell s/a} + e^{-\ell s/a}}.$$

Multiplying the numerator and the denominator by the factor

$$e^{-\ell s/a} - e^{-3\ell s/a}$$

and rearranging we get

$$\left(1 - e^{-4\ell s/a}\right) U(x, s)$$

$$= \frac{af_0}{Es^2} \left\{ e^{-(\ell-x)s/a} - e^{-(\ell+x)s/a} - e^{-(3\ell-x)s/a} + e^{-(3\ell+x)s/a} \right\} .$$

Since the denominator has a term of the type

$$\left(1 - e^{-4\ell s/a}\right) ,$$

the inverse transform is a periodic function with period $T = 4\ell/a$; we can then write

$$u(x, t) = \frac{af_0}{E} \begin{cases} 0 & 0 < t < (\ell - x)/a, \\ t - (\ell - x)/a & (\ell - x)/a < t < (\ell + x)/a, \\ 2x/a & (\ell + x)/a < t < (3\ell - x)/a, \\ -t + (3\ell + x)/a & (3\ell - x) < t < (3\ell + x)/a, \\ 0 & (3\ell + x) < t < 4\ell/a, \end{cases}$$

which is the desired result.

SE 9.5 Using separation of variables, obtain the temperature on an infinite circular cylinder of radius r_0, with the condition that its initial temperature is given by

$$u(r, 0) = u_0 \left(1 - \frac{r^2}{r_0^2}\right) ,$$

where u_0 is a positive constant and the temperature on the surface $r = r_0$ is kept equal to zero.

Solution: We must solve the heat equation written in cylindrical coordinates (r, ϕ, z). Due to its symmetry, the solution is independent of coordinates ϕ and z, so that our partial differential equation reduces to

$$\frac{\partial}{\partial t} u(r, t) = a^2 \left(\frac{\partial^2}{\partial r^2} u(r, t) + \frac{1}{r} \frac{\partial}{\partial r} u(r, t) \right) ,$$

where a^2 is a positive constant.

   We will apply the method of separation of variables. To this end, we introduce a function with the form of a product,

$$u(r, t) = R(r)T(t) ,$$

obtaining the following linear ordinary differential equation in variable r:

$$\frac{d^2}{dr^2}R(r) + \frac{1}{r}\frac{d}{dr}R(r) + \lambda^2 R(r) = 0,$$

where λ^2 is a separation constant. The general solution of this differential equation is given by

$$R_n(r) = A\,\mathcal{J}_0(\lambda r) + B\,Y_0(\lambda r),$$

with A and B constants. $\mathcal{J}_0(x)$ and $Y_0(x)$ are the zero order Bessel functions of the first and second kind, respectively; $\mathcal{J}_0(x)$ is regular at $x = 0$ while $Y_0(x)$ is regular at $x = \infty$, so that we must leave aside $Y_0(x)$.

In order to obtain nontrivial solutions, we must have

$$\mathcal{J}_0(\lambda r_0) = 0.$$

This is a transcendental equation. It will be true whenever $\lambda r_0 = \mu_n$, where μ_n is any positive root of the zero order Bessel function $\mathcal{J}_0(x)$; we can then write

$$R(r) = A\,\mathcal{J}_0\left(\mu_n \frac{r}{r_0}\right).$$

As for the ordinary differential equation in variable t, we have

$$\frac{d}{dt}T(t) + a^2\lambda^2 T(t) = 0,$$

whose solution is given by

$$T(t) = B\,e^{-a^2\lambda^2 t},$$

where B is a constant. Combining the solutions of both linear ordinary differential equations, we can write, using the superposition principle, the following result:

$$u(r, t) = \sum_{n=0}^{\infty} A_n\, e^{-a^2\lambda^2 t}\, \mathcal{J}_0\left(\mu_n \frac{r}{r_0}\right),$$

where A_n still has to be determined.

Applying the initial condition, we get

$$u(r, 0) = \sum_{n=0}^{\infty} A_n \mathcal{J}_0\left(\mu_n \frac{r}{r_0}\right) = u_0\left(1 - \frac{r^2}{r_0^2}\right).$$

This is a Fourier-Bessel series, whose coefficients A_n are given by the integral

$$A_n = \frac{2/r_0^2}{[\mathcal{J}_1(\mu_n)]^2} \int_0^{r_0} r\,\mathcal{J}_0\left(\mu_n \frac{r}{r_0}\right) u_0 \left(1 - \frac{r^2}{r_0^2}\right) dr \,.$$

This integral can be found in tables of integrals [2]; it permits us to write

$$A_n = \frac{4u_0}{\mu_n^2} \frac{\mathcal{J}_2(\mu_n)}{[\mathcal{J}_1(\mu_n)]^2} \,.$$

We have thus obtained the solution of our problem, that is,

$$u(r, t) = 4u_0 \sum_{n=0}^{\infty} \frac{1}{\mu_n^2} \frac{\mathcal{J}_2(\mu_n)}{[\mathcal{J}_1(\mu_n)]^2} \, e^{-a^2 \mu_n^2 t/r_0^2} \, \mathcal{J}_0\left(\mu_n \frac{r}{r_0}\right) \,.$$

9.5 Proposed Exercises

PE 9.1 Let $u(x, y) = x^2 - y^2 + 1$. Show that this function satisfies the two-dimensional Laplace equation written in Cartesian coordinates.

PE 9.2 Is the function $u(x, y) = x^2 - y^2 + 2xy$ a solution of the two-dimensional Laplace equation in Cartesian coordinates?

PE 9.3 Let $f(x \pm t)$ be two differentiable functions. Show that $u(x, t) = f(x + t) + f(x - t)$ is a solution of the one-dimensional wave equation.

PE 9.4 Show that $u(x, t) = \frac{1}{\sqrt{t}} \exp(-x^2/4t)$, with $t > 0$, is a solution of the one-dimensional homogeneous heat equation.

PE 9.5 Let $x, y \in \mathbb{R}^*$. Consider the partial differential equation

$$y \frac{\partial u}{\partial x} - x \frac{\partial u}{\partial y} = 0\,,$$

with $u = u(x, y)$. Is the function

$$u(x, y) = \exp\left[-\frac{\lambda}{2}(x^2 + y^2)\right],$$

with λ a positive constant, a solution of this partial differential equation?

PE 9.6 Let $u = u(r, \theta)$ be the solution of the Laplace equation written in polar coordinates:

$$\frac{\partial^2 u}{\partial r^2} + \frac{1}{r} \frac{\partial u}{\partial r} + \frac{1}{r^2} \frac{\partial^2 u}{\partial \theta^2} = 0\,.$$

Consider $u(r, \theta) = r^\lambda T(\theta)$, with λ is a nonnegative parameter. Obtain an ordinary differential equation for $T(\theta)$ and solve it.

PE 9.7 Let $u(r, \theta, \phi)$ be a solution of the Laplace equation in spherical coordinates (r, θ, ϕ), with $r > 0$, $0 < \theta < \pi$ and $0 < \phi < 2\pi$. Suppose that u depends only on the radial variable, that is, $u = u(r)$. Find u.

PE 9.8 Let $u = u(x, y)$. Classify the partial differential equation

$$\frac{\partial^2 u}{\partial x^2} + 2\frac{\partial^2 u}{\partial x \partial y} - 3\frac{\partial^2 u}{\partial y^2} = 2022 \,.$$

PE 9.9 Let $a \in \mathbb{R}^*$ and $u = u(x, y)$. Classify the partial differential equation

$$a\frac{\partial^2 u}{\partial x^2} + 2a^2\frac{\partial^2 u}{\partial x \partial y} \pm a^3\frac{\partial^2 u}{\partial y^2} + u = 0 \,.$$

PE 9.10 Let $u(r) = A + B \ln r$, with A and B arbitrary constants, be the solution of the radial Laplace equation in polar coordinates, (r, θ). Use separation of variables to get the solution of the angular ordinary differential equation.

PE 9.11 Classify the following partial differential equations as linear or nonlinear.

$$\text{(a)} \quad \frac{\partial^2 u}{\partial x^2} + \frac{\partial^2 u}{\partial y^2} + \frac{\partial^2 u}{\partial z^2} = 0, \ u = u(x, y, z);$$

$$\text{(b)} \quad i\hbar \frac{\partial}{\partial t} \Psi(x, t) = -\frac{\hbar^2}{2m}\frac{\partial^2}{\partial x^2} \Psi(x, t),$$

where $\hbar$ and m are constants;

$$\text{(c)} \quad \left(\frac{\partial u}{\partial t}\right)^2 + \frac{\partial^2 u}{\partial x^2} = f(x, t), \ u = u(x, t);$$

$$\text{(d)} \quad u\frac{\partial u}{\partial t} = \frac{\partial^3 u}{\partial x^3}, \ u = u(x, t).$$

PE 9.12 Classify with respect to homogeneity the following partial differential equations:

$$\text{(a)} \quad \frac{\partial u}{\partial t} + \frac{\partial^2 u}{\partial x^2} + u = 0, \ u = u(x, t);$$

$$\text{(b)} \quad \frac{\partial^2 u}{\partial x^2} + \frac{\partial^2 u}{\partial y^2} = -\rho(x, y), \ u = u(x, y);$$

$$\text{(c)} \quad \frac{1}{c^2}\frac{\partial^2 u}{\partial t^2} - \frac{\partial^2 u}{\partial x^2} - \frac{\partial^2 u}{\partial y^2} = f(x, t), \ u = u(t, x, y), \ c = \text{constant.}$$

PE 9.13 Classify according to their order the equations given in **PE 9.11** and **PE 9.12**.

PE 9.14 The two-dimensional Laplace equation in Cartesian coordinates is given by

$$\nabla^2 u(x, y) = \frac{\partial^2}{\partial x^2} u(x, y) + \frac{\partial^2}{\partial y^2} u(x, y) = 0.$$

Verify that the functions below are solutions of the Laplace equation:

$$\text{(a)} \quad u(x, y) = x^2 - y^2;$$

$$\text{(b)} \quad u(x, y) = e^x \cos y;$$

$$\text{(c)} \quad u(x, y) = \arctan \frac{y}{x}.$$

PE 9.15 Classify according to linearity, order, and homogeneity and solve the linear partial differential equations

$$\text{(a)} \quad \frac{\partial u}{\partial x} + x \frac{\partial u}{\partial y} = 0,$$

$$\text{(b)} \quad \frac{\partial u}{\partial x} + \frac{\partial u}{\partial y} = 0,$$

where $u = u(x, y)$.

PE 9.16 Show that:

(a) The Laplace equation $\dfrac{\partial^2 u}{\partial x^2} + \dfrac{\partial^2 u}{\partial y^2} = 0$, with $u = u(x, y)$, is elliptic.

(b) The heat equation $\dfrac{\partial u}{\partial t} = \dfrac{\partial^2 u}{\partial x^2}$, with $u = u(x, t)$, is parabolic.

(c) The wave equation $\dfrac{\partial^2 u}{\partial t^2} = \dfrac{\partial^2 u}{\partial x^2}$, with $u = u(x, t)$, is hyperbolic.

(d) The Tricomi equation $\dfrac{\partial^2 u}{\partial x^2} + y \dfrac{\partial^2 u}{\partial y^2} = 0$, with $u = u(x, y)$, is elliptic on the upper half-plane and hyperbolic on the lower half-plane.

PE 9.17 For $u = u(x, y)$, find the general solutions of the following partial differential equations:

$$\text{(a)}\quad \frac{\partial u}{\partial y} = 0 \qquad \text{(b)}\quad \frac{\partial u}{\partial x} = 0$$

$$\text{(c)}\quad \frac{\partial^2 u}{\partial x \partial y} = \frac{\partial u}{\partial x} \qquad \text{(d)}\quad \frac{\partial^2 u}{\partial x \partial y} = 0$$

$$\text{(e)}\quad \frac{\partial^2 u}{\partial x^2} = 0 \qquad \text{(f)}\quad \frac{\partial^2 u}{\partial y^2} = 0$$

PE 9.18   Using the transformations $v = x + y$ and $z = 3x + y$, solve the equation

$$\frac{\partial^2 u}{\partial x^2} - 4\frac{\partial^2 u}{\partial x \partial y} + 3\frac{\partial^2 u}{\partial y^2} = 0,$$

where $u = u(x, y)$

PE 9.19   Separate the Laplace equation in polar coordinates

$$\frac{\partial^2 u}{\partial r^2} + \frac{1}{r}\frac{\partial u}{\partial r} + \frac{1}{r^2}\frac{\partial^2 u}{\partial \theta^2} = 0,$$

where $u = u(r, \theta)$.

PE 9.20   Spherical coordinates r, θ, and ϕ are related to Cartesian coordinates x, y, and z by

$$x = r \sin\theta \cos\phi \,;$$

$$y = r \sin\theta \sin\phi \,;$$

$$z = r \cos\theta \,,$$

with $r \geq 0, 0 < \theta \leq \pi$ and $0 < \phi \leq 2\pi$. Write and separate the Laplace equation written in spherical coordinates.

PE 9.21   Using the same coordinates as in the preceding problem, write and separate the d'Alembert [1717 – Jean-le-Rond D'Alembert – 1783] equation

$$\frac{\partial^2 u}{\partial t^2} - \frac{\partial^2 u}{\partial x^2} - \frac{\partial^2 u}{\partial y^2} - \frac{\partial^2 u}{\partial z^2} = \frac{\partial^2 u}{\partial t^2} - \nabla^2 u \equiv \Box u = 0,$$

where $u = u(t, x, y, z)$. In this equation, there appears the Laplace operator, or Laplacian,

$$\Delta \equiv \nabla^2 = \frac{\partial^2}{\partial x^2} + \frac{\partial^2}{\partial y^2} + \frac{\partial^2}{\partial z^2} \,,$$

and the so-called d'Alembert operator or Dalembertian

$$\Box = \frac{\partial^2}{\partial t^2} - \frac{\partial^2}{\partial x^2} - \frac{\partial^2}{\partial y^2} - \frac{\partial^2}{\partial z^2} \, .$$

PE 9.22 Proceed as in **PE 9.20** for the cylindrical coordinates

$$x = r \sin\theta, \quad y = r \cos\theta, \quad z = z,$$

with $r \geq 0$, $0 < \theta \leq 2\pi$, and $-\infty < z < +\infty$.

PE 9.23 Proceed as in **PE 9.21** for the cylindrical coordinates given in the preceding exercise.

PE 9.24 Let $u(x_1, x_2, x_3, x_4) = u$. Write and separate the four-dimensional Laplace equation,

$$\nabla^2 u(x_1, x_2, x_3, x_4) = \frac{\partial^2 u}{\partial x_1^2} + \frac{\partial^2 u}{\partial x_2^2} + \frac{\partial^2 u}{\partial x_3^2} + \frac{\partial^2 u}{\partial x_4^2} = 0,$$

using the polar coordinates r, θ, ϕ, and ψ defined by the relations

$$\begin{aligned}
x_1 &= r \sin\theta \sin\phi \cos\psi\,, \\
x_2 &= r \sin\theta \sin\phi \sin\psi\,, \\
x_3 &= r \sin\theta \cos\phi\,, \\
x_4 &= r \cos\theta\,,
\end{aligned}$$

with $r \geq 0$, $0 < \theta \leq \pi$, $0 < \phi \leq 2\pi$, and $0 < \psi \leq 2\pi$.

PE 9.25 Discuss in detail the Sturm-Liouville problems presented in **SE 9.1**.

PE 9.26 Solve the vibrating string problem **SE 9.1** with the initial conditions $u(x, 0) = 2$ and $\left.\dfrac{\partial u}{\partial t}\right|_{t=0} = 0$.

PE 9.27 (Heat Conduction on a Rod)   Using the method of separation of variables, solve the partial differential equation

$$\frac{\partial u}{\partial t} = K \frac{\partial^2 u}{\partial x^2},$$

with $0 < x < l$ and $t > 0$, where $u = u(x, t)$ and K is a constant and $u(x, t)$ satisfies the boundary conditions $u(0, t) = 0$ and $u(l, t) = u_0 = $ constant and the initial condition $u(x, 0) = f(x)$, $0 \leq x \leq l$.

PE 9.28 Proceed as in **PE 9.27** to solve a nonhomogeneous equation whose nonhomogeneous term is time independent, that is,

$$\frac{\partial^2}{\partial t^2}u(x,t) = c^2 \frac{\partial^2}{\partial x^2}u(x,t) + F(x),$$

where c^2 is a positive constant and $F(x)$ is independent of t. Suppose that the initial conditions are given by

$$u(x,0) = \phi(x), \ 0 \le x \le l;$$

$$\frac{\partial}{\partial t}u(x,t)\Big|_{t=0} = \psi(x), \ 0 \le x \le l;$$

with the boundary conditions $u(0,t) = u(l,t) = 0$.

PE 9.29 Proceed as in the preceding exercise, with boundary conditions given by $u(0,t) = \alpha$ and $u(l,t) = \beta, t > 0$, where α and β are constants.

PE 9.30 Let $u(x,t) = u$. Solve the following partial differential equation:

$$\frac{\partial^2 u}{\partial t^2} - c^2 \frac{\partial^2 u}{\partial x^2} = 0, \ 0 < x < l \ \text{ and } \ t > 0,$$

with initial conditions

$$u(x,0) = \frac{\partial}{\partial t}u(x,t)\Big|_{t=0} = 0,$$

and boundary conditions

$$u(0,t) = t, \ u(l,t) = 1.$$

To this end, introduce a change of dependent variable that reduces the nonhomogeneous boundary conditions to homogeneous conditions.

PE 9.31 Let $u(x,t) = u$. Consider the problem

$$\frac{\partial u}{\partial t} - k\frac{\partial^2 u}{\partial x^2} = 0, \ 0 < x < l \ \text{and} \ t > 0,$$

with $u(x,0) = 1$, $u(0,t) = 0$, and $u(l,t) = e^{-t}$. Determine the conditions associated with the parameters k and l in such a way that there exists a solution of the heat equation with the form $u(x,t) = v(x)\exp(-t)$ that satisfies the above conditions.

PE 9.32 Let $u(r,\theta) = u$. Show that the solution of the partial differential equation

$$\frac{\partial^2 u}{\partial r^2} + \frac{1}{r}\frac{\partial u}{\partial r} + \frac{1}{r^2}\frac{\partial^2 u}{\partial \theta^2} = 0,$$

for $0 < \theta < \pi$ and $0 < r < a$, with the conditions $u(a, \theta) = u_0 = $ constant and $u(r, 0) = u(r, \pi) = 0$, is

$$u(r, \theta) = \frac{2u_0}{\pi} \sum_{k=1}^{\infty} \frac{1 - (-1)^k}{k} \left(\frac{r}{a}\right)^k \sin k\theta.$$

PE 9.33 Let $u(x, t) = u$. Obtain a solution of the heat equation

$$\frac{\partial u}{\partial t} - k\frac{\partial^2 u}{\partial x^2} = 0, \quad 0 < x < l \quad \text{and} \quad t > 0,$$

with the boundary conditions $u(0, t) = 0$, $u(l, t) = t$, and $t > 0$, in the form of a polynomial in variables x and t, with k a constant. The initial condition is $u(x, 0) = 0$.

PE 9.34 Using the result obtained in the preceding exercise, show that the solution can be used to solve the same equation with the same boundary conditions but with initial condition given by $u(x, 0) = f(x)$ for $0 < x < l$.

In the following three exercises, consider the partial differential equation

$$\frac{\partial u}{\partial t} - k\frac{\partial^2 u}{\partial x^2} = 0, \quad 0 < x < l \quad \text{and} \quad t > 0,$$

with $u(x, t) = u$ and the initial and boundary conditions

$$u(x, 0) = 0$$

and

$$u(0, t) = 0, \quad u(l, t) = 1,$$

respectively, where k is a positive constant.

PE 9.35 Solve the homogeneous equation obtained when we eliminate the nonhomogeneous term appearing in the boundary conditions.

PE 9.36 Use the finite Fourier sine transform to solve the complete problem.

PE 9.37 Compare the result obtained in **PE 9.36** with **PE 9.35**.

PE 9.38 Let $u(x, y) = u$. Use the finite Fourier sine transform to solve the Dirichlet problem for the Laplace equation on a rectangle $0 < x < a, 0 < y < b$,

$$\frac{\partial^2 u}{\partial x^2} + \frac{\partial^2 u}{\partial y^2} = 0,$$

with the conditions $u(x, 0) = f(x)$, $u(x, b) = g(x)$, $u(0, y) = p(y)$, $u(a, y) = q(y)$, where $f(x)$, $g(x)$, $p(y)$, and $q(y)$ are known functions.

PE 9.39 Let $u(x, y) = u$. Apply the finite Fourier cosine transform to solve the so-called Neumann problem for the Poisson equation on a rectangle $0 < x < a$, $0 < y < b$:

$$\frac{\partial^2 u}{\partial x^2} + \frac{\partial^2 u}{\partial y^2} = -F(x, y),$$

$$\frac{\partial}{\partial x} u(x, y)\bigg|_{x=0} = h(y), \quad \frac{\partial}{\partial x} u(x, y)\bigg|_{x=a} = i(y),$$

$$\frac{\partial}{\partial y} u(x, y)\bigg|_{y=0} = f(x), \quad \frac{\partial}{\partial y} u(x, y)\bigg|_{y=b} = g(x).$$

PE 9.40 Let $u(x, y) = u$. Use the Laplace transform to solve the partial differential equation

$$\frac{\partial^2 u}{\partial t^2} - c^2 \frac{\partial^2 u}{\partial x^2} = 0, \quad 0 < x < \infty, \quad t > 0,$$

satisfying the initial conditions

$$u(x, 0) = 0 = \frac{\partial}{\partial t} u(x, t)\bigg|_{t=0}$$

and the boundary condition $u(0, t) = f(t)$.

In the following two exercises, consider the Cauchy problem for $u = u(x, t)$:

$$x \frac{\partial^2 u}{\partial x^2} - \frac{\partial^2 u}{\partial t^2} + \frac{\partial u}{\partial x} = 0, \quad x > 0, \tag{9.14}$$

$$u(x, 0) = \varphi_0(x),$$

$$\frac{\partial}{\partial t} u(x, t)\bigg|_{t=0} = \varphi_1(x).$$

PE 9.41 Let $\omega(\xi, \eta) = \omega$. Convert Eq. (9.14) to the corresponding canonical form

$$\frac{\partial^2 \omega}{\partial \xi \partial \eta} - \frac{1}{4} \frac{\omega}{(\xi - \eta)^2} = 0 \, ,$$

using the transformations $\xi = t/2 + \sqrt{x}$, $\eta = t/2 - \sqrt{x}$ and $\omega = u\sqrt{\xi - \eta}$.

PE 9.42 Verify that

$$\omega \equiv G(\lambda) = {}_2F_1\left(\frac{1}{2}, \frac{1}{2}; 1; \lambda\right) ,$$

where ${}_2F_1(a, b; c; x)$ is a hypergeometric function and

$$\lambda = \frac{(\xi - \xi_0)(\eta - \eta_0)}{(\xi_0 - \eta_0)(\xi - \eta)}$$

is a solution of **PE 9.41**.

PE 9.43 Use the Fourier method to solve the problem of the elastic circular vibrating membrane, given by the partial differential equation

$$\frac{\partial^2 u}{\partial t^2} = c^2 \left[\frac{\partial^2 u}{\partial r^2} + \frac{1}{r}\frac{\partial u}{\partial r} + \frac{1}{r^2}\frac{\partial^2 u}{\partial \theta^2} \right] ,$$

for the elongation $u = u(r, \theta, t)$, $0 < r < R$. To simplify, consider solutions with radial symmetry, that is, independent of θ, with the boundary condition $u(R, t) = 0$ for all $t \geq 0$, and the initial conditions $u(r, 0) = f(r)$ and $\left.\dfrac{\partial u}{\partial t}\right|_{t=0} = g(r)$. Note that the boundary condition at $r = R$ must be imposed in such a way that we do not have infinities in our solution, i.e., $u(r, t) < \infty$ for all r on this interval.

PE 9.44 Find $u = u(x, t)$ satisfying the partial differential equation

$$\frac{\partial^2 u}{\partial t^2} - \frac{\partial^2 u}{\partial x^2} = x - t \, ,$$

with $u(x, 0) = x$, $(\partial u/\partial t)_{t=0} = x^2/2$, $u(0, t) = t$ and $u(1, t) = t^2/2$.

PE 9.45 Find the stationary state temperature on a straight circular cylinder of radius 2 and height 4, that is, solve the partial differential equation, with $u = u(r, z)$,

$$\frac{\partial^2 u}{\partial r^2} + \frac{1}{r}\frac{\partial u}{\partial r} + \frac{\partial^2 u}{\partial z^2} = 0, \quad 0 < r < 2, \quad 0 < z < 4 \, ,$$

satisfying the conditions $u(2, z) = 0$, $u(r, 0) = u_0$, and $u(r, 4) = 0$.

PE 9.46 Show that the stationary state temperature on a sphere, that is, a solution of the partial differential equation

$$\frac{\partial^2 u}{\partial r^2} + \frac{2}{r}\frac{\partial u}{\partial r} + \frac{1}{r^2}\frac{\partial^2 u}{\partial \theta^2} + \frac{\cot\theta}{r^2}\frac{\partial u}{\partial \theta} = 0,$$

with $u = u(r, \theta)$, for $0 < r < a$ and $0 < \theta < \pi$, with $u(a, \theta) = f(\theta)$, is given by

$$u(r, \theta) = \sum_{k=0}^{\infty}\left[\frac{2k+1}{2}\int_0^{\pi} f(\theta')P_k(\cos\theta')\sin\theta'\,d\theta'\right]\left(\frac{r}{a}\right)^k P_k(\cos\theta),$$

where $P_k(x)$ are the Legendre polynomials of order k.

PE 9.47 Let $u = u(x, t)$. Find *one* stationary state solution $\psi(x)$ for the equation

$$k\frac{\partial^2 u}{\partial x^2} = \frac{\partial u}{\partial t}, \quad 0 < x < \pi \text{ and } t > 0,$$

satisfying the boundary and initial conditions

$$u(0, t) = u_0, \quad -\frac{\partial}{\partial x}u(x, t)\bigg|_{x=\pi} = u(\pi, t) - u_1, \ t > 0,$$

$$u(x, 0) = 0, \quad 0 < x < \pi,$$

where u_0 and u_1 are constants.

PE 9.48 (Poisson Formula) Solve the Laplace equation on the upper half-plane $y > 0$ using the Fourier transform, i.e., find a function $u(x, y)$ such that

$$\frac{\partial^2 u}{\partial x^2} + \frac{\partial^2 u}{\partial y^2} = 0, \quad -\infty < x < \infty,$$

with $u(x, 0) = f(x)$, $|u(x, y)| \le M$ for $-\infty < x < \infty$, and where $f(x)$ is a smooth by parts function such that

$$\int_{-\infty}^{\infty} |f(x)|\,dx < \infty.$$

PE 9.49 Let $u(\rho, \phi, t) = u$. Find the solution of the heat equation on an infinite cylinder of radius R, that is, solve the partial differential equation

$$\frac{\partial u}{\partial t} = k\Delta u + \Omega,$$

for $t > 0$, $0 \le \rho < R$, $-\pi \le \phi \le \pi$, with $u(R, \phi, t) = T_a$ and $u(R, \phi, 0) = T_b$, where Ω, k, T_a and T_b are positive constants. Note that ρ and ϕ are cylindrical coordinates and that u is independent of z.

PE 9.50 Solve the Laplace equation in spherical coordinates for the following case: Determine the potential between two concentric spheres with radii a and b, kept at constant and distinct potentials, $u(a) = A$ and $u(b) = B$, where A and B are constants.

References

1. P.R. Garabedian, *Partial Differential Equations* (AMS Chelsea Publishing, Rhode Island, 1964)
2. I.S. Gradshteyn, I.M. Ryzhik, *Table of Integrals, Series and Products* (Academic Press, New York, 1980)
3. S. Hassani, *Foundations of Mathematical Physics* Prentice-Hall, New York, 1991)
4. M.A. Pinsky, *Partial Differential Equations and Boundary Value Problems with Applications* (McGraw-Hill, New York, 1991)
5. I. Stakgold, *Boundary Value Problems of Mathematical Physics* (Macmillan, New York, 1967)
6. E. Zauderer, *Partial Differential Equations of Applied Mathematics* (Wiley, New York, 1983)

Chapter 10
Fractional Calculus

Nothing takes place in the world whose meaning is not that of some maximum or minimum.

1707 – Leonhard Euler – 1783

Fractional calculus, a popular name for the calculus of arbitrary order, is a branch of mathematical analysis which basically studies several different ways of defining a derivative. With these different definitions, it is possible, for instance, to build and solve fractional differential equations. We present here just an introduction to the subject. Specifically, we discuss two types of fractional derivatives, the Riemann-Liouville and the Caputo [1927 – Michele Caputo –] fractional derivatives. We also introduce the one-parameter Mittag-Leffler [1846 – Magnus Gösta Mittag-Leffler – 1927] function, the queen of special functions of fractional calculus, together with two generalizations, also called Mittag-Leffler functions, with two and three parameters. We use integral transforms to solve simple fractional ordinary differential equations and simple fractional integral equations.

It is important to say that one reason for choosing this theme for the last chapter before the applications was that it provides a way to use some results learned in previous chapter, namely, ordinary derivatives, which appeared in Chaps. 1 and 2; analytic functions and the residue theorem, discussed in Chap. 3; gamma function, beta function, and the confluent hypergeometric function, presented in Chap. 4; and the Laplace transform studied in Chap. 6. Another, not less important reason, is just that this theme is beginning to appear in many undergraduate courses, due to its being a topic on the rise, a hot topic [1–12].

10.1 Fractional Derivatives

Unlike the case with classical calculus, which involves derivatives and integrals of integer orders only and in which we normally start from the definition of derivative and then introduce the concept of integral, it may be more intuitive to begin the study of fractional calculus with the concept of integral and then, later, to address the idea of fractional derivative. This is indeed more natural since an integral is associated with an area, a more primitive notion than the concept of rate of change associated with the derivative.

10.1.1　Fractional Integral

In order to introduce the concept of a fractional order integral, we start by showing that an integral of order n, with $n \in \mathbb{N}$, of a function $f(x)$ with $x \in \mathbb{R}$, can be seen as the Laplace convolution product of function $f(x)$ and the Gel'fand-Shilov [1913 – Israel Moiseevich Gel'fand – 2009]-[1917 – Georgi Evgen'evich Shilov – 1975] function $\phi_n(x)$ of order n. This integer order integral is also called multiple integral or iterated integral. With the help of the generalization of the factorial concept provided by the gamma function, we then introduce the concept of noninteger order integral or fractional order integral.

Definition 10.1.1 (Integer Order Integral)　Let $t > 0$ and let $f(t)$ be a real and integrable function. The integral operator of integer order, denoted by $\mathcal{J}$, acting on function $f(t)$, is given by

$$\mathcal{J}f(t) = \int_0^t f(t_1)\, \mathrm{d}t_1 .$$

From Definition 10.1.1 and iterating, we obtain

$$\mathcal{J}^2 f(t) = \mathcal{J}[\mathcal{J}f(t)] = \int_0^t \int_0^{t_1} f(t_2)\, \mathrm{d}t_2\, \mathrm{d}t_1 ,$$

$$\mathcal{J}^3 f(t) = \mathcal{J}[\mathcal{J}^2 f(t)] = \int_0^t \int_0^{t_1} \int_0^{t_2} f(t_3)\, \mathrm{d}t_3\, \mathrm{d}t_2\, \mathrm{d}t_1 .$$

Thus, the integral of order $n \in \mathbb{N}$ is given by

$$\mathcal{J}^n f(t) = \int_0^t \int_0^{t_1} \int_0^{t_2} \cdots \int_0^{t_{n-2}} \int_0^{t_{n-1}} f(t_n)\, \mathrm{d}t_n\, \mathrm{d}t_{n-1} \cdots \mathrm{d}t_3\, \mathrm{d}t_2\, \mathrm{d}t_1 . \tag{10.1}$$

We can express the integral of order n, with $n \in \mathbb{N}$, through a theorem involving the Gel'fand-Shilov function and the Laplace convolution product.

Definition 10.1.2 (Gel'fand-Shilov Function)　Let $n \in \mathbb{N}$. We define the Gel'fand-Shilov *function*, denoted by $\phi_n(t)$, as

$$\phi_n(t) := \begin{cases} \dfrac{t^{n-1}}{(n-1)!} & \text{for } t \geq 0, \\[2em] 0 & \text{for } t < 0. \end{cases}$$

Theorem 10.1 (Integral of order n) *Let $n \in \mathbb{N}$, $t \in \mathbb{R}_+$ and $f(t)$ an integrable real function. The integral of order n is given by*

$$\begin{aligned} \mathcal{J}^n f(t) &= \phi_n(t) \star f(t) \\ &:= \int_0^t \phi_n(t - \tau) f(\tau)\, d\tau \\ &= \int_0^t \frac{(t - \tau)^{n-1}}{(n-1)!} f(\tau)\, d\tau, \end{aligned}$$

where $\star$ denotes Laplace's convolution product.

Proof We prove the theorem by induction on parameter n. Using the definition of the Gel'fand-Shilov function with $n = 1$, we have

$$\mathcal{J} f(t) = \int_0^t \frac{(t - \tau)^{1-1}}{(1-1)!} f(\tau)\, d\tau = \phi_1(t) \star f(t) .$$

We then show that, if $\mathcal{J}^n f(t) = \phi_n(t) \star f(t)$, then $\mathcal{J}^{n+1} f(t) = \phi_{n+1}(t) \star f(t)$. By the induction hypothesis, we can write

$$\begin{aligned} \mathcal{J}^{n+1} f(t) &= \mathcal{J}[\mathcal{J}^n f(t)] = \mathcal{J}[\phi_n(t) \star f(t)] \\ &= \int_0^t \phi_n(u) \star f(u)\, du \\ &= \int_0^t \int_0^u \frac{(u - \tau)^{n-1}}{(n-1)!} f(\tau)\, d\tau\, du. \end{aligned}$$

By Goursat's [1858 – Édouard Goursat – 1936] theorem, it is possible to change the order of integration. It follows that

$$\mathcal{J}^{n+1} f(t) = \int_0^t \left[\int_\tau^t \frac{(u - \tau)^{n-1}}{(n-1)!} du \right] f(\tau)\, d\tau.$$

Evaluating the integral inside the square brackets, we finally have

$$\mathcal{J}^{n-1} f(t) = \int_0^t \frac{(t - \tau)^n}{n!} f(\tau)\, d\tau = \phi_{n+1}(t) \star f(t),$$

which is the desired result. $\qquad\square$

Using Theorem 10.1 and the concept of gamma function, a generalization of the factorial, as already discussed, we can generalize the order of the integral for an arbitrary number $v \in \mathbb{R}$.

Definition 10.1.3 (Integral of Order v)
Let $f(t)$ be an integrable function. We define the integral of order $v \in \mathbb{R}$ of function $f(t)$, denoted $\mathcal{J}^v f(t)$, by means of the expression

$$\mathcal{J}^v f(t) = \phi_v(t) \star f(t) = \int_0^t \frac{(t - \tau)^{v-1}}{\Gamma(v)} f(\tau) \, \mathrm{d}\tau. \tag{10.2}$$

When the parameter defining the order of the integral, v, is such that $v = n + 1$, with $n \in \mathbb{N}$, we recover the result for integer order.

10.1.2   Riemann-Liouville Fractional Integral

Fractional integrals can be construed on the left and right of a point. Nevertheless, we present here only the formulation to the left, as this is just an introduction to the subject. The respective formulation on the right, with simple modifications, follows the same steps as the formulation on the left. With these caveats in mind, we start by defining the Riemann-Liouville fractional integral on the left. From now on, we shall omit the nomenclature "on the left." We present here the definition of the Riemann-Liouville fractional integral. There are other formulations, in particular the Hadamard [1865 – Jacques Salomon Hadamard – 1963] fractional integral, whose kernel contains a logarithm [2, 7].

Definition 10.1.4 (Riemann-Liouville Fractional Integral) Let $t \in \mathbb{R}$ with $\mathrm{Re}(v) > 0$. The Riemann-Liouville fractional integral of order v, acting on $f \in L^p[a, b]$, $1 \leq p < +\infty$, $-\infty < a < b < +\infty$, for $t \in [a, b]$, is defined by

$$(\mathcal{J}^v f)(t) \equiv \mathcal{J}^v f(t) := \frac{1}{\Gamma(v)} \int_a^t \frac{f(\tau)}{(t - \tau)^{1-v}} \mathrm{d}\tau, \qquad t > a. \tag{10.3}$$

In the particular case $v = 0$, we have $(\mathcal{J}^0 f)(t) = f(t)$.

Riemann-Liouville fractional integrals are characterized by the particular class of functions on which this operator acts, as well as by the respective integration interval to be considered. For example, in the case in which $a = 0$ in Eq. (10.3), we obtain the definition of fractional integral proposed by Riemann, but without the so-called complementary function [7]. It should be noted that the expression in Eq. (10.3) will be used to introduce the concept of fractional derivative in the Riemann-Liouville sense (an integer order derivative of a fractional order integral) and in the Caputo sense (a fractional order integral of an integer order derivative). This difference in the orders of integration and differentiation plays an important role, especially when we use the integral transform methodology to solve a problem involving a differential equation with the imposition of initial and/or boundary conditions of integer orders.

10.2   Fractional Derivatives

We now present the definitions of the Riemann-Liouville and the Caputo fractional derivatives. The Riemann-Liouville and the Caputo formulations are stated with the help of the Riemann-Liouville integral, making clear the nonlocal character of fractional derivatives, differently from integer order derivatives, which have local character.

10.2.1   Riemann-Liouville Fractional Derivative

Before defining the Riemann-Liouville and the Caputo fractional derivatives, we mention that the space of functions used here is the space of absolutely continuous functions on the interval $[a, b]$, denoted by $AC[a, b]$, where $[a, b]$ is any finite interval with $-\infty < a < b < +\infty$. Also, we use for the integer order derivative operator the notation $\dfrac{\mathrm{d}^n}{\mathrm{d}t^n} := \mathsf{D}^n$.

Definition 10.2.1 (Riemann-Liouville Fractional Derivative) Let $t \in \mathbb{R}$, $\mathrm{Re}(v) > 0$, $n = [\mathrm{Re}(v)] + 1$ where $[\mathrm{Re}(v)]$ is the integer part of $\mathrm{Re}(v)$, $n \in \mathbb{N}$ and $v \notin \mathbb{N}$. The Riemann-Liouville fractional derivative of order v, acting on a function $f \in AC[a, b]$, with $-\infty < a < b < +\infty$, for $t \in [a, b]$, is given by

$$(\mathcal{D}^v f)(t) \equiv \mathcal{D}^v f(t) := \mathsf{D}^n \left(\mathcal{J}^{n-v} f \right)(t)$$

$$= \frac{1}{\Gamma(n-v)} \mathsf{D}^n \int_a^t \frac{f(\tau)}{(t-\tau)^{v-n+1}} \mathrm{d}\tau . \tag{10.4}$$

For $v = n$, we define $\mathcal{D}^v f(t) = \mathsf{D}^n f(t)$.

Equation (10.4) tells us that a derivative of arbitrary order, in the Riemann-Liouville formulation, is equivalent to the derivative of integer order of an integral of arbitrary order. Note that what has been said about integrals on the left and on the right is also valid for derivatives, i.e., we do have both types of derivatives but we deal here only with the formulation on the left.

10.2.2   Caputo Fractional Derivative

Definition 10.2.2 (Caputo Fractional Derivative) Let $t \in \mathbb{R}$ with $\mathrm{Re}(\nu) > 0$, $n = [\mathrm{Re}(\nu)] + 1$, where $[\mathrm{Re}(\nu)]$ is the integer part of $\mathrm{Re}(\nu)$, $n \in \mathbb{N}$ and $\nu \notin \mathbb{N}$. Let $f(t) \in AC[a, b]$. The Caputo fractional derivative of order ν, acting on a function f, with $-\infty < a < b < +\infty$ for $t \in [a, b]$, is given by

$$(^{\mathrm{C}}\mathcal{D}^{\nu} f)(t) \equiv {}^{\mathrm{C}}\mathcal{D}^{\nu} f(t) := \left(\mathcal{J}^{n-\nu} \mathsf{D}^{n} f \right)(t)$$

$$= \frac{1}{\Gamma(n - \nu)} \int_{a}^{t} \frac{\mathsf{D}^{n} f(\tau)}{(t - \tau)^{\nu - n + 1}} \mathrm{d}\tau . \tag{10.5}$$

For $\nu = n$, we define $\mathcal{D}^{\nu} f(t) = \mathsf{D}^{n} f(t)$.

As with the formulation of the Riemann-Liouville fractional derivative, Eq. (10.5) allows us to say that a derivative of arbitrary order, according to Caputo's formulation, is equivalent to an arbitrary order integral of an integer order derivative.

10.2.3   Riemann-Liouville × Caputo

The relation between the Riemann-Liouville derivative, Eq. (10.4), and the Caputo derivative, Eq. (10.5), is given by

$$(^{\mathrm{C}}\mathcal{D}^{\nu} f)(t) = (\mathcal{D}^{\nu} f)(t) - \sum_{k=0}^{\infty} f^{(k)}(0^{+}) \frac{t^{k-\nu}}{\Gamma(k - \nu + 1)} . \tag{10.6}$$

Before we discuss the calculation of the Laplace transform of specific fractional derivatives, we present an expression that can help us move from one formulation to another. For this sake we define the following integrodifferential expression:

Definition 10.2.3 (Fractional Derivative) Let $m \leq p < m + 1$, $m \in \mathbb{N}$, and let $f(x)$ be a real function $m + 1$-times continuously differentiable, with p being the order of the derivative. We define the following integrodifferential expression:

$$_{a}\mathcal{D}_{x}^{p} f(x) = \left(\frac{\mathrm{d}}{\mathrm{d}x} \right)^{m+1} \int_{a}^{x} (x - \xi)^{m-p} f(\xi) \, \mathrm{d}\xi$$

with $x > a$.

Just to mention that, from this expression, we can obtain both the Riemann-Liouville formulation and the Caputo formulation of the fractional derivative using integration by parts and integer order derivatives.

Due to their importance in the solution of fractional differential equations, we evaluate in the exercises the integral Laplace transforms of the Riemann-Liouville fractional integral and of the Caputo fractional derivative. We also obtain the Laplace transform (and the corresponding inverse Laplace transform) of the Mittag-Leffler function with three parameters. The Laplace transform of the Mittag-Leffler function with two parameters and the Laplace transform of the classical (one parameter) Mittag-Leffler function are left as proposed exercises.

10.3 Mittag-Leffler Functions

The most important special function of fractional calculus is the Mittag-Leffler function, a generalization of the exponential function. The function, denoted $\mathbb{E}_\alpha(\cdot)$, involves a parameter $\alpha \in \mathbb{C}$ and is given by the series

$$\mathbb{E}_\alpha(z) = \sum_{k=0}^{\infty} \frac{z^k}{\Gamma(\alpha k + 1)}, \tag{10.7}$$

for $z, \alpha \in \mathbb{C}$ and $\mathrm{Re}(\alpha) > 0$.

This function was introduced in 1903 by Mittag-Leffler and is also called classical Mittag-Leffler function because, since its appearance, several other similar functions have been proposed, with more than one parameter or more than one independent variable or both. Here, we work only with $z = x \in \mathbb{R}$ and the Mittag-Leffler functions with two and three parameters.

Example 10.1 Discuss the Mittag-Leffler function in the case $\alpha = 1$.

Taking $\alpha = 1$ in Eq. (10.7) we obtain

$$\mathbb{E}_1(z) = \sum_{k=0}^{\infty} \frac{z^k}{\Gamma(k + 1)} = \sum_{k=0}^{\infty} \frac{z^k}{k!},$$

where, in the second equality, we have used $\Gamma(k + 1) = k!$ because k is zero or is a positive integer. We then see that the last sum is exactly the exponential function,

$$E_1(z) = \exp(z),$$

which is the desired result. This is the reason why we can say that the classical Mittag-Leffler function is a generalization of the exponential function. $\qquad\square$

The Mittag-Leffler function with two parameters was introduced by Agarwal [1925 – Ratan Prakash Agarwal – 2008] in 1953 and is defined by the series

$$\mathbb{E}_{\alpha,\beta}(z) = \sum_{k=0}^{\infty} \frac{z^{\alpha}}{\Gamma(\alpha k + \beta)}, \qquad (10.8)$$

with $z \in \mathbb{C}$ and $\alpha, \beta \in \mathbb{C}$ with $\mathrm{Re}(\alpha) > 0$.

For $\beta = 1$ we have $\mathbb{E}_{\alpha,1}(\cdot) = \mathbb{E}_{\alpha}(\cdot)$, so that one can interpret the Mittag-Leffler function with two parameters as a generalization of the classical Mittag-Leffler function. Also, taking $\alpha = \beta = 1$ we recover the exponential function.

Example 10.2 Let $z \in \mathbb{C}$. Express the Mittag-Leffler function with two parameters,

$$\mathbb{E}_{2,2}(z^2),$$

in terms of a hyperbolic function.

Considering $\alpha = \beta = 2$ in Eq. (10.8), we get

$$\mathbb{E}_{2,2}(z^2) = \sum_{k=0}^{\infty} \frac{z^{2k}}{\Gamma(2k + 2)}.$$

Since the arguments of the gamma functions in this series are integer, we can write them in terms of factorials. Then, multiplying and dividing the series by z we arrive at

$$\mathbb{E}_{2,2}(z^2) = \frac{1}{z} \sum_{k=0}^{\infty} \frac{z^{2k+1}}{(2k + 1)!}.$$

The remaining sum is exactly the hyperbolic sine, so

$$\mathbb{E}_{2,2}(z^2) = \frac{\sinh z}{z},$$

which is the desired result. $\qquad\qquad\square$

Another type of Mittag-Leffler function was introduced in 1970 by Prabhakar, the so-called Mittag-Leffler function with three parameters, defined by the series

$$\mathbb{E}_{\alpha,\beta}^{\gamma}(z) = \sum_{k=0}^{\infty} \frac{(\gamma)_k}{\Gamma(\alpha k + \beta)} \frac{z^k}{k!}, \qquad (10.9)$$

with $z \in \mathbb{C}$ and $\alpha, \beta, \gamma \in \mathbb{C}$ with $\mathrm{Re}(\alpha) > 0$. In this expression, $(\gamma)_k$ is the Pochhammer symbol, defined in Chap. 4 as the quotient of two gamma functions,

$$(\gamma)_k = \frac{\Gamma(\gamma + k)}{\Gamma(\gamma)}.$$

Example 10.3 Let $k \in \mathbb{N}$. Show that the k-th derivative of the Mittag-Leffler function with two parameters can be written as a product of $k!$ by a Mittag-Leffler function with three parameters.

To show this result we consider, without loss of generality, $\alpha > 0$ and $\beta > 0$ in $\mathbb{E}_{\alpha,\beta}(x)$. We then evaluate its k-th derivative, denoting it by I_M,

$$I_M = \left(\frac{\mathrm{d}}{\mathrm{d}x} \right)^k \mathbb{E}_{\alpha,\beta}(x).$$

Using Eq. (10.8) and interchanging the derivative with the sum, assuming this change to be valid, we obtain

$$\sum_{n=0}^{\infty} \frac{1}{\Gamma(\alpha n + \beta)} \left(\frac{\mathrm{d}}{\mathrm{d}x} \right)^k x^n .$$

As k and n are positive integers, we have

$$I_M = \left(\frac{\mathrm{d}}{\mathrm{d}x} \right)^k x^n = \frac{n!}{(n-k)!} x^{n-k},$$

for $n \geq k$, since if $n < k$ the derivative is null. This means that the resulting sum starts at $n = k$ and we can write

$$I_M = \sum_{n=k}^{\infty} \frac{(1)_n}{\Gamma(\alpha n + \beta)} \frac{x^{n-k}}{(n-k)!}.$$

Introducing the change $n \to n + k$ and manipulating the sum, we have

$$I_M = \sum_{k=0}^{\infty} \frac{(1)_{n+k}}{\Gamma(\alpha n + \alpha k + \beta)} \frac{x^n}{n!} .$$

Finally, using the property of the Pochhammer symbol

$$(\rho)_{k+m} = (\rho)_k (\rho + k)_m,$$

with $\rho = 1$ and $m = n$ in the expression above and simplifying, we get

$$I_M = k! \sum_{k=0}^{\infty} \frac{(k+1)_n}{\Gamma(\alpha n + \alpha k + \beta)} \frac{x^n}{n!}.$$

The remaining sum can be easily expressed as a Mittag-Leffler function with three parameters, thus

$$I_M = k! \, \mathbb{E}^{k+1}_{\alpha,\alpha k+\beta}(x) \,,$$

that is,

$$\left(\frac{\mathrm{d}}{\mathrm{d}x}\right)^k \mathbb{E}_{\alpha,\beta}(x) = k! \, \mathbb{E}^{k+1}_{\alpha,\alpha k+\beta}(x) \,, \tag{10.10}$$

which is the desired result. $\square$

This last result guarantees that working with the k-th derivative of the Mittag-Leffler function with two parameters is the same as working with the Mittag-Leffler function with three parameters.

As already mentioned, there are several types of Mittag-Leffler functions, though we restricted our presentation to functions with one, two, and three parameters. Besides, we discuss a few simple fractional differential equations in the section of solved exercises. For other types of the Mittag-Leffler functions, we mention [4].

10.4   Solved Exercises

SE 10.1 Evaluate the integral of order $\frac{1}{2}$ of the function $f(t) = t$, i.e., $\mathcal{J}^{\frac{1}{2}} t$.
Solution: Substituting $v = 1/2$ and $f(t) = t$ into Eq. (10.3), we get

$$\mathcal{J}^{\frac{1}{2}} t = \frac{1}{\Gamma(1/2)} \int_a^t \frac{\tau}{(t-\tau)^{\frac{1}{2}}} \mathrm{d}\tau \,, \qquad t > a \,.$$

Introducing the change of variable $\xi = t - \tau$, we obtain

$$\mathcal{J}^{\frac{1}{2}} t = \frac{1}{\sqrt{\pi}} \int_0^{t-a} \frac{t-\xi}{\sqrt{\xi}} \, \mathrm{d}\xi \,,$$

where we used the result $\Gamma(1/2) = \sqrt{\pi}$. With another change, $\xi = (t-a)u$, the integral takes the form

$$\mathcal{J}^{\frac{1}{2}} t = \frac{\sqrt{t-a}}{\sqrt{\pi}} \int_0^1 \frac{t-(t-a)u}{\sqrt{u}} \, \mathrm{d}u \,.$$

Separating in two integrals and integrating, we obtain

$$\mathcal{J}^{\frac{1}{2}} t = 2t \frac{\sqrt{t-a}}{\sqrt{\pi}} - \frac{2}{3\sqrt{\pi}} \sqrt{(t-a)^3} \,.$$

If we take $a = 0$ and disconsider terms before zero (the so-called memory effect) and what Riemann calls complementary function, we can write, after

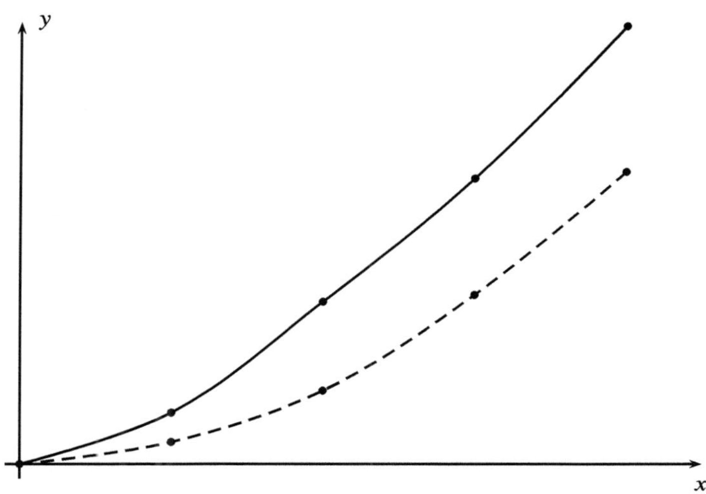

Fig. 10.1 Sketch of curves $y(x) = \frac{x^2}{2}$ and $y(x) = \frac{4x\sqrt{x}}{3\sqrt{\pi}}$

simplifications,

$$\mathcal{J}^{\frac{1}{2}} t = \frac{4}{3\sqrt{\pi}}\sqrt{t^3}.$$

Figure 10.1 sketches the graphs of $y(x) = \frac{x^2}{2}$ (a parabola) and of $y(x) = \frac{4x\sqrt{x}}{3\sqrt{\pi}}$.

SE 10.2 Using the Riemann-Liouville formulation with $a = 0$, evaluate the derivative of order $1/2$ of function $f(t) = t$.

Solution: First, note that $n = 1$ because the integer part of $\nu = 1/2$ is equal to zero. So, substituting $\nu = 1/2$ and $f(t) = t$ into Eq. (10.4), we have

$$\mathcal{D}^{\frac{1}{2}} t = \frac{1}{\Gamma(1/2)}\mathsf{D}\int_0^t \frac{\tau}{\sqrt{t-\tau}}\mathrm{d}\tau .$$

Introducing the change of variable $\xi = t - \tau$, we get

$$\mathcal{D}^{\frac{1}{2}} t = -\frac{1}{\Gamma(1/2)}\mathsf{D}\int_0^t \left(\xi^{\frac{1}{2}} - t\xi^{-\frac{1}{2}}\right)\mathrm{d}\xi,$$

whose integration yields

$$\mathcal{D}^{\frac{1}{2}} t = -\frac{1}{\Gamma(1/2)}\mathsf{D}\left(-\frac{4}{3}t^{\frac{3}{2}}\right).$$

Evaluating the derivative and simplifying, we can write

$$\mathcal{D}^{\frac{1}{2}} t = 2\sqrt{\frac{x}{\pi}}\,, \tag{10.11}$$

which is the desired result.

SE 10.3 For $a = 0$, calculate the derivative of order $1/2$ of function $f(t) = t$ using Caputo's formulation.

Solution: First, note that since $v = 1/2$, we have $n = 1$. Then, considering $a = 0$ and substituting $v = 1/2$ and $f(t) = t$ into Eq. (10.5) we can write

$$^{C}\!\mathcal{D}^{\frac{1}{2}} t = \frac{1}{\Gamma(1/2)} \int_{0}^{t} \frac{\mathsf{D}(\tau)}{\sqrt{t-\tau}} d\tau \,.$$

Evaluating the derivative of order 1, introducing the change of variable $\xi = t - \tau$ and simplifying, we get

$$^{C}\!\mathcal{D}^{\frac{1}{2}} t = \frac{1}{\sqrt{\pi}} \int_{0}^{t} \xi^{-\frac{1}{2}} d\xi \,,$$

whose integration yields

$$^{C}\!\mathcal{D}^{\frac{1}{2}} t = 2\sqrt{\frac{x}{\pi}}\,,$$

exactly as in the Riemann-Liouville formulation, Eq. (10.11).

SE 10.4 Consider $n \in \mathbb{N}$ and suppose that the order μ of a Riemann-Liouville derivative is such that $n - 1 \leq \mu < n$. Let $f(x)$ be a $(n + 1)$-times continuously differentiable function and let p be the parameter of the Laplace transform, that is, $\mathscr{L}[f(x)] = F(p)$. If the Riemann-Liouville fractional derivative is denoted by $^{RL}_{0}\mathbb{D}^{\mu}_{x}$, show that

$$\mathscr{L}\left[^{RL}_{0}\mathbb{D}^{\mu}_{x} f(x); p \right] = p^{\mu} F(p) - \sum_{k=0}^{n-1} p^{k} \left[^{RL}_{0}\mathbb{D}^{\mu-k-1}_{x} f(x) \right]_{x=0} . \tag{10.12}$$

Solution: The Riemann-Liouville fractional derivative is given by the expression

$$^{RL}_{0}\mathbb{D}^{\mu}_{x} f(x) = \mathsf{D}^{n}\left(\mathcal{J}^{n-\mu} f(x) \right) ,$$

where $\mathcal{J}^{v}(\cdot)$ is the Riemann-Liouville integral operator. Taking the Laplace transform on both sides of the previous expression, we get

$$\mathscr{L}\left[^{RL}_{0}\mathbb{D}^{\mu}_{x} f(x); p \right] = \mathscr{L}\left[\mathsf{D}^{n} g(x); p \right] , \tag{10.13}$$

with $g(x) = \mathcal{J}^{n-\mu} f(x)$. As the Laplace transform of the integer order derivative, n, is given by

$$\mathcal{L}\left[\mathrm{D}^n g(x)\right] = p^n \mathcal{L}[g(x)] - \sum_{k=0}^{n-1} p^k g^{(n-k-1)}(0),$$

we can evaluate the Laplace transform of the Riemann-Liouville fractional integral

$$\mathcal{L}[g(x)] = \mathcal{L}\left[\mathcal{J}^{n-\mu} f(x)\right]$$

$$= \mathcal{L}\left[\frac{1}{\Gamma(n-\mu)} \int_0^x (x-\xi)^{n-\mu-1} f(\xi)\, \mathrm{d}\xi\right] \tag{10.14}$$

$$= \mathcal{L}\left[\frac{x^{n-\mu-1}}{\Gamma(n-\mu)}\right] F(p),$$

where the convolution product was used.

Using the result of the **PE 10.23** in Eq. (10.14), we obtain

$$\mathcal{L}[g(x)] = \frac{F(p)}{p^{n-\mu}}.$$

On the other hand, we must evaluate the derivative of integer order $(n-k-1)$ of function $g(x)$, and evaluate it at $x = 0$. Then, we can write for the derivative

$$\frac{\mathrm{d}^{n-k-1}}{\mathrm{d}x^{n-k-1}} g(x) = \frac{\mathrm{d}^{n-k-1}}{\mathrm{d}x^{n-k-1}} \mathcal{J}^{n-\mu} f(x)$$

$$= \mathrm{D}^{n-k-1} \mathbb{D}^{-(n-\mu)} f(x)$$
$$= {}^{\mathrm{RL}}_{0}\mathbb{D}_x^{\mu-k-1} f(x).$$

Returning to Eq. (10.14) and simplifying, we finally obtain

$$\mathcal{L}\left[{}^{\mathrm{RL}}_{0}\mathbb{D}_x^{\mu} f(x); p\right] = p^{\mu} F(p) - \sum_{k=0}^{n-1} p^k \left[{}^{\mathrm{RL}}_{0}\mathbb{D}_x^{\mu-k-1} f(x)\right]_{x=0},$$

which is the desired result.

SE 10.5 Obtain the expression for the Laplace transform of the Caputo fractional derivative.

Solution: The process is similar to the case of the Riemann-Liouville fractional derivative. We start with the Caputo fractional derivative written as

$$^{\mathrm{C}}_{0}\mathbb{D}_x^{\mu} f(x) = \mathcal{J}^{n-\mu} \left(\mathrm{D}^n f(x)\right),$$

where $\mathcal{J}^{\nu}(\cdot)$ is the Riemann-Liouville fractional integral operator. Taking the Laplace transform on both sides of this expression, we have

$$\mathcal{L}\left[{}^{C}_{0}\mathbb{D}^{\mu}_{x} f(x);\, p\right] = \mathcal{L}\left[\mathcal{J}^{n-\mu} g(x);\, p\right],$$

with $g(x) = \mathbb{D}^{n} f(x)$. With the help of Eq. (10.14), we get

$$\mathcal{L}\left[\mathcal{J}^{n-\mu} g(x)\right] = \mathcal{L}\left[\frac{1}{\Gamma(n-\mu)} \int_{0}^{x} (x-\xi)^{n-\mu-1} f(\xi)\, d\xi\right]$$

$$= \mathcal{L}\left[\frac{x^{n-\mu-1}}{\Gamma(n-\mu)}\right] F(p) = \frac{F(p)}{p^{n-\mu}},$$

where we used the convolution product and the notation $F(p) = \mathcal{L}[f(x)]$. Returning to the expression for the Laplace transform and using the expression for the Laplace transform of an integer order derivative, we can write, after simplification,

$$\mathcal{L}\left[{}^{C}_{0}\mathbb{D}^{\mu}_{x} f(x);\, p\right] = p^{\mu} F(p) - \sum_{k=0}^{n-1} p^{\mu-k-1} f^{(k)}(0) \tag{10.15}$$

with the notation $f^{(k)}(0) = \mathbb{D}^{k} f(x)|_{x=0}$, which is the desired result.

SE 10.6 Let $\lambda \in \mathbb{C}$, $\text{Re}(\alpha) > 0$, $\beta > 0$ and $\mu > 0$. Prove the relation

$$\frac{1}{\Gamma(\mu)} \int_{0}^{z} (z-\xi)^{\mu-1} \mathbb{E}_{\alpha,\beta}(\lambda \xi^{\alpha}) \xi^{\beta-1} d\xi = z^{\mu+\beta-1} E_{\alpha,\mu+\beta}(\lambda z^{\alpha}),$$

where $\mathbb{E}_{\alpha,\beta}(\cdot)$ is the Mittag-Leffler with two parameters.
Solution: Denote the first member by Λ, that is,

$$\Lambda = \frac{1}{\Gamma(\mu)} \int_{0}^{z} (z-\xi)^{\mu-1} \mathbb{E}_{\alpha,\beta}(\lambda \xi^{\alpha}) \xi^{\beta-1} d\xi.$$

Introducing the series expansion for the Mittag-Leffler function with two parameters and changing the order of the integral and the sum, we can write

$$\Lambda = \frac{1}{\Gamma(\mu)} \sum_{k=0}^{\infty} \frac{\lambda^{k}}{\Gamma(\alpha k + \beta)} \int_{0}^{z} (z-\xi)^{\mu-1} \xi^{\alpha k+\beta-1}\, d\xi.$$

Let $\xi = zt$ be a new variable. Substituting into the integral and simplifying, we get

$$\Lambda = \frac{1}{\Gamma(\mu)} \sum_{k=0}^{\infty} \frac{\lambda^{k} z^{\alpha k+\beta+\mu-1}}{\Gamma(\alpha k + \beta)} \int_{0}^{1} (1-t)^{\mu-1} t^{\alpha k+\beta-1}\, dt.$$

The remaining integral is one of the representations of the beta function. Expressing this beta function in terms of the gamma function yields

$$\Lambda = \frac{z^{\mu+\beta-1}}{\Gamma(\mu)} \sum_{k=0}^{\infty} \frac{(\lambda z^{\alpha})^k}{\Gamma(\alpha k + \beta)} \frac{\Gamma(\mu)\Gamma(\alpha k + \beta)}{\Gamma(\alpha k + \mu + \beta)} \, .$$

With further simplification, we find that

$$\Lambda = z^{\mu+\beta-1} \sum_{k=0}^{\infty} \frac{(\lambda z^{\alpha})^k}{\Gamma(\alpha k + \beta + \mu)} \, .$$

Comparing the series on the right-hand side of this equation with the definition of the Mittag-Leffler function with two parameters, we see that

$$\Lambda = z^{\mu+\beta-1} \mathbb{E}_{\alpha,\beta+\mu}(\lambda z^{\alpha}) \, ,$$

which is the desired result.

SE 10.7 Let $\mathrm{Re}(\alpha) > 0$ and $\beta > 0$. Prove the recurrence relation

$$\mathbb{E}_{\alpha,\beta}(z) - z\mathbb{E}_{\alpha,\alpha+\beta}(z) = \frac{1}{\Gamma(\beta)},$$

where $\mathbb{E}_{\alpha,\beta}(\cdot)$ is the Mittag-Leffler with two parameters.

Solution: Denote the first member of this equality by Ω. Introducing the series expansion for the Mittag-Leffler function with two parameters, we obtain

$$\Omega = \sum_{k=0}^{\infty} \frac{z^k}{\Gamma(\alpha k + \beta)} - z \sum_{k=0}^{\infty} \frac{z^k}{\Gamma(\alpha k + \alpha + \beta)}$$

$$= \sum_{k=0}^{\infty} \frac{z^k}{\Gamma(\alpha k + \beta)} - \sum_{k=0}^{\infty} \frac{z^{k+1}}{\Gamma(\alpha k + \alpha + \beta)} \, .$$

Change the summation index $k \to k - 1$ in the second sum. We then have

$$\Omega = \sum_{k=0}^{\infty} \frac{z^k}{\Gamma(\alpha k + \beta)} - \sum_{k=1}^{\infty} \frac{z^k}{\Gamma(\alpha k + \beta)} \, .$$

The term $k = 0$ in the first sum is equal to $1/\Gamma(\beta)$ and the remaining terms cancel the second sum. We have thus proved that

$$\Omega = \frac{1}{\Gamma(\beta)} \, ,$$

which is the desired result.

SE 10.8 (Prabhakar Function) Let $x \in \mathbb{R}$ and $\alpha, \beta, \gamma \in \mathbb{C}$ with $\mathrm{Re}(\alpha) > 0$. Evaluate the Laplace transform of

$$\mathcal{E}^{\gamma}_{\alpha,\beta}(x) =: x^{\beta-1} \mathbb{E}^{\gamma}_{\alpha,\beta}(x^{\alpha}),$$

a function known as Prabhakar function, with $\mathbb{E}^{\gamma}_{\alpha,\beta}(\cdot)$ the Mittag-Leffler function with three parameters.

Solution: Let $s \in \mathbb{C}$ with $\mathrm{Re}(s) > 0$ be the parameter of the Laplace transform. We must evaluate the following integral:

$$\mathscr{L}\left[\mathcal{E}^{\gamma}_{\alpha,\beta}(x)\right] = \int_{0}^{\infty} e^{-sx} x^{\beta-1} \mathbb{E}^{\gamma}_{\alpha,\beta}(x^{\alpha}) \, dx.$$

Introducing the series expansion of the Mittag-Leffler with three parameters and changing the order of the integral and the sum, we can write

$$\mathscr{L}\left[\mathcal{E}^{\gamma}_{\alpha,\beta}(x)\right] = \sum_{k=0}^{\infty} \frac{(\gamma)_k}{k! \, \Gamma(\alpha k + \beta)} \int_{0}^{\infty} e^{-sx} x^{\beta-1} x^{\alpha k} \, dx.$$

Let J be the integral

$$J = \int_{0}^{\infty} e^{-sx} x^{\alpha k + \beta - 1} \, dx \, .$$

Introducing in this last integral the change of variable $sx = t$ and rearranging, we obtain

$$J = \frac{1}{s^{\alpha k + \beta}} \int_{0}^{\infty} e^{-t} t^{\alpha k + \beta - 1} \, dt \, .$$

Using the definition of the gamma function, we see that

$$J = \frac{\Gamma(\alpha k + \beta)}{s^{\alpha k + \beta}}.$$

Substituting this last result into the expression for the Laplace transform and simplifying, we get

$$\mathscr{L}\left[\mathcal{E}^{\gamma}_{\alpha,\beta}(x)\right] = \sum_{k=0}^{\infty} \frac{(\gamma)_k}{k! \, s^{\alpha k + \beta}} \, .$$

Expressing the Pochhammer symbol in terms of binomial coefficients and substituting into the above expression, we can rewrite the Laplace transform as

$$\mathscr{L}\left[\mathcal{E}^{\gamma}_{\alpha,\beta}(x)\right] = \frac{1}{s^{\beta}} \sum_{k=0}^{\infty} \binom{\gamma + k}{k} (s^{-\alpha})^{k} \, .$$

Assuming that $|s^{-\alpha}| < 1$, we have a geometric series with infinite terms. Evaluating its sum and substituting into the last equality, we find that

$$\mathscr{L}\left[\mathcal{E}^{\gamma}_{\alpha,\beta}(x)\right] = \frac{1}{s^{\beta}}\left(1 - s^{-\alpha}\right)^{-\gamma} .$$

This expression can be put in the form

$$\mathscr{L}\left[\mathcal{E}^{\gamma}_{\alpha,\beta}(x)\right] = \frac{s^{\alpha\gamma-\beta}}{(s^{\alpha}-1)^{\gamma}} ,$$

which is the desired result.

SE 10.9 We now present a recent result, involving two Mittag-Leffler functions [3], which uses an expansion in a Maclaurin series, discussed in Chap. 2, and the relation obtained in Eq. (10.10). Let $a, b \in \mathbb{R}$, $x \in \mathbb{R}$, and $\alpha, \beta \in \mathbb{C}$ with $\mathrm{Re}(\alpha) > 0$. We will show that

$$\sum_{k=0}^{\infty}(bx^{\alpha})^{k}\mathbb{E}_{\alpha,\alpha k+\beta}(ax^{\alpha}) = \mathbb{E}_{\alpha,\beta}((a+b)x^{\alpha}), \tag{10.16}$$

where $\mathbb{E}_{\alpha,\beta}(\cdot)$ is the Mittag-Leffler function with two parameters.

Solution: We first consider the Maclaurin series expansion of $\mathbb{E}_{\alpha,\beta}(x)$, that is,

$$\mathbb{E}_{\alpha,\beta}(x) = \frac{1}{\Gamma(\beta)} + x\,\mathbb{E}^{2}_{\alpha,\alpha+\beta}(0) + x^{2}\,\mathbb{E}^{3}_{\alpha,2\alpha+\beta}(0) + \cdots$$

which can be rewritten as

$$\mathbb{E}_{\alpha,\beta}(x) = \frac{1}{\Gamma(\beta)} + \frac{x}{\Gamma(\alpha+\beta)} + \frac{x^{2}}{\Gamma(2\alpha+\beta)} + \cdots$$

Thus, the right-hand side of Eq. (10.16) can be written as

$$\mathbb{E}_{\alpha,\beta}((a+b)x^{\alpha}) = \frac{1}{\Gamma(\beta)} + \frac{(a+b)x^{\alpha}}{\Gamma(\alpha+\beta)} + \frac{(a+b)^{2}x^{2\alpha}}{\Gamma(2\alpha+\beta)} + \cdots$$

Remark that in the above expression, each bracket $(a+b)^{k}$ has $k+1$ terms in the k-th row of a Pascal [1623 – Blaise Pascal – 1662] triangle. By adding together the terms in b^{k} in the Maclaurin series expansion, we find that the sum of the first diagonal is $\sum_{j=0}^{\infty} a^{j}$; the second diagonal yields $b\sum_{j=0}^{\infty}\binom{j+1}{1}a^{j}$; the third diagonal gives $b^{2}\sum_{j=0}^{\infty}\binom{j+2}{2}a^{j}, \ldots$ The Maclaurin series can thus be rearranged in terms of powers of b. We then get

$$\mathbb{E}_{\alpha,\beta}((a+b)x^{\alpha}) = \sum_{k=0}^{\infty} b^k \sum_{j=0}^{\infty} \frac{a^j \binom{j+k}{k} x^{\alpha(j+k)}}{\Gamma(\alpha j + \alpha k + \beta)}$$

$$= \sum_{k=0}^{\infty} \frac{(bx^{\alpha})^k}{k!} \sum_{j=0}^{\infty} \frac{(ax^{\alpha})^j \Gamma(j+k+1)}{\Gamma(\alpha j + \alpha k + \beta)}$$

$$= \sum_{k=0}^{\infty} (bx^{\alpha})^k \, \mathbb{E}_{\alpha,\alpha k+\beta}^{k+1}(ax^{\alpha}) ,$$

which is the desired result. It is important to note that the result is also valid if a and b are two commuting matrices [3].

SE 10.10 (Relaxation and Oscillation) Classical relaxation and oscillation problems are characterized by first-order and second-order ordinary differential equations, respectively. Here, we approach both problems by means of a single fractional ordinary differential equation using Caputo's formulation for the derivative, namely,

$$\mathscr{D}^{\alpha} x(t) + x(t) = q(t) , \tag{10.17}$$

where α is the order of the derivative, $n - 1 < \alpha \le n$ with $n \in \mathbb{N}^*$, $\mathscr{D} \equiv \frac{\mathrm{d}}{\mathrm{d}t}$ is the differential operator, and $q(t)$ is the nonhomogeneous term.

We have, for specific values of α,

$$n = 1 \longrightarrow 0 < \alpha \le 1 \longrightarrow \text{relaxation,}$$
$$n = 2 \longrightarrow 1 < \alpha \le 2 \longrightarrow \text{oscillation.}$$

Note that the integer (classical) case for relaxation is recovered when $\alpha = 1$; for oscillation, when $\alpha = 2$. Solve the fractional differential equation, Eq. (10.17).

Solution: To discuss Eq. (10.17), a fractional differential equation, we use the Laplace transform methodology. Taking the Laplace transform on both sides and remembering the expression for the Laplace transform of the Caputo fractional derivative, we have

$$s^{\alpha} \mathscr{L}[x(t)] - \sum_{k=0}^{n-1} s^{\alpha-k-1} \mathscr{D}^k x(t) \Big|_{t=0} + \mathscr{L}[x(t)] = \mathscr{L}[q(t)] ,$$

where s is the parameter of the Laplace transform.

Introducing the notations

$$Q(s) = \mathscr{L}[q(t)] \qquad \text{and} \qquad \mathscr{D}^k x(t) \Big|_{t=0} = x^{(k)}(0)$$

and rearranging, we can write the following expression:

$$\mathscr{L}[x(t)] = \sum_{k=0}^{n-1} \frac{s^{\alpha-k-1}}{s^{\alpha}+1} x^{(k)}(0) + \frac{Q(s)}{s^{\alpha}+1},$$

where $x^{(k)}(0)$ denotes the k-th integer order derivative of function $x(t)$, evaluated at $t = 0$.

In order to find the solution of Eq. (10.17), we take the inverse Laplace transform (a linear operator), obtaining

$$x(t) = \sum_{k=0}^{n-1} \mathscr{L}^{-1}\left[\frac{s^{\alpha-k-1}}{s^{\alpha}+1}\right] x^{(k)}(0) + \mathscr{L}^{-1}\left[\frac{Q(s)}{s^{\alpha}+1}\right].$$

Using the result of **PE 10.32**, it follows that

$$x(t) = \sum_{k=0}^{n-1} t^{k} \mathbb{E}_{\alpha,k+1}(-t^{\alpha}) x^{(k)}(0) + \mathscr{L}^{-1}\left[\frac{Q(s)}{s^{\alpha}+1}\right].$$

In order to evaluate the remaining inverse Laplace transform, we use the convolution theorem to identify

$$\frac{Q(s)}{s^{\alpha}+1} = Q(s) \cdot (s^{\alpha}+1)^{-1} \equiv Q(s)F(s),$$

where $F(s)$ is known. Thus,

$$f(t) = \mathscr{L}^{-1}\left[\frac{1}{s^{\alpha}+1}\right] = t^{s-1} \mathbb{E}_{\alpha,\alpha}(-t^{\alpha}).$$

Returning with these results, we obtain the solution of the fractional ordinary differential equation,

$$x(t) = \sum_{k=0}^{n-1} t^{k} \mathbb{E}_{\alpha,k+1}(-t^{\alpha}) x^{(k)}(0) - q(t) \star \mathbb{E}'_{\alpha}(-t^{\alpha})$$

where $\star$ denotes the convolution product.

Before we move on to another fractional differential equation, we emphasize that an analogous treatment can be given to the differential equations associated with the problems of radioactive decay and population growth/decay. The corresponding fractional ordinary differential equations are, respectively,

$$\mathscr{D}^{\alpha} m(t) = -km(t), \qquad \text{and} \qquad \mathscr{D}^{\alpha} P(t) = kP(t),$$

where k is a positive constant. The corresponding solutions are

$$m(t) = m_0\, \mathbb{E}_\alpha(-kt^\alpha) \qquad \text{and} \qquad P(t) = P_0\, \mathbb{E}_\alpha(kt^\alpha)\,,$$

where m_0 and P_0 positive constants and $\mathbb{E}_\alpha(\cdot)$ is the classical Mittag-Leffler function.

SE 10.11 (Heat Transfer) We discuss a fractional version of Newton's law of cooling, given by the fractional ordinary differential equation

$$\mathscr{D}^\alpha T(t) = -k[T(t) - T]$$

with initial condition $T(0) = T_0$, where $0 < \alpha \le 1$, $k \in \mathbb{R}^*_+$ and where $\mathscr{D}^\alpha$ is the Caputo fractional derivative of order α. If T is the environment temperature, we must have

$$\lim_{t\to\infty} T(t) = T.$$

Solution: We take the Laplace transform on both sides of the differential equation. Using the expression for the Laplace transform of the Caputo fractional derivative, we obtain an algebraic equation whose solution is given by the expression

$$F(s) = T_0 \frac{s^{\alpha-1}}{s^\alpha + k} + kT \frac{s^{-1}}{s^\alpha + k}\,,$$

where $F(s) = \mathscr{L}[T(t)]$ and s is the parameter of the Laplace transform. Calculating the corresponding inverse Laplace transform, we obtain

$$T(t) = T_0 \mathscr{L}^{-1}\left[\frac{s^{\alpha-1}}{s^\alpha + k}\right] + kT\,\mathscr{L}^{-1}\left[\frac{s^{-1}}{s^\alpha + k}\right].$$

Using the result of **PE 10.36**, we obtain the solution for the cooling problem,

$$T(t) = T_0 \mathbb{E}_\alpha(-kt^\alpha) + kT t^\alpha \mathbb{E}_{\alpha,\alpha+1}(-kt^\alpha)\,,$$

where $\mathbb{E}_\alpha(\cdot)$ and $\mathbb{E}_{\alpha,\beta}(\cdot)$ are Mittag-Leffler functions with one and two parameters, respectively.

10.5   Proposed Exercises

PE 10.1 Justify the permutation of the order of integrations in Theorem 10.1.

PE 10.2 Let $n \in \mathbb{N}$ and $\mu \in \mathbb{R}_+$. Use Eq. (10.3) with $a = 0$ and the definition of the beta function to evaluate the integral of order μ of $f(x) = x^n$.

PE 10.3 Let $\nu = \frac{1}{2}$ and $f(x) = \sqrt{x}$. Use Eq. (10.3) with $a = 0$ to evaluate the fractional integral of order $1/2$ of $f(x) = \sqrt{x}$.

PE 10.4 Let $\nu, \mu \in \mathbb{C}$. Using Eq. (10.4) with $a = 0$ and the definition of the beta function, evaluate the derivative of order μ of $f(x) = x^\nu$. Impose, if necessary, restrictions on parameters μ and ν.

PE 10.5 Use the result of **PE 10.4**, considering $\mu = 1/2$ and $\nu = 1$, to recover the result obtained in **SE 10.2**.

PE 10.6 Let $\mu \in \mathbb{R}_+$. Justify the impossibility of calculating the derivative of order μ of function $f(x) = \frac{1}{x}$ in the Riemann-Liouville sense with $a = 0$. Discuss the integer order derivative of that same function.

PE 10.7 Using the Riemann-Liouville formulation, in the case in which the lower extreme is $a = 0$, calculate the derivative of order $1/2$ of $f(x) = e^x$, expressing it in terms of an error function.

PE 10.8 Let $\nu, \mu \in \mathbb{C}$. Use Eq. (10.5) with $a = 0$ and the definition of beta function to evaluate the derivative of order μ of $f(x) = x^\nu$. Impose, if necessary, restrictions on the parameters μ and ν.

PE 10.9 Let $\mu \in \mathbb{R}_+$. Justify the impossibility of evaluating the derivative of order μ of function $f(x) = \frac{1}{x}$ in Caputo's sense with $a = 0$. Discuss the derivative of integer order for the same function.

PE 10.10 Using the Caputo formulation in the case in which the lower extreme is $a = 0$, evaluate the fractional derivative of order $1/2$ of $f(x) = e^x$.

PE 10.11 Evaluate the derivative of order $1/2$ in Caputo's sense, with $a = 0$, for the function $f(x) = x^{\frac{1}{2}}$.

PE 10.12 Consider $n = 1$ in Eqs. (10.4) and (10.5), both with $a = 0$ and $0 < \nu < 1$. Use integration by parts to prove the relation

$$({}^C\mathcal{D}^\nu f)(x) = (\mathcal{D}^\nu f)(x) - f(0^+)\frac{x^{-\nu}}{\Gamma(1 - \nu)}.$$

This result is a particular case of Eq. (10.6).

PE 10.13 Let $x > 0$ and $f(x)$ a function that admits the fractional derivatives of Riemann-Liouville and Caputo. Determine the conditions under which the two derivatives are equivalent.

PE 10.14 Let f be a function defined for every $x > 0$. Show that the Laplace transform of the Riemann-Liouville fractional integral of order μ of f is given by

$$\mathscr{L}[\mathcal{J}^\mu f(x)] = \frac{\mathscr{L}[f(x)]}{s^\mu}.$$

PE 10.15 Let $v > 0$, $\alpha > 0$ and $\gamma > 0$. Evaluate the Riemann-Liouville fractional integral of order v, with $a = 0$, of the Mittag-Leffler function with three parameters $\mathbb{E}^{\gamma}_{\alpha,1}(-x^{\alpha})$ and recover the particular case in which $\gamma = 1$, that is, the classical Mittag-Leffler function.

PE 10.16 Let $v > 0$, $\lambda \in \mathbb{R}$ and $x \geq 0$. Show that $y(x) = \mathbb{E}_v(\lambda x^v)$, where $\mathbb{E}_{\alpha}(\cdot)$ is the Mittag-Leffler function with one parameter, is a solution of the fractional differential equation

$$^{C}\mathcal{D}^{v} y(x) = \lambda y(x) \,,$$

where the derivative is taken in the Caputo sense.

PE 10.17 Let $m \in \mathbb{R}$. Evaluate the Caputo fractional derivative of order v, with $n - 1 < v \leq n, n \in \mathbb{N}$, for the function

$$f(x) = (x + 1)^{m} \,.$$

PE 10.18 Let $v > 0$ and $t > a$. Denoting by $_a\mathcal{D}^{v}_t$ and $_a\mathcal{J}^{v}_t$ the Riemann-Liouville fractional derivative and the Riemann-Liouville fractional integral, respectively, both of order v and initialized at $x = a$, show that

$$_a\mathcal{D}^{v}_t \left[_a\mathcal{J}^{v}_t f(x) \right] = f(x) \,.$$

This relation can be read as follows: The Riemann-Liouville fractional derivative is the inverse, to the left, of the Riemann-Liouville fractional integral.

PE 10.19 Let C be a positive constant. Show that the derivative of $f(t) = C$ is zero when we use the Caputo fractional derivative and is different from zero when we use the Riemann-Liouville derivative.

PE 10.20 Let $0 \leq \alpha < 1$. Consider the following integral equation, in which the unknown function $y(x)$ appears inside the integral symbol:

$$y(x) = \frac{x^{1-\alpha}}{\Gamma(2 - \alpha)} - \frac{1}{\Gamma(1 - \alpha)} \int_0^x (x - \xi)^{-\alpha} y(\xi) \, d\xi \,.$$

Verify that it admits a solution

$$y(x) = 1 - \mathbb{E}_{1-\alpha}(-x^{1-\alpha}) \,,$$

where $\mathbb{E}_v(\cdot)$ is the classical Mittag-Leffler function.

PE 10.21 Let $0 < \alpha < 1$ and $f(x)$ a continuous function. Solve the integral equation

$$\int_0^x (x - \tau)^{-\alpha} f(\tau)\, d\tau = 1 .$$

PE 10.22 Let $\mu > -1$ and $\nu > 0$. Use the definition of the Caputo fractional derivative with $a = 0$ to get

$$^C\mathcal{D}_x^\nu x^\mu = \frac{\Gamma(\mu + 1)}{\Gamma(\mu - \nu + 1)} x^{\mu - \nu} .$$

PE 10.23 Let $n \in \mathbb{N}$ and $\mu \in \mathbb{R}$. Show that

$$\mathcal{L}\left[\frac{x^{n-\mu-1}}{\Gamma(n - \mu)}\right] = \frac{1}{s^{n-\mu}}$$

where s is the parameter of the Laplace transform, imposing conditions on the parameters.

PE 10.24 Use Eq. (10.12) to find the Laplace transforms in the particular cases in which $0 \leq \mu < 1$ and $1 \leq \mu < 2$.

PE 10.25 Use Eq. (10.15) to find the Laplace transforms in the particular cases in which $0 \leq \mu < 1$ and $1 \leq \mu < 2$.

PE 10.26 Let $\mathrm{Re}(\alpha) > 0$, $\beta > 0$ and $|z| < 1$. Show that

$$\int_0^\infty e^{-t} \mathbb{E}_{\alpha,\beta}(zt^\alpha) t^{\beta-1}\, dt = \frac{1}{1 - z} .$$

Note that the second member is independent of both parameters. Justify.

PE 10.27 For $n \in \mathbb{N}$, show that

$$\frac{d^n}{dz^n}\left[z^{\beta-1}\mathbb{E}_{\alpha,\beta}(z^\alpha)\right] = z^{\beta-n-1}\mathbb{E}_{\alpha,\beta-n}(z^\alpha).$$

PE 10.28 Use the result of **PE 10.27** to discuss the particular case in which $n = 1$.

PE 10.29 Let $\alpha > 0$, $\beta > 0$, $k \in \mathbb{N}$ and $\gamma > 0$. Show that

$$\left(\frac{d}{dx}\right)^k \mathbb{E}_{\alpha,\beta}^\gamma(x) = (\gamma)_k \mathbb{E}_{\alpha,\beta+\alpha k}^{\gamma+k}(x) ,$$

where $\mathbb{E}_{\alpha,\beta}^\gamma(\cdot)$ is the Mittag-Leffler function with three parameters and $(\rho)_k$ the Pochhammer symbol.

PE 10.30 With the same procedure presented in **SE 10.9**, show that

$$\sum_{k=0}^{\infty} \frac{\Gamma(\gamma + k)}{\Gamma(\gamma)k!} (bt^{\alpha})^k \, \mathbb{E}_{\alpha,\alpha k+\beta}^{\gamma+k}(ax^{\alpha}) = \mathbb{E}_{\alpha,\beta}^{\gamma}((a+b)x^{\alpha}) \,.$$

PE 10.31 Let $x > 0$ and $\alpha, \beta \in \mathbb{C}$ with $\mathrm{Re}(\alpha) > 0$. Evaluate the integral

$$\int_0^x (x-\xi)^{\beta-1} \mathbb{E}_{\alpha}(\xi^{\alpha}) \, \mathrm{d}\xi \,,$$

where $\mathbb{E}_{\alpha}(\cdot)$ is the classical Mittag-Leffler function.

PE 10.32 Let $\alpha, \beta \in \mathbb{R}$ with $\alpha > 0$; let $\mathbb{E}_{\alpha,\beta}(\cdot)$ be a Mittag-Leffler function with two parameters. Show that

$$\mathscr{L}\left[t^k \mathbb{E}_{\alpha,k+1}(-t^{\alpha})\right] = \frac{s^{\alpha-k-1}}{s^{\alpha}+1} \,.$$

Applying the inverse transform to both sides, we find that

$$\mathscr{L}^{-1}\left[\frac{s^{\alpha-k-1}}{s^{\alpha}+1}\right] = t^k \mathbb{E}_{\alpha,k+1}(-t^{\alpha}) \,.$$

PE 10.33 Let $\alpha \in \mathbb{R}_+^*$. Prove the relation

$$\frac{\mathrm{d}}{\mathrm{d}x}\mathbb{E}_{\alpha}(x^{\alpha}) = x^{\alpha-1}\mathbb{E}_{\alpha,\alpha}(x^{\alpha}) \,.$$

PE 10.34 Consider the results presented in **SE 10.10**. Find the particular solutions for the relaxation problem with $n = 1$ and for the oscillation problem with $n = 2$.

PE 10.35 Use the Laplace transform methodology to solve the equations for $m(t)$, radioactive decay, and, for $P(t)$, population growth/decay, obtaining the results presented in the text.

PE 10.36 Let $\alpha, \beta \in \mathbb{R}$ with $\alpha > 0$. Show that

$$\mathscr{L}^{-1}\left[\frac{s^{\alpha-\beta}}{s^{\alpha}+\lambda}\right] = t^{\beta-1}\mathbb{E}_{\alpha,\beta}(-\lambda t^{\alpha}) \,,$$

where λ is a positive constant and $\mathbb{E}_{\alpha,\beta}(\cdot)$ is a Mittag-Leffler function with two parameters.

PE 10.37 Let $\alpha \in \mathbb{R}$ with $\alpha > 0$. Show that

$$\mathbb{E}_{\alpha}(x) = x\,\mathbb{E}_{\alpha,\alpha+1}(x) + 1 \,.$$

PE 10.38 Discuss the limiting case $\alpha = 1$ in Newton's law of cooling.

PE 10.39 Let $z \in \mathbb{C}$, α a complex parameter such that $\mathrm{Re}(\alpha) > 0$, and $k = 0, 1, 2, \ldots$ Show that

$$\left(\frac{\mathrm{d}}{\mathrm{d}z}\right)^k \mathbb{E}_k(z^k) = \mathbb{E}_k(z^k) \,.$$

PE 10.40 Let $z \in \mathbb{C}$ and α a complex parameter with $\mathrm{Re}(\alpha) > 0$. Prove the so-called duplication formula

$$\mathbb{E}_\alpha(z) + \mathbb{E}_\alpha(-z) = 2\mathbb{E}_{2\alpha}(z^2) \,.$$

PE 10.41 Let $z \in \mathbb{C}$ and let α and β be two complex parameters with $\mathrm{Re}(\alpha) > 0$. Prove the recurrence relation for the first derivative of the Mittag-Leffler function with two parameters:

$$\frac{\mathrm{d}}{\mathrm{d}z}\mathbb{E}_{\alpha,\beta}(z) = \left(z\frac{\mathrm{d}}{\mathrm{d}z} + 1\right)\mathbb{E}_{\alpha,\alpha+\beta}(z) \,.$$

PE 10.42 Let $z \in \mathbb{C}$. Express the quotient

$$\Lambda = \frac{e^z - z - 1}{z^2}$$

as a Mittag-Leffler function with two parameters.

PE 10.43 Let $z \in \mathbb{C}$ and α a complex parameter with $\mathrm{Re}(\alpha) > 0$. Denote the confluent hypergeometric function by ${}_1F_1(1; \alpha; z)$. Prove the following relation between the Mittag-Leffler function with two parameters and the confluent hypergeometric function:

$$\Gamma(\alpha)\mathbb{E}_{1,\alpha}(z) = {}_1F_1(1; \alpha; z) \,.$$

Note that both the confluent hypergeometric function and the Mittag-Leffler function have a parameter which is a constant.

PE 10.44 Let $\beta, \gamma \in \mathbb{C}$, and $z \in \mathbb{C}$. Generalize the relation obtained in **PE 10.43** to obtain

$${}_1F_1(\gamma, \beta; z) = \Gamma(\beta)\mathbb{E}^\gamma_{1,\beta}(z) \,,$$

where $\Gamma(\cdot)$ is the gamma function.

PE 10.45 Let $\lambda \in \mathbb{C}$, $\mathrm{Re}(\alpha) > 0$, $\beta > 0$, and $\mu > 0$. Prove the relation between two Mittag-Leffler functions with two parameters given by the following integral:

$$\frac{1}{\Gamma(\mu)} \int_0^z (z - \xi)^{\mu-1}\mathbb{E}_{\alpha,\beta}(\lambda\xi^\alpha)\xi^{\beta-1}\mathrm{d}\xi = z^{\mu+\beta-1}\mathbb{E}_{\alpha,\mu+\beta}(\lambda z^\alpha) \,.$$

PE 10.46 Let $\mathrm{Re}(\alpha) > 0$ and $\beta > 0$. Show that the difference

$$\mathbb{E}_{\alpha,\beta}(z) - z\mathbb{E}_{\alpha,\alpha+\beta}(z)\,,$$

where $\mathbb{E}_{\alpha,\beta}(z)$ and $\mathbb{E}_{\alpha,\alpha+\beta}(z)$ are Mittag-Leffler functions with two parameters, is independent of parameter α.

PE 10.47 Let $\alpha,\beta,\gamma \in \mathbb{C}$ with $\mathrm{Re}(\alpha) > 0$ and $z \in \mathbb{C}$. For $k = 1,2,\ldots$, and $\mathrm{Re}(\beta) > k$, show that

$$\left(\frac{\mathrm{d}}{\mathrm{d}z}\right)^k \left[z^{\beta-1}\mathbb{E}_{\alpha,\beta}^{\gamma}(z^{\alpha})\right] = z^{\beta-k-1}\mathbb{E}_{\alpha,\beta-k}^{\gamma}(z^{\alpha})\,,$$

where $\mathbb{E}_{\alpha,\beta}^{\gamma}(\cdot)$ is a Mittag-Leffler function with three parameters.

PE 10.48 Let $n,k \in \mathbb{N}$ and $\mathbb{E}_{n,k}(\cdot)$ the Mittag-Leffler function with two parameters. Show that

$$\sum_{k=1}^{n} z^{k-1}\mathbb{E}_{n,k}(z^n) = e^z\,,$$

independent of n.

PE 10.49 (Christoffel-Darboux Formula) Let $\alpha \in \mathbb{C}$ with $\mathrm{Re}(\alpha) > 0$; $x, y \in \mathbb{R}$ with $x \neq y$; $\beta > 0$, $\gamma > 0$ and $\mathbb{E}_{\alpha,\beta}(\cdot)$ the Mittag-Leffler function with two parameters. Show the so-called Christoffel-Darboux formula for Mittag-Leffler functions:

$$\int_0^z \xi^{\gamma-1}\mathbb{E}_{\alpha,\gamma}(x\xi^{\alpha})(z-\xi)^{\beta-1}\mathbb{E}_{\alpha,\beta}[y(z-\xi)^{\alpha}]\,\mathrm{d}\xi$$

$$= \frac{x\mathbb{E}_{\alpha,\beta+\gamma}(x\,z^{\alpha}) - y\mathbb{E}_{\alpha,\beta+\gamma}(yz^{\alpha})}{x-y}z^{\beta+\gamma-1}\,.$$

PE 10.50 Using the result obtained in **PE 10.49** and the l'Hôpital rule, discuss the particular case $y \to x$.

References

1. E. Capelas de Oliveira, *Funções Especiais com Aplicações* (Special Functions with Applications), Editora Livraria da Física, São Paulo (2005)
2. E. Capelas de Oliveira, *Solved Exercises in Fractional Calculus* (Springer, Switzerland, 2019)
3. S.A. Deif, E. Capelas de Oliveira, System of Cauchy fractional differential equations and new properties of Mittag-Leffler functions with matrix argument. J. Comput. Appl. Math. **406**, 113977 (2022). http://doi.org/10.1016/j.com.2021.113977

4. R. Gorenflo, A.A. Kilbas, F. Mainardi, S.V. Rogosin, *Mittag-Leffler Functions, Related Topics and Applications* (Springer, Heidelberg, 2014)
5. R. Hilfer (ed.) *Applications of Fractional Calculus in Physics* (World Scientific, Singapore, 2000)
6. F. Mainardi, *Fractional Calculus and Waves in Linear Viscoelasticity* (Imperial College Press, London, 2010)
7. K.S. Miller, B. Ross, *An Introduction to the Fractional Calculus and Fractional Differential Equations* (Wiley, New York, 1993)
8. M.D. Ortigueira, *Fractional Calculus for Scientists and Engineers* (Springer, Netherlands, 2011)
9. I. Podlubny, *Fractional Differential Equations*. Mathematics in Science and Engineering, Vol. 198 (Academic Press, San Diego, 1999)
10. J.A. Tenreiro Machado, V. Kiryakova, Historical survey: The chronicles of fractional calculus. Fract. Calc. Appl. Anal. **20**, 307–336 (2017)
11. J.A. Tenreiro Machado, V. Kiryakova, F. Mainardi, A poster about the recent history of fractional calculus. Fract. Calc. Appl. Anal. **13**, 329–334 (2010)
12. J.A. Tenreiro Machado, V. Kiryakova, F. Mainardi, A poster about the old hystory of fractional calculus. Fract. Calc. Appl. Anal. **13**, 447–454 (2010)

Chapter 11
Applications

Mathematics is a language.

1839 – Josiah Willard Gibbs – 1903

In this chapter, we present and solve several exercises involving techniques studied in the preceding chapters. We took the care of presenting the exercises in the same sequence as the techniques were presented in the chapters, so that the applications associated with a particular chapter do not involve formalisms and/or techniques presented in subsequent chapters [1–10]. Differently from what we did in Chaps. 1–10, we present the proposed exercises at the end of each section instead of collecting them at the end of the chapter.

The titles of the ten following sections correspond to the titles of the ten chapters already presented; in each section, we discuss three applications in the first three subsections, leaving some exercises to be solved by the reader.

11.1 Ordinary Differential Equations

We believe that the most canonical applications of homogeneous and nonhomogeneous, linear ordinary differential equations are the mechanical mass-spring problem and its electrical analog, the RLC circuit. As we have already presented these problems in Chap. 1, we discuss here three other applications.

11.1.1 Newton's Heat Transfer Law

SE 11.1 The differential equation describing heat transfer, due to Newton, appears in several other contexts such as solute diffusion, where it is known as Fick's [1829 – Adolf Eugen Fick – 1901] law. Here we will consider a body with no internal heat sources, at an initial temperature T_0, placed in a certain environment whose

© The Author(s), under exclusive license to Springer Nature Switzerland AG 2025

E. Capelas de Oliveira, J. E. Maiorino, *Analytical Methods in Applied Mathematics*, Problem Books in Mathematics, https://doi.org/10.1007/978-3-031-74794-6_11

temperature is $T < T_0$. A concrete example would be a cup of hot coffee whose temperature decreases until it reaches room temperature.

We know that, due to the exchange of heat between the body and the environment, the body temperature, $T(t)$, changes over time until it becomes equal to the ambient temperature T, supposed constant. This is expressed by the limit $\lim_{t\to\infty} T(t) = T$.

Newton's heat transfer law states that as $T(t)$ approaches T, the velocity with which $T(t)$ tends to T gradually decreases. Supposing that this velocity $\frac{d}{dt}T(t)$ is proportional to $T(t) - T$, the difference between the temperatures, the body temperature can then be described by the ordinary differential equation

$$\frac{d}{dt}T(t) = -k[T(t) - T],$$

where $k > 0$ is a proportionality constant. Find $T(t)$.

Solution: This is an ordinary, homogeneous, first-order and separable linear differential equation, which can be rewritten as

$$\frac{dT(t)}{T(t) - T} = -k\,dt.$$

Integrating both members, we obtain

$$\ln[T(t) - T] = -kt + \ln C,$$

where $C > 0$ is a constant. Solving for $T(t)$, we have

$$T(t) = C\exp(-kt) + T.$$

To obtain C we use the condition $T(0) = T_0$. We thus obtain

$$T(t) = (T_0 - T)e^{-kt} + T,$$

which is the desired result. Note that, as expected, $\lim_{t\to\infty} T(t) = T$.

11.1.2 Vertical Launch of a Body

SE 11.2 (a) Write the equation of motion for a mass m that is thrown vertically upward, knowing that the resistance of the air is proportional to its speed. (b) Solve the linear ordinary differential equation obtained in (a), assuming that the mass was launched with initial velocity V_0. (c) Obtain the maximum height attained by the mass.

Fig. 11.1 Diagram of the forces acting on the body launched vertically upward, near the Earth's surface, subject to gravity and to the resistance of the air

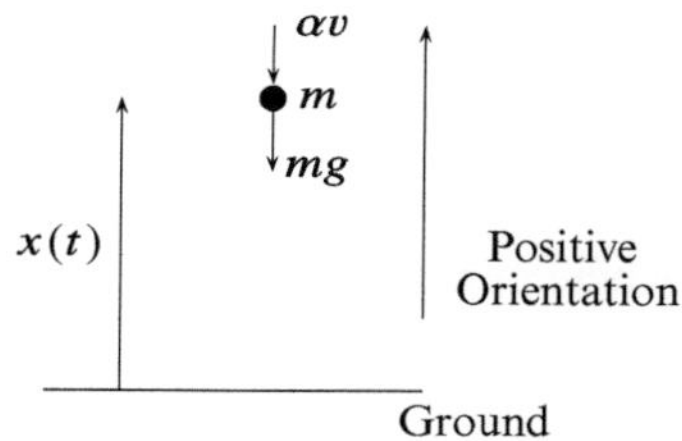

Solution: (a) We consider the launch point, the ground, as the origin of our coordinate system. The trajectory is oriented upward and we take the coordinate $x(t)$ as the distance between the mass and the origin, at time t, as in Fig. 11.1. Upon the mass m, there act two forces: the gravitational force, of constant magnitude $-mg$, where g is the gravity acceleration, and the resistance of the air, with magnitude $-\alpha v$, where v is the velocity of the mass in the direction of motion x and α is a proportionality constant. As the body is launched upwardly, both forces are directed toward the ground, as we can see in Fig. 11.1.

According to Newton's second law, we have *mass $\times$ acceleration = resultant*. That is,

$$m\frac{\mathrm{d}^2}{\mathrm{d}t^2}x(t) = -mg - \alpha v$$

$$= -mg - \alpha\frac{\mathrm{d}}{\mathrm{d}t}x(t). \tag{11.1}$$

Dividing both members of Eq. (11.1) by m, we obtain

$$\frac{\mathrm{d}^2}{\mathrm{d}t^2}x(t) + k\frac{\mathrm{d}}{\mathrm{d}t}x(t) = -g, \tag{11.2}$$

where we have introduced $k = \alpha/m$; this is the desired equation.

(b) As shown in Chap. 1, this is a nonhomogeneous second order linear ordinary differential equation with constant coefficients. To solve this differential equation, we can use the method of variation of parameters, but as the coefficients are constant, it is more convenient to apply the method of undetermined coefficients.

First, we solve the corresponding homogeneous ordinary differential equation, i.e.,

$$\frac{\mathrm{d}^2}{\mathrm{d}t^2}x(t) + k\frac{\mathrm{d}}{\mathrm{d}t}x(t) = 0.$$

Then, putting $\dfrac{\mathrm{d}x}{\mathrm{d}t} \equiv v(t)$ we have

$$\frac{\mathrm{d}}{\mathrm{d}t}v(t) + kv(t) = 0,$$

which is a separable ordinary differential equation, with solution

$$v(t) = c_1 \exp(-kt),$$

where c_1 is an arbitrary constant. Another integration yields

$$x(t) = -\frac{c_1}{k} \exp(-kt) + c_2 ,$$

where c_2 is another arbitrary constant. This is the general solution of the homogeneous linear, ordinary differential equation, since it contains two arbitrary constants.

Now, to obtain a particular solution of the corresponding nonhomogeneous linear ordinary differential equation, we use the method of undetermined coefficients. The nonhomogeneous term is a constant function, that is, a polynomial of degree zero. We then look for a particular solution with the form

$$x_p(t) = d_1 t + d_2 , \tag{11.3}$$

where d_1 and d_2 must be determined. Substituting Eq. (11.3) into the linear ordinary differential equation, Eq. (11.2), we get

$$kd_1 = -g \qquad \Rightarrow \qquad d_1 = \frac{-g}{k} ,$$

and this yields a particular solution given by

$$x_p(t) = -\frac{g}{k}t + d_2,$$

where d_2 is a constant. Finally, the general solution of the nonhomogeneous differential equation is given by the sum of the general solution of the homogeneous differential equation and a particular solution of the corresponding nonhomogeneous differential equation

$$x(t) = -\frac{c_1}{k} \exp(-kt) - \frac{g}{k}t + d_3 ,$$

where $d_3 = c_2 + d_2$ is a constant.

Using the known conditions, we can determine the values of constants c_1 and d_3. We know that $x(0) = 0$, because we considered the launch point the origin of the coordinate system; in the same way, choosing the origin of times on the instant when the body is launched, the initial velocity is $v(0) = v_0$. Introducing these conditions into the general solution, we get

$$x(0) = -\frac{c_1}{k} + d_3 = 0;$$

and

$$v(0) = c_1 - \frac{g}{k} = v_0 \,.$$

Solving this linear system for c_1 and d_3 and substituting the result into the general solution, we finally obtain

$$x(t) = -\frac{1}{k^2}(g + kv_0)[1 - \exp(-kt)] - \frac{g}{k}t \,, \qquad (11.4)$$

which is the desired solution.

(c) To determine the maximum height, we must first obtain the ascension time, i.e., the time the body takes to attain a velocity equal to zero, when there occurs the inversion in the sense of motion. Differentiating the solution in Eq. (11.4) with respect to t, we obtain an expression for the velocity:

$$v(t) = \frac{1}{k}(g + kv_0)\exp(-kt) - \frac{g}{k} \,.$$

We must obtain the value of t for which

$$\frac{1}{k}(g + kv_0)\exp(-kt) - \frac{g}{k} = 0 \,,$$

which is given by

$$t = \frac{1}{k}\ln\left(\frac{g + kv_0}{g}\right) \,.$$

Substituting this value of t into the equation for $x(t)$, Eq. (11.4), we get the value of the maximum height, i.e.,

$$\begin{aligned}
x_{\max} &= \frac{1}{k^2}(g + kv_0)\left(1 - \frac{g}{g + kv_0}\right) - \frac{g}{k}\frac{1}{k}\ln\left(\frac{g + kv_0}{g}\right) \\
&= \frac{1}{k}\left[v_0 - \frac{g}{k}\ln\left(\frac{g + kv_0}{g}\right)\right] \,,
\end{aligned}$$

which is the desired result.

11.1.3 Falling Body with Air Resistance

SE 11.3 (a) Determine the equation of motion of a body of mass m that falls with initial velocity equal to zero, in the atmosphere, considering the air resistance

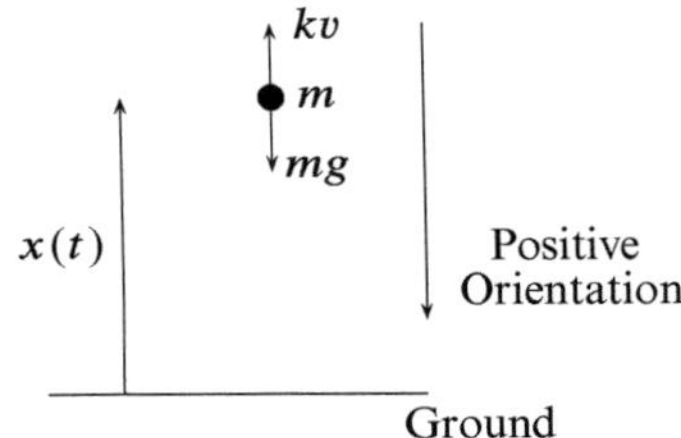

Fig. 11.2 Forces acting on a body in free fall, subject to air resistance, with the same system of coordinates used in the resolution of **SE 11.2**

proportional to the square of its velocity. (b) Show that, when the air resistance is equal to zero, the law of motion is independent of the mass of the body.

Solution: (a) Consider that a body is abandoned from a height h, with initial velocity $v_0 = 0$. Take $x(t)$ to be the space traversed by the body since the moment it was abandoned until time t. Choose a downward orientation for the trajectory (cf. Fig. 11.2), so that the weight force mg has a positive sign and the resistance force, oriented upwardly, has negative sign, with k the constant of proportionality between the resistance force and the square of velocity.

Using Newton's second law, we obtain the following ordinary differential equation:

$$mg - kv^2(t) = m\frac{\mathrm{d}^2}{\mathrm{d}t^2}x(t)\,. \tag{11.5}$$

Substituting $v(t) = \dfrac{\mathrm{d}}{\mathrm{d}t}x(t)$ into Eq. (11.5), without making explicit the dependence on variable t, we obtain

$$m\frac{\mathrm{d}^2 x}{\mathrm{d}t^2} = mg - k\left(\frac{\mathrm{d}x}{\mathrm{d}t}\right)^2,$$

which is a nonlinear ordinary differential equation. To solve this nonlinear second-order ordinary differential equation, we first transform it into a linear ordinary differential equation.

Indeed, recalling that $\mathrm{d}x/\mathrm{d}t = v(t)$, we can rewrite the nonlinear ordinary differential equation in the form

$$m\frac{\mathrm{d}v}{\mathrm{d}t} = mg - kv^2,$$

or

$$\frac{\mathrm{d}v}{\mathrm{d}t} = g - \frac{k}{m}v^2 = g\left(1 - \frac{k}{mg}v^2\right)\,.$$

This is a separable first-order ordinary differential equation. To integrate it we write

$$\frac{dv}{1 - \frac{k}{mg}v^2} = g\,dt\,.$$

The left-hand side of the above ordinary differential equation can be expanded by means of partial fractions, yielding

$$\frac{dv}{1 - \sqrt{\frac{k}{mg}}\,v} + \frac{dv}{1 + \sqrt{\frac{k}{mg}}\,v} = 2g\,dt\,. \tag{11.6}$$

Integrating both sides of Eq. (11.6), we get

$$\sqrt{\frac{mg}{k}}\ln\left(1 + \sqrt{\frac{k}{mg}}\,v\right) - \sqrt{\frac{mg}{k}}\ln\left(1 - \sqrt{\frac{k}{mg}}\,v\right) = 2gt + C\,,$$

where C is an arbitrary constant of integration. Solving it for variable v, we have

$$\sqrt{\frac{k}{mg}}\,v = \frac{\exp\left[\sqrt{\frac{k}{mg}}(2gt + C)\right] - 1}{\exp\left[\sqrt{\frac{k}{mg}}(2gt + C)\right] + 1}$$

$$= \frac{\exp\left[\sqrt{\frac{k}{mg}}\left(gt + \frac{C}{2}\right)\right] - \exp\left[-\sqrt{\frac{k}{mg}}\left(gt + \frac{C}{2}\right)\right]}{\exp\left[\sqrt{\frac{k}{mg}}\left(gt + \frac{C}{2}\right)\right] + \exp\left[-\sqrt{\frac{k}{mg}}\left(gt + \frac{C}{2}\right)\right]}\,.$$

Recalling the hyperbolic relation

$$\frac{\sinh u}{\cosh u} = \tanh u = \frac{e^u - e^{-u}}{e^u + e^{-u}}\,,$$

we finally get

$$v(t) = \sqrt{\frac{mg}{k}}\tanh\left[\sqrt{\frac{k}{mg}}\left(gt + \frac{C}{2}\right)\right]\,. \tag{11.7}$$

To obtain $x(t)$ we integrate the last expression

$$x(t) = \sqrt{\frac{mg}{k}}\int \frac{\sinh\left[\sqrt{\frac{k}{mg}}\left(gt + \frac{C}{2}\right)\right]}{\cosh\left[\sqrt{\frac{k}{mg}}\left(gt + \frac{C}{2}\right)\right]}\,dt$$

$$= \sqrt{mgk}\ln\left\{\cosh\left[\sqrt{\frac{k}{mg}}\left(gt + \frac{C}{2}\right)\right]\right\}\left(\sqrt{\frac{k}{mg}}\,g\right)^{-1} + D\,,$$

which can be written as

$$x(t) = \frac{m}{k} \ln \left\{ \cosh \left[\sqrt{\frac{k}{mg}} \left(gt + \frac{C}{2} \right) \right] \right\} + D \, ,$$

where D is another arbitrary integration constant. Imposing the condition that $x(0) = 0$ at the origin of times, we have

$$D + \frac{m}{k} \ln \left\{ \cosh \left(\frac{D}{2} \sqrt{\frac{k}{mg}} \right) \right\} = 0 \, .$$

As the body was *abandoned*, $v_0 = 0$, and Eq. (11.7) then implies that $C = 0$. This result implies that $D = 0$ and we can then write

$$x(t) = \frac{m}{k} \ln \left[\cosh \left(\sqrt{\frac{kg}{m}} t \right) \right] \, . \tag{11.8}$$

(b) To show this item, we must substitute $k = 0$ into the solution given by Eq.(11.8); the result will be an indetermination. So, in order to obtain the value of the limit $\overline{x}(t)$ of $x(t)$ when k goes to zero, we use the l'Hôpital rule, which yields

$$\overline{x}(t) = m \lim_{k \to 0} \frac{1}{k} \ln \left[\cosh \left(\sqrt{\frac{kg}{m}} t \right) \right]$$

$$= m \lim_{k \to 0} \frac{\sinh \sqrt{\frac{kg}{m}} t}{\cosh \sqrt{\frac{kg}{m}} t} \sqrt{\frac{g}{m}} t \frac{1}{2} \frac{1}{\sqrt{k}}$$

$$= \frac{t}{2} \sqrt{gm} \lim_{k \to 0} \frac{\sinh \sqrt{\frac{kg}{m}} t}{\cosh \sqrt{\frac{kg}{m}} t} \frac{\sqrt{\frac{g}{m}} t}{\sqrt{\frac{g}{m}} t}$$

$$= \frac{g}{2} t^2 \lim_{k \to 0} \frac{\sinh \sqrt{\frac{kg}{m}} t}{\sqrt{\frac{kg}{m}} t} \frac{1}{\cosh \sqrt{\frac{kg}{m}} t}$$

$$= \frac{g}{2} t^2 \, ,$$

i.e., we have a free fall, a result independent of the mass.

11.1.4 Proposed Exercises

PE 11.1 Suppose that a body with mass m falls freely on the Earth, which has mass M. According to Newton's law of gravitation, the force acting on the body is given by

$$F = G \frac{mM}{r^2},$$

where G is a constant of proportionality and r is the distance from the center of the body to the center of the Earth. Note that when $r = R$, the radius of the Earth, F, is equal to the weight of the body. (a) Write a nonlinear ordinary differential equation for the motion of the body with the form

$$m \frac{\mathrm{d}^2}{\mathrm{d}t^2} r(t) = -gm \frac{R^2}{r^2(t)},$$

where g is the acceleration of gravity on the surface of the Earth. (b) Show that the time taken by the body to fall on the surface of the Earth is given by

$$t = \sqrt{\frac{r_0}{2gR^2}} \left[\sqrt{r_0 R - R^2} + \frac{r_0}{2} \arcsin\left(\frac{2R - r_0}{r_0}\right) + \frac{\pi r_0}{4} \right],$$

where r_0 is the initial distance from the body to the center of the Earth.

PE 11.2 As we have already mentioned, the mechanical system composed of a mass attached to a spring is analogous to an RLC electrical circuit, in the sense that they obey the same linear, second-order ordinary differential equation, as shown in Table 11.1. Using the notation shown in that table, (a) show that, as R, L, and C are positive constants, all solutions of the homogeneous ordinary differential equation go to zero when $t \to \infty$. (b) Obtain a particular solution for the case $E(t) = E_0 \sin \omega t$, where E_0 and ω are positive constants. (c) Since m, c, and k are positive constants, can we say about the spring-mass system the same we said in item (a)? (d) An electrical system (or the corresponding mechanical system) is said *superdamped*, *critically damped*, or *subdamped*, if $R^2 - 4L/C > 0$, $R^2 - 4L/C = 0$ or $R^2 - 4L/C < 0$, respectively. Find the corresponding solutions for each of these three cases for the homogeneous ordinary differential equation—that with $E(t) = 0$— and for the problem considered in item (b).

PE 11.3 In several problems in which central forces play a fundamental rule, as in the Kepler problem, the study of limited and periodic motions is done in terms of the linear second-order ordinary differential equation

$$\frac{U'(r)}{3u'(r) + rU''(r)} = C,$$

Table 11.1 Relation between the spring-mass system and the RLC electrical circuit

Mechanical spring-mass system			RLC Electric circuit		
$m\ddot{s} + c\dot{s} + ks = F(t)$			$L\ddot{Q} + R\dot{Q} + (1/C)Q = E(t)$		
s	$\rightarrow$	displacement	Q	$\rightarrow$	charge
m	$\rightarrow$	mass	L	$\rightarrow$	inductance
c	$\rightarrow$	dumping constant	R	$\rightarrow$	resistance
k	$\rightarrow$	elastic constant	$1/C$	$\rightarrow$	elastance
$\dot{s}$	$\rightarrow$	velocity	$\dot{Q}$	$\rightarrow$	current
$F(t)$	$\rightarrow$	imposed force	$E(t)$	$\rightarrow$	imposed voltage

where $c > 0$ is a constant and $U(r)$ is the potential, which depends only on the radial coordinate that characterizes the central force. Introducing another constant $A = 3 - 1/C$, we obtain the linear second-order ordinary differential equation:

$$rU''(r) + Au'(r) = 0.$$

(a) Given a real constant a, look for solutions of this homogeneous second-order linear ordinary differential equation with the form

$$u_1(r) = a\,r^\alpha$$

where α is another real constant. (b) Discuss the two particular cases $\alpha = -1$ (Kepler potential) and $\alpha = 2$ (harmonic oscillator). (c) Look for solutions with the form $U_2(r) = b \ln r$, where b is a real constant. (d) Obtain the general solution, that is, a solution containing two arbitrary constants.

11.2   Power Series and the Frobenius Method

The Taylor series and the Frobenius method are useful to solve homogeneous linear ordinary differential equations. Here we discuss, as concrete applications, the resolution of a first-order ordinary differential equation using the Taylor (MacLaurin) series and of a linear second-order ordinary differential equation by means of the Frobenius method. Also, an exercise involving the method of variation of parameters is discussed in detail.

11.2.1   Linear First-Order Ordinary Differential Equation

SE 11.4 Consider the linear first-order ordinary differential equation

$$\frac{\mathrm{d}}{\mathrm{d}x}y(x) = x^2 y(x)\,.$$

(a) Find a solution in power series. (b) Determine the radius of convergence of this series. (c) Write the result in terms of an elementary function.

Solution: (a) Consider a power series in variable x with the following form:

$$y(x) = \sum_{k=0}^{\infty} a_k x^k\,,\tag{11.9}$$

where a_k are coefficients to be determined and where we assume that $a_0 \neq 0$.

Differentiating formally this series with respect to x, we have

$$\frac{\mathrm{d}}{\mathrm{d}x}y(x) = \sum_{k=1}^{\infty} k a_k x^{k-1}\,.\tag{11.10}$$

Substituting Eqs. (11.9) and (11.10) into the ordinary differential equation, we get

$$\sum_{k=1}^{\infty} k a_k x^{k-1} = x^2 \sum_{k=0}^{\infty} a_k x^k = \sum_{k=0}^{\infty} a_k x^{k+2}\,.$$

Let us introduce a change of indices in the third sum, $k \to k - 3$. Then, the last equation can be rewritten as

$$\sum_{k=1}^{\infty} k a_k x^{k-1} = \sum_{k=3}^{\infty} a_{k-3} x^{k-1}\,,$$

or, equating the coefficients of equal powers,

$$a_1 + 2a_2 x + \sum_{k=3}^{\infty} k a_k x^{k-1} = \sum_{k=3}^{\infty} a_{k-3} x^{k-1}\,.$$

From this equality we conclude that $a_1 = a_2 = 0$ and we obtain the recurrence formula

$$k a_k = a_{k-3} \quad \text{for } k \geq 3\,.$$

Since $a_0 \neq 0$ we have $a_3 = a_0/3$ and $a_4 = a_5 = 0$. Evaluating explicitly other terms, we find $a_6 = a_3/6 = a_0/3!\,3$ and $a_7 = a_8 = 0$. Thus,

$$y(x) = a_0 \left(1 + \frac{x^3}{3} + \frac{x^6}{2!\,3^2} + \frac{x^9}{3!\,3^3} + \cdots\right) = a_0 \sum_{k=0}^{\infty} \frac{x^{3k}}{k!\,3^k}\,.$$

(b) To find the radius of convergence we use the *ratio criterion*. By this criterion, the radius of convergence of the series obtained above is

$$R = \lim_{k\to\infty}\left|\frac{a_k}{a_{k+1}}\right| = \lim_{k\to\infty}\frac{x^{3k}}{k!3^k}\cdot\frac{(k+1)!3^{k+1}}{x^{3k+3}} = \lim_{k\to\infty}\frac{(k+1)3}{x^3} = \infty.$$

Therefore, the series converges for all values of x.

(c) Finally, we identify this series with the series for an exponential function, given by

$$y(x) = a_0\sum_{k=0}^{\infty}\frac{1}{k!}\left(\frac{x^3}{3}\right)^k = a_0\exp\left(\frac{x^3}{3}\right),$$

where a_0 is an arbitrary constant.

11.2.2   Schrödinger Equation for the Harmonic Oscillator

SE 11.5 Discuss the Schrödinger ordinary differential equation for the quantum harmonic oscillator,

$$-\frac{\hbar^2}{2m}\frac{d^2}{dx^2}\psi(x) + \frac{kx^2}{2}\psi(x) = E\psi(x),$$

where $\hbar$, m and k are positive constants, E is the energy, and $\psi(x)$ is the so-called wave function of a particle, using the Frobenius method.

Solution: Let $\psi(x) = \psi$. Before we apply the Frobenius series, we simplify this linear ordinary differential equation, writing it as

$$\frac{d^2}{dx^2}\psi + \left(\frac{2mE}{\hbar^2} - \frac{mk}{\hbar^2}x^2\right)\psi = 0. \tag{11.11}$$

To simplify future calculations, we also introduce the parameters

$$\beta = \frac{2E}{\hbar\omega} \qquad \text{and} \qquad \alpha^2 = \frac{mk}{\hbar^2} = \frac{m^2\omega^2}{\hbar^2},$$

so that

$$\alpha\beta = \frac{2Em}{\hbar^2},$$

and Eq. (11.11) is reduced to

$$\frac{d^2}{dx^2}\psi + (\alpha\beta - \alpha^2 x^2)\psi = 0.$$

Introducing the change of independent variable $\xi = \sqrt{\alpha}x$, we finally obtain a dimensionless ordinary differential equation equivalent to the original one:

$$\frac{d^2}{d\xi^2}\psi + (\beta - \xi^2)\psi = 0.$$

Usually, this type of ordinary differential equation is solved directly by means of the Frobenius method. Here, before we apply this method, we use the physical fact that the wave function ψ must tend to zero when $\xi \to \pm\infty$. To take advantage of this fact, we introduce the function

$$\psi(\xi) = e^{-\xi^2/2}\, H(\xi)$$

in the ordinary differential equation above, and we obtain the following ordinary differential equation for the dependent variable $H(\xi)$:

$$\frac{d^2}{d\xi^2}H(\xi) - 2\xi\frac{d}{d\xi}H(\xi) + (\beta - 1)H(\xi) = 0. \tag{11.12}$$

This ordinary differential equation is known as *Hermite equation*.

Now, applying the Frobenius method, we assume for $H(\xi)$ a solution with the form

$$H(\xi) = \sum_{m=0}^{\infty} c_m\xi^m = c_0 + c_1\xi + c_2\xi^2 + c_3\xi^3 + \cdots$$

Introducing this series into the ordinary differential equation Eq. (11.12), differentiating term by term and factoring terms with the same power, we find

$$\sum_{m=0}^{\infty} [(m+1)(m+2)c_{m+2} + (\beta - 1 - 2m)c_m]\xi^m = 0.$$

Therefore, in order for $H(\xi)$ to be a solution of the given differential equation, we must have

$$c_{m+2} = \frac{2m + 1 - \beta}{(m+1)(m+2)}c_m,$$

which permits us to calculate the coefficients with $m \geq 2$ in terms of the coefficients c_0 and c_1, which are arbitrary and must be determined from the initial conditions.

We note that our result is comprised of two power series, an *odd* series, involving the coefficient a_0, and an *even* series, involving the coefficient a_1. In order to obtain solutions which are regular at the origin, we look for polynomial solutions. For this we must have

$$\frac{\beta - 1}{2} = n,$$

where $n = 0, 1, 2, \ldots$. From this relation and from the definition of the parameter β, we get

$$E_n = (n + 1/2)\hbar\omega,$$

which are the possible values of the energy of the quantum harmonic oscillator (discrete values), different from the continuous spectrum predicted by classical mechanics.

Imposing this last condition, we obtain for the recurrence relation

$$c_{m-2} = -\frac{m(m-1)}{2(n-m+2)}c_m,$$

for $m \le n$. Choosing now $c_n = 2^n$ and proceeding iteratively, we obtain

$$c_{n-2} = -2^{n-2}\frac{n(n-1)}{1!};$$

$$c_{n-4} = 2^{n-4}\frac{n(n-1)(n-2)(n-3)}{2!};$$

or, substituting into the expression for the series,

$$H_n(\xi) = (2\xi)^n - \frac{n(n-1)}{1!}(2\xi)^{n-2} + \frac{n(n-1)(n-2)(n-3)}{2!}(2\xi)^{n-4} + \cdots$$

to which we must add $c_1\xi$ if n is odd and c_0 if n is even. We thus have a family of solutions, one for each value of n:

$$H_0(\xi) = 1, \quad H_1(\xi) = 2\xi, \quad H_2(\xi) = 4\xi^2 - 2, \ldots$$

These are the classical *Hermite polynomials*, which are normalized to unity by means of the formula

$$\int_{-\infty}^{\infty} e^{-x^2}[H_n(x)]^2 \, dx = 2^n n! \sqrt{\pi}.$$

11.2.3   Variation of Parameters

SE 11.6 Let $-\pi/2 < x < \pi/2$. Obtain a particular solution for the nonhomogeneous second-order linear ordinary differential equation

$$\frac{d^2}{dx^2} y(x) + y(x) = \sec x$$

using the method of variation of parameters.

Solution: We know that the general solution of the corresponding homogeneous second-order linear ordinary differential equation is given by

$$y(x) = A \sin x + B \cos x \,,$$

where A and B are two arbitrary constants. We look for a particular solution $y_P(x)$ of the nonhomogeneous second-order linear ordinary differential equation by imposing that

$$y_P(x) = A(x) \sin x + B(x) \cos x$$

where $A(x)$ and $B(x)$ are to be determined. To this end, we start evaluating the first derivative, that is,

$$y_P'(x) = A'(x) \sin x + A(x) \cos x + B'(x) \cos x - B(x) \sin x \,.$$

As we are free to impose another condition (the ordinary differential equation is a second-order equation), we consider (for simplicity)

$$A'(x) \sin x + B'(x) \cos x = 0 \,.$$

This allows us to work with the first derivative involving only $A(x)$ and $B(x)$. So, evaluating the second derivative, introducing it into the nonhomogeneous second-order linear ordinary differential equation and simplifying, we get

$$A'(x) \cos x - B'(x) \sin x = \sec x \,.$$

Using the two last equations, we obtain a linear system involving $A'(x)$ and $B'(x)$, that is,

$$\begin{cases} A'(x) \sin x + B'(x) \cos x = 0 \\ A'(x) \cos x - B'(x) \sin x = \sec x \end{cases}$$

whose solution can be written as

$$A'(x) = 1 \qquad \text{and} \qquad B'(x) = -\tan x \,.$$

We now integrate these first-order ordinary differential equations, obtaining

$$A(x) = x + \alpha_1 \qquad \text{and} \qquad B(x) = \ln(\cos x) + \alpha_2 \,,$$

where α_1 and α_2 are arbitrary constants.

Using these two values for $A(x)$ and $B(x)$, we finally obtain the particular solution of the nonhomogeneous second-order linear ordinary differential equation:

$$y_P(x) = x \sin x + \cos x \, \ln(\cos x) \,.$$

Note that, as this is a particular solution of the nonhomogeneous equation, it does not contain any arbitrary constants, which appear only in general solutions of homogeneous equations.

11.2.4 Proposed Exercises

PE 11.4 (a) Using power series, solve the ordinary differential equation

$$(x - 1)\frac{\mathrm{d}}{\mathrm{d}x} y(x) + 2y(x) = 0 \,.$$

(b) Find the radius of convergence of the resulting power series and express it in terms of elementary functions.

PE 11.5 Using the Frobenius method, discuss the Schrödinger differential equation treated above without making physical considerations, i.e., substitute the series directly into the ordinary differential equation

$$\frac{\mathrm{d}^2}{\mathrm{d}\xi^2}\psi(\xi) + (\beta - \xi^2)\psi(\xi) = 0 \,,$$

where $\beta = 2E/\hbar\omega$. Compare the results.

PE 11.6 Obtain a particular solution for the nonhomogeneous second-order linear ordinary differential equation with constant coefficients:

$$\frac{\mathrm{d}^2}{\mathrm{d}x^2}y(x) + 6\frac{\mathrm{d}}{\mathrm{d}x}y(x) + 13y(x) = 60\cos x + 26 \,.$$

11.3   Laurent Series and Residues

In this section we present an application of the residue theorem (contour integrals) to calculate the sum of a series, together with two applications to the calculation of real integrals.

11.3.1   Summing Series by Means of Contour Integrals

SE 11.7 (a) Evaluate the following sum:

$$\sum_{k=1}^{\infty} \frac{2x}{x^2 + k^2 \pi^2} .$$

(b) Using the result obtained in the preceding item, express $(\sin x)/x$ as an infinite product.

Solution: Before we solve the proposed exercise, we prove a theorem concerning the properties of the two functions $\pi \cotg \pi z$ and $\pi \cosec \pi z$. Both functions have simple poles at the zeros of the function $\sin \pi z$, i.e., for $z = k = 0, \ \pm 1, \ \pm 2, \ldots$, with residues given, respectively, by

$$\frac{\pi \cos k\pi}{[d(\sin \pi z)/dz]_{z=k}} = 1 \text{ for } \pi \cotg \pi z$$

and

$$\frac{\pi}{[d(\sin \pi z)/dz]_{z=k}} = (-1)^k \text{ for } \pi \cosec \pi z .$$

From these properties we can prove the following theorem.

Theorem 11.1 *Suppose that $f(z)$ is a meromorphic function and Γ is a contour that encircles the zeros $m, \ m + 1, \ldots k$ of $\sin \pi z$. Denote by $\sum \operatorname{Res} f$ the sum of the residues of the integrand, in expressions* (a) *or* (b) *below, at the poles of $f(z)$ in Γ. Then:*

$$\text{(a)} \qquad \sum_{r=m}^{k} f(r) = \frac{1}{2\pi i} \int_{\Gamma} \pi \cotg \pi z f(z) \, dz - \sum \operatorname{Res} f;$$

$$\text{(b)} \quad \sum_{r=m}^{k} (-1)^k f(r) = \frac{1}{2\pi i} \int_{\Gamma} \pi \cosec \pi z f(z) \, dz - \sum \operatorname{Res} f .$$

Proof: Using the definition of residue, we can write

(a) $\quad \int_{\Gamma} \pi \cot g \, \pi z f(z) \, dz = 2\pi i \sum \left[\text{Residues at the poles of the integrand}\right]$

$$= 2\pi i \left[\sum \text{Res} \, f + \sum_{r=m}^{k} f(r)\right],$$

since the residue of the integrand at the poles $z = r$ of $\pi \cot \pi z$ is $f(r)$.

In the same way we obtain

(b) $\quad \int_{\Gamma} \pi \csc \pi z f(z) \, dz = 2\pi i \left[\sum \text{Res} \, f + \sum_{r=m}^{k} (-1)^r f(r)\right].$

We now turn to the solution of the proposed exercise. Consider the function

$$f(z) = \frac{2x}{x^2 + \pi^2 z^2},$$

for which we can write

$$\sum_{k=-N}^{N} \frac{2x}{x^2 + k^2\pi^2} = \frac{1}{2\pi i} \int_{\gamma} (\pi \cot g \, \pi z) f(z) \, dz -$$

$$- \sum \text{Residues of the integrand at the poles of } f(z),$$

where Γ is a contour that goes to infinity, in all directions, as $N \to \infty$, and which does not pass on any of the singularities of the integrand. Thus, the integral goes to zero when $N \to \infty$. Then, as the residue of the integrand in each pole $z = \pm ix/\pi$ is the same and the series is absolutely convergent, we have

$$\lim_{N \to \infty} \sum_{k=-N}^{N} = \sum_{k=-\infty}^{\infty},$$

so that we can write

$$\sum_{k=-\infty}^{\infty} \frac{2x}{x^2 + k^2\pi^2} = -2\pi \cot g \left(\frac{\pi i x}{\pi}\right) \frac{2x}{2\pi^2 i x/\pi},$$

or, equivalently,

$$\sum_{k=-\infty}^{\infty} \frac{2x}{x^2 + k^2\pi^2} = 2 \coth x.$$

Finally, we have for the sum

$$\sum_{k=1}^{\infty} \frac{2x}{x^2 + k^2\pi^2} = \coth x - \frac{1}{x}.$$

Second, we integrate the above result in variable x, obtaining

$$\sum_{k=1}^{\infty} \ln(x^2 + k^2\pi^2) = \ln\left[\frac{A \sinh x}{x}\right],$$

where A is an integration constant, determined by means of the limit $x \rightarrow 0$, namely,

$$\ln A = \sum_{k=1}^{\infty} \ln(k^2\pi^2).$$

Thus, we get

$$\frac{\sinh x}{x} = \prod_{k=1}^{\infty} \left(1 + \frac{x^2}{k^2\pi^2}\right),$$

which is a convergent product for all real or complex x. Taking $x \Rightarrow ix$, we finally have

$$\frac{\sin x}{x} = \prod_{k=1}^{\infty} \left(1 - \frac{x^2}{k^2\pi^2}\right)$$

which is the desired result.

11.3.2   Real Integral

SE 11.8 Use complex variables to evaluate the integral

$$\int_0^{\infty} \frac{dx}{1 + x^2}.$$

Solution: In order to evaluate this real integral with the help of complex variables, let us consider the following integral on the complex plane:

$$\oint_C \frac{dz}{1 + z^2}, \tag{11.13}$$

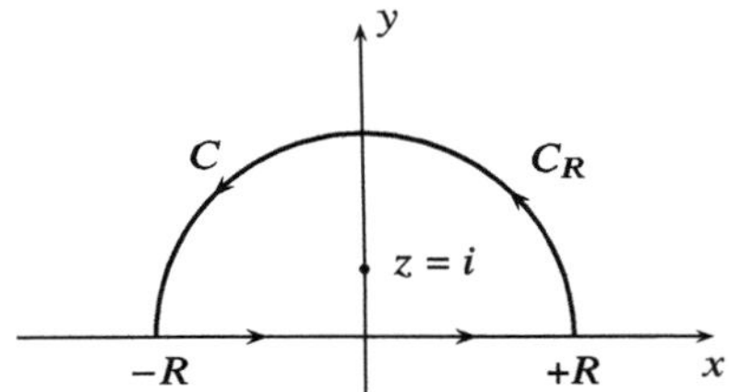

Fig. 11.3 Contour for the integral in Eq. (11.13)

where $z = x + iy$ and $x, y \in \mathbb{R}$. The contour C, oriented in the positive sense (counterclockwise), is formed by a semicircumference C_R of radius $R > 1$ centered at the origin, on the upper half-plane, and a line segment $[-R, R]$, as in Fig. 11.3.

The singularities of the integrand are simple poles located at $z = i$ and $z = -i$. Since only the point $z = i$ lies within C, we have, by the residue theorem

$$\int_{-R}^{R} \frac{dx}{1 + x^2} + \int_{C_R} \frac{dz}{1 + z^2} = 2\pi i \operatorname{Res}(z = i). \tag{11.14}$$

The same result will be obtained if we consider a semicircumference on the lower half-plane; in this case, the pole contributing to the residues will be the one at $z = -i$.

Using Jordan's lemma, we can show that the integral on C_R goes to zero as $R \to \infty$. To this end, we consider a parameterization $z = R\,e^{i\theta}$, with $0 < \theta < \pi$, that allows us to write

$$|1 + z^2| \leq 1 + |z^2| = 1 + R^2.$$

We can then write for the modulus of the integral:

$$\left| \int_{C_R} \frac{dz}{1 + z^2} \right| \leq \int_0^{\pi} \frac{R\,d\theta}{1 + R^2} = \frac{\pi R}{1 + R^2}.$$

It is then clear that, in the limit $R \to \infty$, the integral on C_R goes to zero. Now, taking the limit $R \to \infty$ in Eq. (11.14) and considering the expression for the limit, we can write

$$\int_{-\infty}^{\infty} \frac{dx}{1 + x^2} = 2\pi i \lim_{z \to i} \left[(z - i)\frac{1}{1 + z^2} \right] = 2\pi i \lim_{z \to i} \left(\frac{1}{i + z} \right) = \pi.$$

Since the integrand in the previous equation is an even function and the integration range is symmetric, we can write

$$\int_{-\infty}^{\infty} \frac{dx}{1 + x^2} = 2 \int_0^{\infty} \frac{dx}{1 + x^2} = \pi,$$

so that

$$\int_0^\infty \frac{dx}{1+x^2} = \frac{\pi}{2},$$

which is the desired result. Note that an identical result can be obtained using the trigonometric substitution $x = \tan\theta$.

11.3.3   Meromorphic Function

SE 11.9 Integrate the function $f(z) = z/(\lambda - e^{-iz})$ along a rectangular contour with vertices at $\pm\pi$, $\pm\pi + iR$ (see Fig. 11.4) to show that

$$\frac{\lambda}{\pi} \int_0^\pi \frac{x \sin x}{\lambda^2 - 2\lambda \cos x + 1}\, dx = \begin{cases} \ln(1 + \lambda) & \text{if } 0 < \lambda < 1, \\ \ln\left(1 + \dfrac{1}{\lambda}\right) & \text{if } \lambda > 1, \end{cases}$$

where λ is a positive real number.
Solution: The function $f(z)$ has poles at

$$\lambda - e^{-iz} = 0,$$

that is, at $z = i \ln \lambda$. Then: (a) if $\lambda > 1$ we have a pole at $z = i \ln \lambda$, inside the contour and (b) if $0 < \lambda < 1$ we have a pole at $z = -i \ln \lambda$, outside the contour.

Integrating along the contour, we have

$$\oint_\Gamma \frac{z\, dz}{\lambda - e^{-iz}} = \int_{-\pi}^\pi \frac{x}{\lambda - e^{-ix}}\, dx + \int_0^R \frac{(\pi + iy)}{\lambda - e^{-i(\pi + iy)}} i\, dy +$$

$$+ \int_\pi^{-\pi} \frac{(x + iR)}{\lambda - e^{-i(x+iR)}}\, dx + \int_R^0 \frac{(-\pi + iy)}{\lambda - e^{-i(-\pi + iy)}} i\, dy$$

$$= \int_{-\pi}^\pi \frac{x\, dx}{\lambda - e^{-ix}} + \int_0^R \frac{\pi i\, dy}{\lambda + e^y} + \int_\pi^{-\pi} \frac{x\, dx}{\lambda - e^{-ix+R}} +$$

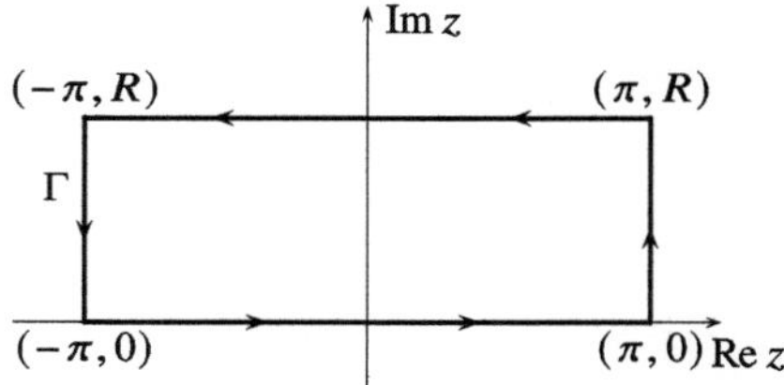
Fig. 11.4 Contour of integration used in **SE 11.9**

$$+ \int_{\pi}^{-\pi} \frac{i R \, dx}{\lambda - e^{-ix+R}} + \int_0^R \frac{\pi i \, dy}{\lambda + e^y} \,.$$

In the limit $R \to \infty$ these integrals reduce to

$$\oint_\Gamma \frac{z \, dz}{\lambda - e^{-iz}} = \int_{-\pi}^{\pi} \frac{x \, dx}{\lambda - e^{-ix}} + 2\pi i \int_0^\infty \frac{dy}{\lambda + e^y} \,,$$

which is equivalent to

$$\int_{-\pi}^{\pi} \frac{x(\lambda - \cos x)}{\lambda^2 - 2\lambda \cos x + 1} \, dx = -i \int_{-\pi}^{\pi} \frac{x \sin x}{\lambda^2 - 2\lambda \cos x + 1} \, dx + 2\pi i \int_0^\infty \frac{dy}{\lambda + e^y} \,. \tag{11.15}$$

The integral on the left-hand side is null, since the integrand is odd and the interval is symmetric. The same argument, applied to the second integral, permits us to rewrite Eq. (11.15) in the form

$$0 = -2i \int_0^{\pi} \frac{x \sin x}{\lambda^2 - 2\lambda \cos x + 1} dx + 2\pi i \int_0^\infty \frac{dy}{\lambda + e^y} \,.$$

We then use the residue theorem. We will first consider the case $\lambda > 1$; then

$$\int_0^{\pi} \frac{x \sin x}{\lambda^2 - 2\lambda \cos x + 1} \, dx = \frac{\pi}{\lambda} \ln(\lambda + 1) - \frac{\pi}{\lambda} \ln \lambda,$$

where the first term on the second member of the equality is obtained by evaluating the integral, while the second term comes from the residue. Therefore,

$$\frac{\lambda}{\pi} \int_0^{\pi} \frac{x \sin x}{\lambda^2 - 2\lambda \cos x + 1} \, dx = \ln\left(1 + \frac{1}{\lambda}\right) \,.$$

In the case $0 < \lambda < 1$, no term contributes to the residue; hence

$$\frac{\lambda}{\pi} \int_0^{\pi} \frac{x \sin x}{\lambda^2 - 2\lambda \cos x + 1} \, dx = \ln(1 + \lambda) \,,$$

which is desired result.

11.3.4   Proposed Exercises

PE 11.7 Show that, for $a > b > 0$,

$$\int_0^{2\pi} \frac{d\theta}{a^2 \cos^2 \theta + b^2 \sin^2 \theta} = \frac{2\pi}{ab} \,.$$

PE 11.8 For p and q positive integers such that $p > q + 1$, show that

$$\int_0^\infty \frac{x^q}{x^p + 1}\, dx = \frac{\pi}{p}\, \frac{1}{\sin\left[\pi\left(\frac{p+q}{p}\right)\right]}.$$

Recover as a particular case the result discussed in **SE 11.8**.

PE 11.9 If $|x| \le \pi$, show that

$$\sum_{k=0}^\infty (-1)^k \frac{\cos kx}{k^2 + t^2} = \frac{1}{2t^2} + \frac{\pi}{2t}\, \frac{\cosh tx}{\sinh \pi t}.$$

PE 11.10 For $-\pi < \theta < \pi$ and α noninteger and nonnull, show that

$$\sum_{k=-\infty}^\infty (-1)^k \frac{\cos(k + \alpha)\theta}{k + \alpha} = \frac{\pi}{\sin \pi \alpha}.$$

11.4 Special Functions

As there are many real problems involving applications of special functions other than the ones we have already studied, we present in this section two more families of such functions, the Hermite polynomials and the confluent hypergeometric function, emphasizing some of their properties but without concerning ourselves with specific applications. A particular application involving a hypergeometric function is also discussed.

11.4.1 *Hermite Polynomials $H_n(x)$*

SE 11.10 We call *generating function* of a certain family of functions, a function which, when expanded in a power series, yields all the functions belonging to that family in the form of the *coefficients* of this expansion. The generating function $F(x, t)$ associated with the Hermite polynomials $H_n(x)$, $n = 0,\ 1,\ 2,\ \dots$ is given by

$$F(x, t) = e^{x^2 - (t-x)^2} = \sum_{k=0}^\infty H_k(x) \frac{t^k}{k!},$$

where $t,\ x \in \mathbb{R}$ and the series on the right-hand side of the equation is obtained by expanding the function $F(x, t)$ in a power series on variable t around the point

$t = 0$. (a) Express $H_n(x)$ as a contour integral and (b) show that $H_n(x)$ satisfies the second-order ordinary differential equation (Hermite equation)

$$\frac{d^2}{dx^2} H_n(x) - 2x \frac{d}{dx} H_n(x) + 2n H_n(x) = 0 \,.$$

(c) Prove the recurrence relation

$$\frac{d}{dx} H_n(x) = 2n H_{n-1}(x) \,.$$

Solution: (a) We know that

$$\oint \frac{dx}{z^n} = \begin{cases} 2\pi i & \text{if} \quad n = 1, \\ 0 & \text{if} \quad n \neq 1, \end{cases}$$

where the integration is carried along a simple closed contour on the complex plane containing the origin and where $n \in \mathbb{Z}$.

Dividing the generating function $F(x,t)$ by t^{n+1} and integrating along this closed contour around the origin, we have

$$\frac{1}{2\pi i} \oint \frac{F(x,t)}{t^{n+1}} \, dt = \frac{1}{2\pi i} \oint e^{x^2 - (t-x)^2} \frac{dt}{t^{n+1}}$$

$$= \sum_{k=0}^{\infty} H_k(x) \frac{1}{k!} \frac{1}{2\pi i} \oint \frac{t^k}{t^{n+1}} \, dt$$

$$= \frac{H_n(x)}{n!} \,,$$

so that we can write the desired result

$$H_n(x) = \frac{n!}{2\pi i} \oint \frac{\exp[x^2 - (t-x)^2]}{t^{n+1}} \, dt \,.$$

(b) We will show that the quantity

$$\frac{\partial^2 F}{\partial x^2} + 2t \frac{\partial F}{\partial t} - 2x \frac{\partial F}{\partial x}$$

is identically null. Since

$$F(x,t) = e^{-t^2 + 2tx} \,,$$

we have for the derivatives:

$$\frac{\partial F}{\partial x} = 2t\,F(x, t);$$

$$\frac{\partial F}{\partial t} = (-2t + 2x)F(x, t);$$

$$\frac{\partial^2 F}{\partial x^2} = 4t^2 F(x, t).$$

Thus, substituting and simplifying,

$$\frac{\partial^2 F}{\partial x^2} + 2t\frac{\partial F}{\partial t} - 2x\frac{\partial F}{\partial x} = (4t^2 - 4t^2 + 4tx - 4tx)F(x, t) = 0.$$

Using the expansion for $F(x, t)$ in terms of the polynomials $H_n(x)$, this identity takes the form

$$\sum_{n=0}^{\infty} \frac{t^n}{n!}\left[\frac{\mathrm{d}^2}{\mathrm{d}x^2}H_n(x) - 2x\frac{\mathrm{d}}{\mathrm{d}x}H_n(x) + 2nH_n(x)\right] = 0.$$

This is true for all values of t. Therefore, we can write

$$\frac{\mathrm{d}^2}{\mathrm{d}x^2}H_n(x) - 2x\frac{\mathrm{d}}{\mathrm{d}x}H_n(x) + 2nH_n(x) = 0.$$

(c) Differentiating the integral representation for $H_n(x)$ given in item (a), we have

$$\frac{\mathrm{d}}{\mathrm{d}x}H_n(x) = \frac{2n(n-1)!}{2\pi i}\oint \frac{\exp[x^2 - (t-x)^2]}{t^n}\,\mathrm{d}t$$

$$= 2nH_{n-1}(x).$$

11.4.2 *Confluent Hypergeometric Function*

SE 11.11 The hypergeometric function $_2F_1(a, b; c; x)$ admits the following integral representation [2]:

$$_2F_1(a, b; c; x) = \frac{\Gamma(c)}{\Gamma(a)\Gamma(c-b)}\int_0^1 t^{b-1}(1-t)^{c-b-1}(1-tx)^{-a}\,\mathrm{d}t, \qquad (11.16)$$

valid for $|x| < 1$, $\mathrm{Re}\,a > 0$, $\mathrm{Re}\,b > 0$ and $\mathrm{Re}\,(c-b) > 0$.

Knowing that the confluent hypergeometric function $_1F_1(a; c; x)$ is obtained from the hypergeometric function through the limit

$$\lim_{b\to\infty} {}_2F_1\left(a, b; c; \frac{x}{b}\right) = {}_1F_1(a; c; x),$$

obtain an integral representation for ${}_1F_1(a; c; x)$.

Solution: Introducing into the integral representation, Eq. (11.16), the change of variable $x \to x/b$ and taking the limit $b \to \infty$, we obtain, for $\mathrm{Re}(c) > \mathrm{Re}(a) > 0$,

$$ {}_1F_1(a; c; x) = \frac{\Gamma(c)}{\Gamma(a)\Gamma(c-a)} \int_0^1 (1-t)^{c-a-1} t^{a-1}\, e^{tx}\, dt,$$

where we have used the fact ${}_2F_1(a, b; c; x) = {}_2F_1(b, a; c; x)$, together with the result

$$\lim_{x\to\infty}\left(1+\frac{a}{x}\right)^x = e^a .$$

11.4.3 Two-Dimensional Flow

SE 11.12 The steady flow of a compressible fluid, without viscous efforts, satisfies the so-called hodograph equations

$$\frac{\partial}{\partial \xi}\phi(\xi, \theta) = P(\xi)\frac{\partial}{\partial \theta}\psi(\xi, \theta),$$

$$\frac{\partial}{\partial \theta}\phi(\xi, \theta) = Q(\xi)\frac{\partial}{\partial \xi}\psi(\xi, \theta),$$

where $\phi(\xi, \theta)$ is a velocity potential, $\psi(\xi, \theta)$ is a flow function, θ is the angle formed by the velocity vector with a fixed direction, and $\xi = q^2/q_{\max}^2$, where q is the flow velocity and $q_{\max}$ is the highest velocity that can be attained.

We consider only the case in which we have a homentropic fluid (homentropic fluid is one in which the entropy is constant in time and uniform in space [7]), for which

$$2\xi(1-\xi)^{\gamma/(\gamma-1)}P(\xi) = \frac{\gamma+1}{\gamma-1}\xi - 1 \quad \text{and} \quad (1-\xi)^{1/(\gamma-1)}Q(\xi) = 2\xi,$$

where ξ is the quotient between the specific heats and $1 < \gamma < 2$. Show that a solution of the partial differential equations satisfied by the flow function can be written in the form

$$\psi(\xi, \theta) = \xi^{\nu/2}\exp(-i\nu\theta)\, {}_2F_1(a_\nu, b_\nu; 1+\nu; \xi),$$

where a_ν and b_ν are solutions of the system $a_\nu + b_\nu = \nu - \frac{1}{\gamma-1}$, $a_\nu b_\nu = -\frac{\nu(\nu+1)}{2(\gamma-1)}$ and ${}_2F_1(a, b; c; z)$ is a hypergeometric function.

Solution: First, we obtain a second-order, linear partial differential equation (as we have seen in Chap. 8) satisfied by the flow function, $\psi(\xi, \theta)$, that is,

$$\frac{d}{d\xi} Q(\xi) \frac{\partial}{\partial \xi} \psi(\xi, \theta) + Q(\xi) \frac{\partial^2}{\partial \xi^2} \psi(\xi, \theta) = P(\xi) \frac{\partial^2}{\partial \theta^2} \psi(\xi, \theta),$$

which is a separable partial differential equation.

Introducing $\psi(\xi, \theta) = T(\theta)R(\xi)$ and using the method of separation of variables, we obtain two homogeneous, second-order, linear ordinary differential equations

$$\frac{d^2}{d\theta^2} T(\theta) + \nu^2 T(\theta) = 0$$

and

$$Q(\xi) \frac{d^2}{d\xi^2} R(\xi) + \frac{d}{d\xi} Q(\xi) \frac{d}{d\xi} R(\xi) + \nu^2 P(\xi) R(\xi) = 0,$$

where ν^2 is the separation constant. A formal solution of the first ordinary differential equation can be written as

$$T(\theta) = A \exp(-i\nu\theta) + B \exp(i\nu\theta),$$

where A and B are two arbitrary constants.

We now turn to the second linear ordinary differential equation. We first introduce the functions $P(\xi)$ and $Q(\xi)$ previously defined, in order to obtain the following homogeneous, second-order, linear ordinary differential equation:

$$\xi(1-\xi) \frac{d^2}{d\xi^2} R(\xi) + \left(1 - \xi + \frac{\xi}{\gamma - 1}\right) \frac{d}{d\xi} R(\xi) + \frac{\nu^2}{4\xi} \left(\frac{\gamma+1}{\gamma-1}\xi - 1\right) R(\xi) = 0.$$

We then introduce a change of dependent variable $R(\xi) = \xi^{\nu/2} F(\xi)$ to obtain

$$\xi(1-\xi) \frac{d^2}{d\xi^2} F(\xi) + \left[1 + \nu - \left(1 + \nu - \frac{1}{\gamma - 1}\right)\xi\right] \frac{d}{d\xi} F(\xi) + \frac{\nu(\nu+1)}{2(\gamma - 1)} F(\xi) = 0,$$

which is a hypergeometric equation whose solution, regular at the origin, is given by

$$F(\xi) = C \, {}_2F_1(a_\nu, b_\nu; \nu + 1; \xi),$$

where C is an arbitrary constant and $a_\nu + b_\nu = \nu - \frac{1}{\gamma-1}$, $a_\nu b_\nu = -\frac{\nu(\nu+1)}{2(\gamma-1)}$.

Putting $AC = 1$ and $B = 0$ we obtain the desired result, that is,

$$\psi(\xi, \theta) = \xi^{\nu/2} \exp(-i\nu\theta) \, _2F_1(a_\nu, b_\nu; 1 + \nu; \xi) \, .$$

For certain values of ν, this result can be written in terms of Jacobi polynomials. To obtain the exact solution we must solve the system for a_ν and b_ν written as functions of ν and γ.

11.4.4   Proposed Exercises

PE 11.11 Let $x \in \mathbb{R}$ and $\ell = 0, 1, 2, \ldots$ Knowing that

$$x \frac{\mathrm{d}}{\mathrm{d}x} \mathcal{P}_\ell(x) = \frac{\mathrm{d}}{\mathrm{d}x} \mathcal{P}_{\ell+1}(x) - (\ell + 1)\mathcal{P}_\ell(x)$$

and

$$\frac{\mathrm{d}}{\mathrm{d}x} \mathcal{P}_{\ell+1}(x) \frac{\mathrm{d}}{\mathrm{d}x} \mathcal{P}_{\ell-1}(x) = (2\ell + 1)\mathcal{P}_\ell(x) \, ,$$

find the generating function for the Legendre polynomials, $\mathcal{P}_\ell(x)$, in the form

$$G(x, t) = \frac{1}{(1 - 2xt + t^2)^{1/2}} = \sum_{\ell=0}^{\infty} t^\ell \mathcal{P}_\ell(x) \, .$$

PE 11.12 Use the Laplace transform to solve the following problem: A semi-infinite spring is fixed at its extreme $x = 0$. The spring is initially at rest. The initial displacement is zero, and the displacement is finite for $x \to \infty$. An external force acts on the spring at the (moving) point $x = vt$ and has constant modulus f_0.

PE 11.13 (Incomplete gamma function) Let $\mathrm{Re}(\mu) > 0$ and $x \in \mathbb{R}$. The incomplete gamma function, denoted by $\gamma(\mu, x)$, is defined by the integral

$$\gamma(\mu, x) = \int_0^x e^{-t} t^{\mu-1} \, \mathrm{d}t \, .$$

Represent the integral that defines the incomplete gamma function in terms of a confluent hypergeometric function.

11.5   Fourier-Bessel and Fourier-Legendre Series

In this section, we discuss only the ways we can express a given function as a Fourier-Bessel series or a Fourier-Legendre series, because the applications of these series will appear in a natural way when we solve partial differential equations associated with problems with cylindrical and spherical symmetry, respectively. We also discuss the so-called radial Laplace equation.

11.5.1 Fourier-Bessel Series

SE 11.13 Consider the Bessel differential equation of order n written in the form

$$(xy')' + \left(\alpha^2 x - \frac{n^2}{x}\right) y = 0, \tag{11.17}$$

where α^2 is a parameter and n is an integer. (a) Multiply each member of this expression by $2xy'$ and integrate it term-by-term, using integration by parts, to obtain

$$2\alpha^2 \int_0^a x [\mathcal{J}_n(\alpha x)]^2 \, dx = \alpha^2 a^2 [\mathcal{J}'_n(\alpha a)]^2 + (\alpha^2 a^2 - n^2)[\mathcal{J}_n(\alpha a)]^2 .$$

(b) Using the result of the preceding item, obtain the specific expressions for the case $n = 0$ when: (i) $\gamma_k = \alpha a$ is a root of the equation $\mathcal{J}_0(x) = 0$; (ii) $\gamma_k = \alpha a$ is a root of the equation $\mathcal{J}'_0(x) = 0$; (iii) $\gamma_k = \alpha a$ is a root of the equation $h\mathcal{J}_0(x) + x\mathcal{J}'_0(x) = 0$, where h is a positive constant.

Solution: (a) We know that a solution of the Bessel equation of order n is $y = \mathcal{J}_n(\alpha x)$. Multiplying Eq. (11.17) by $2xy'$, we obtain

$$2xy'(xy')' + 2xy'\left(\alpha^2 x - \frac{n^2}{x}\right) y = 0,$$

which can be written as

$$[(xy')^2]' + (\alpha^2 x^2 - n^2)[y^2]' = 0.$$

Then, integrating from zero to a, we have

$$\int_0^a [(xy')^2]' \, dx + \int_0^a (\alpha^2 x^2 - n^2)[y^2]' \, dx = 0,$$

or

$$[(xy')^2 + (\alpha^2 x^2 - n^2)y^2]_0^a - 2\alpha^2 \int_0^a xy^2 \, dx = 0.$$

But $y = \mathcal{J}_n(x)$ is a solution of the equation and we therefore have

$$2\alpha^2 \int_0^a x[\mathcal{J}_n(x)]^2\, dx = \alpha^2 a^2 [\mathcal{J}_n'(\alpha a)]^2 + (\alpha^2 a^2 - n^2)[\mathcal{J}_n(\alpha a)]^2 \,.$$

(b) Taking $n = 0$ in the last equation we find that

$$\int_0^a x[\mathcal{J}_0(\alpha x)]^2 dx = \frac{a^2}{2}\{[\mathcal{J}_0(\alpha a)^2] + [\mathcal{J}_1(\alpha a)]^2\}\,, \tag{11.18}$$

where we have used the relation $\mathcal{J}_1(\alpha a) = -\mathcal{J}_0'(\alpha a)$.

(i) Knowing that γ_k is a root of the equation $\mathcal{J}_0(x) = 0$ we obtain, using Eq. (11.18),

$$\int_0^a x\left[\mathcal{J}_0\left(\frac{\gamma_k x}{a}\right)\right]^2 dx = \frac{a^2}{2}[\mathcal{J}_1(\gamma_k)]^2 \,.$$

(ii) Now, for γ_k a root of the equation $\mathcal{J}_1(x) = 0$ we get

$$\int_0^a x\left[\mathcal{J}_0\left(\frac{\gamma_k x}{a}\right)\right]^2 dx = \frac{a^2}{2}[\mathcal{J}_0(\gamma_k)]^2 \,.$$

(iii) Finally, if γ_k is a root of the equation $h\mathcal{J}_0(x) + x\mathcal{J}_0'(x) = 0$ we have the identity

$$\int_0^a x\left[\mathcal{J}_0\left(\frac{\gamma_k x}{a}\right)\right]^2 dx = \frac{a^2}{2}\frac{\gamma_k^2 + h^2}{\gamma_k^2}[\mathcal{J}_0(\gamma_k)]^2 \,,$$

which is the desired result.

11.5.2 *Fourier-Legendre Series*

SE 11.14 Let $f(x)$ be a limited function on the closed interval $I : -1 \le x \le 1$ and continuous on I except for a finite number of discontinuities. Suppose also that for every subinterval of I on which $f(x)$ is continuous, the curve $y = f(x)$ is rectifiable, that is, there exists a series of Legendre polynomials $P_n(x)$, $n = 0, 1, 2, \ldots$, with constant coefficients A_n,

$$\sum_{n=0}^{\infty} A_n P_n(x) \,,$$

such that (a) the series converges on the entire interval I; (b) it converges for $f(x)$ on all points of continuity of $f(x)$ on I, and (c) it is such that the series, after multiplication by an arbitrary $P_n(x)$, is term-by-term integrable on I. Show that the coefficients are given by

$$A_n = \frac{2n+1}{2} \int_{-1}^{1} f(x) P_n(x) \, dx \, ,$$

with $n = 0, 1, 2, \ldots$

Solution: This type of development is important, for example, when we discuss the radial Laplace equation. From hypotheses (a)–(c) we have

$$\int_{-1}^{1} f(x) P_n(x) \, dx = \int_{-1}^{1} A_0 P_0(x) P_n(x) \, dx$$

$$+ \int_{-1}^{1} A_1 P_1(x) P_n(x) \, dx + \cdots + \int_{-1}^{1} A_n P_n(x) P_n(x) \, dx + \cdots$$

Thus, due to the orthogonality property of Legendre polynomials, we get

$$\int_{-1}^{1} f(x) P_n(x) \, dx = A_n \int_{-1}^{1} [P_n(x)]^2 \, dx = A_n \frac{2}{2n+1} \, ,$$

which can be written as

$$A_n = \frac{2n+1}{2} \int_{-1}^{1} f(x) P_n(x) \, dx$$

with $n = 0, 1, 2, \ldots$, which is the desired result.

11.5.3   Laplace Equation in Spherical Coordinates

SE 11.15 (a) Obtain the electric potential, denoted by u, of a spherical conductor of radius a in a uniform electric field $\mathbf{E}_0$ in the $\hat{z}$ direction, that is, solve the Laplace equation in spherical coordinates, (r, θ, ϕ), for u independent of coordinate ϕ, imposing boundary conditions $u(a, \theta) = 0$ and $u(r, \theta) \to -E_0 r \cos \theta$ for $r \to \infty$, where E_0 is a positive constant. (b) Determine the components of the electric field E_r and E_θ, that is, in the directions $\hat{e}_r$ and $\hat{e}_\theta$, where $\hat{e}_r$ and $\hat{e}_\theta$ are the unitary vectors written in spherical coordinates. (c) Determine the charge density at the surface of the conductor and show that the total charge on the sphere is zero.

Solution: (a) Since the uniform electric field is in the $\hat{z}$ direction, the potential is independent of the coordinate ϕ and the Laplace equation to be solved is written as

$$\frac{\partial^2 u}{\partial r^2} + \frac{2}{r} \frac{\partial u}{\partial r} + \frac{1}{r^2} \left(\frac{\partial^2 u}{\partial \theta^2} + \cot \theta \frac{\partial u}{\partial \theta} \right) = 0,$$

with $0 \leq r < a$ (interior problem) and $0 \leq \theta < \pi$, where $u = u(r, \theta)$ is the potential.

Using the method of separation of variables, we obtain two homogeneous, linear, second-order ordinary differential equations

$$\sin^2 \theta \frac{d^2}{d\theta^2} T(\theta) + \sin \theta \cos \theta \frac{d}{d\theta} T(\theta) + \lambda \sin^2 \theta \, T(\theta) = 0$$

and

$$r^2 \frac{d^2}{dr^2} R(r) + 2r \frac{d}{dr} R(r) - \lambda R(r) = 0,$$

where λ is the separation constant. If we choose $\lambda = n(n+1)$, with $n = 0, 1, 2, \ldots$, we have for the general solution of the differential equation in variable θ

$$T(\theta) = A_n P_n(\cos \theta) + B_n Q_n(\cos \theta),$$

where A_n and B_n are arbitrary constants; $P_n(\cos \theta)$ and $Q_n(\cos \theta)$ are the Legendre polynomials and the second kind Legendre functions, respectively. We first impose that the solution is not a singular function at $\theta = 0$ and $\theta = \pi$. This condition implies that $B = 0$, so that we have

$$T_n(\theta) = A_n P_n(\cos \theta),$$

which is the solution of the homogeneous linear second-order ordinary differential equation in variable θ satisfying the boundary conditions.

On the other hand, the ordinary differential equation in variable r (radial equation) is an Euler equation, whose general solution is given by

$$R_n(r) = C_n r^n + \frac{D_n}{r^{n+1}},$$

where C_n and D_n are arbitrary constants.

Combining the solutions $T_n(\theta)$ and $T_n(r)$ and using superposition, we obtain for the potential the general solution

$$u(r, \theta) = \sum_{n=0}^{\infty} \left(a_n r^n + \frac{b_n}{r^{n+1}} \right) P_n(\cos \theta),$$

where $a_n = C_n A_n$ and $b_n = D_n A_n$ are arbitrary constants to be determined from the boundary conditions. We first consider the boundary condition when $r \to \infty$, which gives

$$a_0 = 0, \quad a_1 = -E_0, \quad \text{and} \quad a_n = 0 \text{ for } n \geq 2.$$

We thus have for the potential

$$u(r, \theta) = -E_0 r \cos \theta + \sum_{n=0}^{\infty} \frac{b_n}{r^{n+1}} P_n(\cos \theta) \,.$$

On the other hand, applying the boundary condition at $r = a$ (on the surface of the sphere) and using the previous result, we get

$$0 = -E_0 a \cos \theta + \sum_{n=0}^{\infty} \frac{b_n}{a^{n+1}} P_n(\cos \theta) \,,$$

which can be rearranged as a Fourier-Legendre series,

$$\sum_{n=0}^{\infty} \frac{b_n}{a^{n+1}} P_n(\cos \theta) = E_0 a \cos \theta \,.$$

To determine the coefficients, we use the orthogonality of the Legendre polynomials. The result is

$$b_n = \frac{2n + 1}{2} E_0 a^{n+2} \int_{-\pi}^{\pi} \cos \theta \, P_n(\cos \theta) \, d(\cos \theta)$$

$$= E_0 a^3 \delta_{n,1} \,,$$

where $\delta_{n,1}$ is the Kronecker delta function. Thus, by the definition of the Kronecker delta function, only the term with $n = 1$ contributes to the solution, so that the potential is given by

$$u(r, \theta) = -E_0 r \cos \theta + E_0 \frac{a^3}{r^2} \cos \theta \,.$$

(b) With this expression for the electric potential, we can find the electric field at all points. To do that, we just have to evaluate the partial derivatives of $u(a, \theta)$ with respect to r and with respect to θ,

$$E_r \equiv -\frac{\partial u}{\partial r} = E_0 \left(1 + 2\frac{a^3}{r^3} \right) \cos \theta$$

and

$$E_\theta \equiv -\frac{1}{r} \frac{\partial u}{\partial \theta} = -E_0 \left(1 - \frac{a^3}{r^3} \right) \sin \theta \,,$$

both for $r \geq a$.

(c) To find the charge density at the surface, denoted by σ, we recall that

$$\sigma(\theta) = \epsilon_0 E_r|_{r=a} \,,$$

where ϵ_0 is the so-called permittivity of free space. Thus, we have in this specific case

$$\sigma(\theta) = 3\epsilon_0 E_0 \cos\theta \,.$$

Finally, the total charge on the sphere is given by the expression

$$Q = 2\pi a^2 \int_0^\pi \sigma(\theta) \sin\theta \, d\theta \,,$$

which yields, after integration, $Q = 0$. We observe that, under static conditions, the electric field inside a conductor is zero, $\mathbf{E} = 0$. So, in consequence of Gauss law, the total charge is zero [6].

11.5.4   Proposed Exercises

PE 11.14 Suppose that the elements of the set $\{\gamma_m\}_{m=1}^\infty$ are the positive roots of the equation $\mathcal{J}_0'(x) = 0$. Define

$$f(x) = c_0 + \sum_{m=1}^\infty c_m \mathcal{J}_0\left(\frac{\gamma_m x}{a}\right) . \tag{11.19}$$

(a) Multiply each term of this equation by x and integrate it term-by-term from $x = 0$ to $x = a$ to show that

$$c_0 = \frac{2}{a^2} \int_0^a x f(x) \, dx \,.$$

(b) Multiply each term of Eq. (11.19) by $x \mathcal{J}_0(\gamma_m x/a)$ and integrate it term-by-term to show that

$$c_m = \frac{2}{a^2 [\mathcal{J}_0(\gamma_m)]^2} \int_0^a x f(x) \mathcal{J}_0\left(\frac{\gamma_m x}{a}\right) dx \,.$$

PE 11.15 Use the Rodrigues formula (**PE 4.21**) for the Legendre polynomials, integrating it n times by parts to show that

$$\int_{-1}^1 \mathcal{P}_n^2(x) \, dx = \frac{2}{2n + 1}, \quad n = 0, 1, 2\ldots$$

PE 11.16 Consider the function

$$f(z) = \begin{cases} 1 & a < z < b, \quad -1 \le a < b \le 1, \\ 0 & \text{otherwise.} \end{cases}$$

(a) Find the expansion of $f(z)$ in a Fourier-Legendre series. (b) Expand the function $f(z) = 1$ in a series with the form $\displaystyle\sum_{n=0}^{\infty} A_{2n+1} P_{2n+1}(z)$ on the interval $0 < z < 1$.

11.6   Laplace and Fourier Transforms

In this section we use the method of Laplace transform to find the solution of an integral equation of Volterra [1860 – Vito Volterra – 1940] type. As an application of the Fourier transform, we calculate the transform of a Gaussian. Finally, a nonhomogeneous wave equation is solved by means of the Laplace transform.

11.6.1   Volterra Integral Equation

SE 11.16 Let $t > 0$ and $y(t)$ a real function. Use the Laplace transform to solve the integral equation (with the unknown function under the integral)

$$y(t) = 1 - \sinh t + \int_0^t (1 + \tau)\, y(t - \tau)\, d\tau .$$

Solution: Let $s > 0$. Writing

$$\int_0^\infty e^{-st}\, y(t)\, dt \equiv F(s)$$

for the Laplace transform of $y(t)$, we can use the convolution product

$$\mathcal{L}[(f * g)(t)] = \mathcal{L}[f(t)]\mathcal{L}[g(t)]$$

to evaluate $F(s)$, obtaining

$$F(s) = \frac{s}{s^2 - 1} .$$

We then calculate the corresponding inverse Laplace transform, i.e.,

$$\mathcal{L}^{-1}[F(s)] = y(t) = \frac{1}{2\pi i} \int_{\gamma-i\infty}^{\gamma+i\infty} e^{st} \frac{s}{s^2 - 1}\, ds\,.$$

This integral can be calculated using the Bromwich contour and the residue theorem, with the method of partial fractions. We then find

$$y(t) = \cosh t\,,$$

which is the desired solution.

11.6.2 Fourier Transform of a Gaussian

SE 11.17 For a and b positive real numbers, evaluate the Fourier transform of the Gaussian function

$$g(x) = a \exp(-bx^2)\,.$$

Solution: Introducing the function $g(x)$ given above into the expression for the Fourier transform, we get

$$\mathcal{F}[g(x)] = \frac{1}{\sqrt{2\pi}} \int_{-\infty}^{\infty} e^{-ikx}\, a\, e^{-bx^2}\, dx$$

$$= \frac{a}{\sqrt{2\pi}} \int_{-\infty}^{\infty} e^{-b(x^2 + \frac{ikx}{b})}\, dx\,. \tag{11.20}$$

In order to calculate rigorously this integral, we would have to use complex variables. However, it is possible to solve the problem employing only real variables [4]. Using the relation

$$x^2 + \frac{ik}{b}x = \left(x + \frac{ik}{2b}\right)^2 + \frac{k^2}{4b^2}\,,$$

we rewrite Eq. (11.20) as

$$\mathcal{F}[g(x)] = \frac{a}{\sqrt{2\pi}}\, e^{-k^2/4b} \int_{-\infty}^{\infty} e^{-by^2}\, dy\,,$$

where $y = x + ik/2b$. The integral in the second member is equal to $\sqrt{\dfrac{\pi}{b}}$, and the final result is

$$\mathscr{F}[g(x)] = \frac{a}{\sqrt{2b}}\, e^{-k^2/4b}\,,$$

which is also a Gaussian, but with a width different from the width of the original Gaussian.

11.6.3 Nonhomogeneous Wave Equation

SE 11.18 Let $u = u(x,t)$ with $0 < x < 1$ and $t > 0$. Use the Laplace transform to obtain the solution of the nonhomogeneous wave equation

$$\frac{1}{c^2}\frac{\partial^2 u}{\partial t^2} - \frac{\partial^2 u}{\partial x^2} = A\sin(\pi x)\,,$$

satisfying the homogeneous boundary conditions

$$u(0,t) = 0 = u(1,t)\,,\quad t > 0,$$

and the initial conditions

$$u(x,0) = 0 = \left.\frac{\partial}{\partial t}u(x,t)\right|_{t=0}\,,\qquad 0 < x < 1\,,$$

where $A > 0$ (interpreted as an amplitude) and $c > 0$ (interpreted as a velocity of propagation) are two constants.

Solution: Taking the Laplace transform in variable t, denoted by

$$F(x,s) = \int_0^\infty e^{-st} u(x,t)\,dt\,,$$

with $s > 0$, we obtain a nonhomogeneous linear second-order ordinary differential equation

$$\frac{d^2}{dx^2}F(x,s) - \frac{s^2}{c^2}F(x,s) = -\frac{A}{s}\sin(\pi x)\,.$$

From the homogeneous boundary conditions, we find that

$$F(0,s) = 0 = F(1,s)\,.$$

To solve the nonhomogeneous ordinary differential equation, we must use the general solution of the corresponding homogeneous ordinary differential equation, namely,

$$F_H(x, s) = C_1 \sinh\left(\frac{x}{c}s\right) + C_2 \cosh\left(\frac{x}{c}s\right),$$

where C_1 and C_2 are two arbitrary constants. As the second member of the nonhomogeneous ordinary differential equation is a sine function, we may use the method of undetermined coefficients to obtain a particular solution of the nonhomogeneous ordinary differential equation. The result is

$$F_P(x, s) = \frac{A}{s} \frac{\sin(\pi x)}{s^2 + \pi^2 c^2}.$$

Using these last two expressions, we can write the general solution of the transformed nonhomogeneous wave equation, i.e., of the ordinary differential equation

$$F(x, s) = C_1 \sinh\left(\frac{x}{c}s\right) + C_2 \cosh\left(\frac{x}{c}s\right) + \frac{A}{s} \frac{\sin(\pi x)}{s^2 + \pi^2 c^2},$$

where C_1 and C_2 are two arbitrary constants which are to be determined by imposing the transformed homogeneous boundary conditions, yielding $C_1 = 0 = C_2$. Then, the solution of the initial problem (ordinary differential equation + boundary conditions) can be written in the form (partial fractions)

$$F(x, s) = \frac{A}{\pi^2 c^2}\left(\frac{1}{s} - \frac{s}{s^2 + \pi^2 c^2}\right)\sin(\pi x).$$

Finally, we apply the inverse Laplace transform to this function, to obtain the solution of the original nonhomogeneous wave equation satisfying the homogeneous boundary conditions and initial conditions given. We thus find

$$u(x, t) = \mathcal{L}^{-1}[F(x, s)] = \frac{A}{\pi^2 c^2}[1 - \cos(\pi c t)]\sin(\pi x),$$

which is the desired result.

11.6.4   Proposed Exercises

PE 11.17 Let $\alpha \in \mathbb{R}$. Show that, for $\alpha > 0$,

$$\int_{-\infty}^{\infty} e^{-\alpha x^2}\, dx = \sqrt{\frac{\pi}{\alpha}}.$$

PE 11.18 Use the convolution theorem to show that

$$\int_{-\infty}^{\infty} f(x)g(-x)\,dx = \int_{-\infty}^{\infty} \tilde{f}(k)\tilde{g}(k)\,dk,$$

where $\tilde{f}(k)$ is the Fourier transform of $f(x)$ and $\tilde{g}(k)$ is the Fourier transform of $g(x)$.

PE 11.19 Let $t > 0$ and $y(t)$ a real function. Solve the following integral equation:

$$y(t) = t + \frac{t^2}{2} - \int_0^t y(\tau)\,(t - \tau)\,d\tau .$$

PE 11.20 Let $a > 0$ and $k > 0$. Evaluate the Laplace transform of the periodic function

$$f(t) = \begin{cases} k & \text{for } 0 \le t \le a, \\ -k & \text{for } a \le t \le 2a, \end{cases}$$

with $f(t) = f(t + 2a)$, where a is a positive constant.

PE 11.21 Let $t > 0$. Use the Laplace transform to evaluate the integral

$$\Lambda(t) = \int_0^{\infty} \frac{x\sin(xt)}{x^2 + 1}\,dx .$$

PE 11.22 Let $f(x)$ be a real function. Consider the following integral equation:

$$\int_{-\infty}^{\infty} \frac{f(x)}{(t - x)^2 + 1}\,dx = \frac{1}{t^2 + 4} .$$

Solve this integral equation using the Fourier transform and the convolution product.

11.7 Sturm-Liouville Systems

This section provides a basis for the next two sections because, whenever we separate a homogeneous linear second-order partial differential equation, we always get a problem of Sturm-Liouville type. We present here a singular Sturm-Liouville problem involving the Legendre polynomials. We also calculate, step-by-step, a Green's function in the so-called generalized sense, in order to make clear the difference between the classical Green's function and the generalized Green's function.

11.7.1   Bessel Equation

SE 11.19 Let $x > 0$ and $\mu, \nu \in \mathbb{R}$. Put the Bessel equation

$$x^2 \frac{d^2}{dx^2} u(x) + x^2 \frac{d}{dx} u(x) + (\mu^2 x^2 - \nu^2) u(x) = 0,$$

with $u = u(x)$, in the Sturm-Liouville form. Classify this Sturm-Liouville equation.

Solution: Identifying the coefficients of the equation above with the ones in Eq. (7.1), we have

$$a_1(x) = x^2, \quad a_2(x) = x, \quad a_3(x) = \mu^2 x^2, \quad \text{and} \quad \lambda = \nu^2.$$

Then, evaluating the functions $p(x)$, $q(x)$, and $s(x)$ given by Eq. (7.1), we can write

$$p(x) = \exp\left(\int^x \frac{\xi}{\xi^2} \, d\xi\right) = \exp(\ln x) = x,$$

$$q(x) = \frac{\mu^2 x^2}{x^2} \cdot x = \mu^2 x \qquad \text{and} \qquad s(x) = \frac{x}{x^2} = \frac{1}{x}.$$

Thus, the Bessel equation can be written as

$$\frac{d}{dx}\left[x \frac{d}{dx} u(x)\right] + \left(\mu^2 x + \frac{\nu^2}{x}\right) u(x) = 0.$$

Defining the self-adjoint differential operator

$$\mathfrak{L} \equiv \frac{d}{dx}\left(x \frac{d}{dx}\right) + \mu^2 x,$$

we obtain

$$\mathfrak{L}[u(x)] + \frac{\nu^2}{x} u(x) = 0,$$

which is the Sturm-Liouville form of the Bessel equation. As $p(x) = x$, we have a singular Sturm-Liouville equation.

11.7.2   Legendre Polynomials

SE 11.20   Obtain a formal solution of the nonhomogeneous Sturm-Liouville problem

$$-\frac{d}{dx}\left[(1-x^2)\frac{d}{dx}y(x)\right] = \mu y(x) + f(x),$$

with $y(0) = 0$ and $y(x)$ and $y'(x)$ limited as $x \to 1$; $f(x)$ is a continuous known function on the interval $0 \le x \le 1$ and μ is not an eigenvalue of the associated homogeneous problem.

Solution: As we known, this is a singular Sturm-Liouville problem. We first solve the corresponding homogeneous problem, which is exactly the Legendre equation

$$(1-x^2)\frac{d^2}{dx^2}y(x) - 2x\frac{d}{dx}y(x) + \mu y(x) = 0$$

written in self-adjoint form.

The general solution of this homogeneous ordinary differential equation is a linear combination of the Legendre polynomials, for $\mu = l(l+1)$, $l = 0, 1, 2, \ldots$ and of the Legendre functions of the second kind. However, imposing the conditions that the function $y(x)$ and its derivative are limited as $x \to 1$, we are left with only the polynomial part. Moreover, using the other condition (of separate extremes), we conclude that only polynomials of odd order contribute because the polynomials of even order are not null at $x = 0$.

Hence, we conclude from these conditions that the eigenfunctions, denoted by ϕ_ℓ, associated with this problem are the odd Legendre polynomials

$$\phi_\ell(x) = P_{2\ell-1}(x),$$

corresponding to the eigenvalues $\lambda_\ell = 2\ell(2\ell - 1)$, for $\ell = 1, 2, 3 \ldots$ Therefore, a solution of this equation in terms of the eigenfunctions and corresponding eigenvalues is given by

$$y(x) = \sum_{\ell=1}^{\infty} \frac{A_\ell}{\lambda_\ell - \mu} P_{2\ell-1}(x),$$

where A_ℓ are coefficients to be determined. This is a Fourier-Legendre series and thus the coefficients are given by

$$A_\ell = \frac{\displaystyle\int_0^1 f(x)P_{2\ell-1}(x)\,dx}{\displaystyle\int_0^1 P_{2\ell-1}^2(x)\,dx},$$

where the denominator is a normalization factor for Legendre polynomials.

11.7.3   Generalized Green's Function

SE 11.21 Let $x > 0$ and $f(x)$ a real function. Find the Green's function associated with the nonhomogeneous problem given by the ordinary differential equation

$$\frac{d^2}{dx^2}y(x) + y(x) = -f(x)$$

and the conditions (separate extremes)

$$y(0) = 0, \qquad \text{and} \qquad y\left(\frac{\pi}{4}\right) = y'\left(\frac{\pi}{4}\right).$$

Solution: We begin with the general solution of the associated homogeneous problem, which is given by

$$y(x) = A \sin x + B \cos x,$$

where A and B are two arbitrary constants.

From the first condition we conclude that $B = 0$. We are then left with

$$y(x) = A \sin x,$$

a solution that satisfies both the first and the second condition. As we know, if a solution of the homogeneous ordinary differential equation satisfies both conditions on the separate extremes, we need to use the so-called Green's function in the generalized sense, a function which satisfies the nonhomogeneous ordinary differential equation

$$\left(\frac{d^2}{dx^2} + 1\right) G(x|\xi) = C \sin x \sin \xi,$$

except at $x = \xi$; here, C is a constant. $G(x|\xi)$ is a continuous function at $x = \xi$, and its derivative, $G'(x|\xi)$, is continuous on the entire interval, except at $x = \xi$, where it has a jump of magnitude -1. Moreover, this function (a function depending on two points) must also satisfy the homogeneous boundary conditions.

We must solve the corresponding nonhomogeneous differential equation

$$\left(\frac{d^2}{dx^2} + 1\right) u = \alpha \sin x,$$

where $\alpha = C \sin \xi$ and $u = G(x|\xi)$. We will use the method of variation of parameters. The solution of the corresponding homogeneous ordinary differential equation is

$$u = c_1 \sin x + c_2 \cos x,$$

where c_1 and c_2 are two arbitrary constants. For a particular solution we must have

$$\begin{cases} c_1' \sin x + c_2' \cos x = 0, \\ c_1' \cos x - c_2' \sin x = \alpha \sin x, \end{cases}$$

a linear system whose solution allows us to write

$$c_1' = \alpha \sin x \cos x \qquad \text{and} \qquad c_2' = -\alpha \sin^2 x,$$

which yields, after integration,

$$c_1 = -\frac{\alpha}{4} \cos 2x, \quad c_2 = -\frac{\alpha}{2}\left(x - \frac{\sin 2x}{2}\right).$$

Therefore, a particular solution of the nonhomogeneous differential equation is given by

$$\begin{aligned} u_p &= -\frac{\alpha}{4} \sin x \cos 2x - \frac{\alpha}{2} x \cos x + \frac{\alpha}{4} \sin 2x \cos x \\ &= -\frac{\alpha}{2} x \cos x + \frac{\alpha}{4} \sin x \\ &= \frac{C}{4} \sin \xi \sin x - \frac{C}{2} x \sin \xi \cos x. \end{aligned}$$

Then, we can write for the generalized Green's function

$$G(x|\xi) = \begin{cases} A_1 \sin x + B_1 \cos x + \dfrac{C}{4} \sin \xi \sin x - \dfrac{C}{2} x \sin \xi \cos x, \ 0 < x < \xi, \\[2mm] A_2 \sin x + B_2 \cos x + \dfrac{C}{4} \sin \xi \sin x - \dfrac{C}{2} x \sin \xi \cos x, \ \xi < x < \frac{\pi}{4}, \end{cases}$$

where A_1, B_1, A_2, and B_2 must be determined.

Knowing that the function $G(x|\xi)$ satisfies the boundary conditions

$$G(0|\xi) = 0 \quad \text{and} \quad G\left(\frac{\pi}{4}|\xi\right) = \frac{d}{dx} G(x|\xi)|_{x=\frac{\pi}{4}},$$

we have for the constants B_1 and B_2,

$$B_1 = 0 \qquad \text{and} \qquad B_2 = \frac{C}{4} \sin \xi \left(\frac{\pi}{2} - 1\right),$$

respectively.

Thus, the Green's function $G(x|\xi)$ takes the form

$$
G(x|\xi) = \begin{cases} A_1 \sin x + \frac{C}{4}\sin\xi \sin x - \frac{C}{2}x\sin\xi\cos x, & 0 < x < \xi, \\[2mm] A_2 \sin x + \frac{\pi-2}{8}C\sin\xi\cos x + \frac{C}{4}\sin\xi\sin x - \frac{C}{2}x\sin\xi\cos x, & \xi < x < \frac{\pi}{4}. \end{cases}
$$

Using the condition of continuity of the function $G(x|x')$ at $x = x'$, we have the following relation involving the constants A_1 and A_2:

$$
A_1 = A_2 + \frac{\pi - 2}{8}C\cos\xi\,.
$$

Thus, we can write for the generalized Green's function

$$
G(x|\xi) = \begin{cases} A_2 \sin x + \frac{\pi-2}{8}C\cos\xi\sin x + \frac{C}{4}\sin\xi\sin x - \frac{C}{2}x\sin\xi\cos x, & 0 < x < \xi, \\[2mm] A_2 \sin x + \frac{\pi-2}{8}C\sin\xi\cos x + \frac{C}{4}\sin\xi\sin x - \frac{C}{2}x\sin\xi\cos x, & \xi < x < \frac{\pi}{4}\,. \end{cases}
$$

Analogously, from the discontinuity of the function $G'(x|\xi)$ at $x = \xi$, it follows that

$$
\left.\frac{\mathrm{d}}{\mathrm{d}x}G(x|\xi)\right|_{x=\xi^+} - \left.\frac{\mathrm{d}}{\mathrm{d}x}G(x|\xi)\right|_{x=\xi^-} = -1
$$

and, after simplification, we get for constant C,

$$
C = \frac{8}{\pi - 2}\,.
$$

Therefore, the generalized Green's function takes the form

$$
G(x|\xi) = \begin{cases} A_2 \sin x + \cos\xi \sin x + \frac{2}{\pi-2}\sin\xi\sin x - \frac{4x}{\pi-2}\sin\xi\cos x, & 0 < x < \xi, \\[2mm] A_2 \sin x + \sin\xi\cos x + \frac{2}{\pi-2}\sin\xi\sin x - \frac{4x}{\pi-2}\sin\xi\cos x, & \xi < x < \frac{\pi}{4}\,. \end{cases}
$$

Constant A_2 is determined by means of an additional condition,

$$
\int_a^b G(x|\xi)y(x)\,\mathrm{d}x = \int_0^{\pi/4} G(x|\xi)\sin x\,\mathrm{d}x = 0\,.
$$

So, substituting $G(x|\xi)$, evaluating the resultant integrals and simplifying, we obtain

$$
A_2 = \frac{4}{\pi - 2}\left(\frac{\sin\xi}{\pi - 2} - \xi\cos\xi\right)\,.
$$

Finally, the generalized Green's function is

$$G(x|\xi) = \begin{cases} \left(\Xi + \cos\xi + \frac{2}{\pi-2}\sin\xi\right)\sin x - \frac{4x}{\pi-2}\sin\xi\cos x, & 0 < x < \xi, \\ \left(\Xi + \frac{2}{\pi-2}\sin\xi\right)\sin x - \left(\frac{4x}{\pi-2}\sin\xi - \sin\xi\right)\cos x, & \xi < x < \frac{\pi}{4}, \end{cases}$$

where we have introduced the notation $\Xi \equiv \frac{4}{\pi-2}\left(\frac{\sin\xi}{\pi-2} - \xi\cos\xi\right)$. Note that this is a symmetric function, that is, $G(x|\xi) = G(\xi|x)$.

11.7.4   Proposed Exercises

PE 11.23 Find the eigenvalues and the eigenfunctions of the singular Sturm-Liouville (system) problem

$$\frac{d}{dx}\left[x\frac{d}{dx}y(x)\right] + \frac{\lambda}{x}y(x) = 0,$$

with $y(1) = 0$, $y(e^\pi) = 0$ and $\lambda > 0$.

PE 11.24 Use the Green's function technique to solve the system:

$$\frac{d^2}{dx^2}u(x) = f(x),$$

$$u(0) = 0 \qquad \text{and} \qquad u(1) = \left.\frac{d}{dx}u(x)\right|_{x=1}.$$

PE 11.25 Let $\mu \in \mathbb{R}$. Solve the following Sturm-Liouville system, formed by the homogeneous linear second-order ordinary differential equation:

$$\frac{d^2}{dx^2}u(x) + \mu^2 u(x) = 0$$

and the boundary conditions $u'(0) = 0 = u'(1)$.

PE 11.26 Consider the family of functions $u_k(x)$, $k = 0, 1, 2, \ldots$, found in the preceding exercise. Introducing the concept of normalization by means of the integral

$$\int_0^1 u_m(x)u_n(x)\,dx = \delta_{m,n},$$

with $m, n = 0, 1, 2, \ldots$, determine the corresponding normalized eigenfunctions.

PE 11.27 Consider the Sturm-Liouville system formed by the homogeneous linear second-order ordinary differential equation

$$\frac{\mathrm{d}^2}{\mathrm{d}x^2} u(x) = 12x^2$$

and the boundary conditions $u(0) = 0 = u(1)$. (a) Obtain the corresponding Green's function. (b) Using the preceding item, solve the Sturm-Liouville system.

11.8   Partial Differential Equations

In this section we discuss the classification of a partial differential equation with only two variables, the so-called projective d'Alembert equation; we also obtain the general solution of another partial differential equation.

11.8.1   Projective d'Alembert Equation

SE 11.22 The projective d'Alembert equation, in the two-dimensional case, is given by

$$(1 - x^2)\frac{\partial^2}{\partial x^2}\psi_N(x,t) + 2xt\frac{\partial^2}{\partial x \partial t}\psi_N(x,t) - (1 - t^2)\frac{\partial^2}{\partial t^2}\psi_N(x,t)-$$

$$-2(N-1)\left[x\frac{\partial}{\partial x}\psi_N(x,t) + t\frac{\partial}{\partial t}\psi_N(x,t)\right] + N(N-1)\psi_N(x,t) = 0,$$

where N is a parameter independent of t and x. This is a homogeneous linear second-order partial differential equation with two independent variables and nonconstant coefficients. Here, too, we have the differential equation written in dimensionless form, i.e., we consider the transformation $t \to ct/R$ and $x \to x/R$, where c and R are constants associated with the speed of light and the radius of the so-called de Sitter [1872 – Willen de Sitter – 1934] Universe, respectively. Classify the projective d'Alembert equation according to its type.

Solution: As we have already seen, the classification of a linear second-order partial differential equation depends on the coefficients of the second derivatives. Using the notation introduced in Chap. 8, we have

$$A(x,t) = 1 + x^2, \quad B(x,t) = 2xt, \quad C(x,t) = -(1 - t^2).$$

So, the discriminant is given by

$$\Delta = B^2 - 4AC = 4x^2t^2 + 4(1 + x^2)(1 - t^2) = 4(1 + x^2 - t^2).$$

Then, for $1 + x^2 - t^2 > 0$ this equation is of hyperbolic type; for $1 + x^2 - t^2 = 0$ the equation is of parabolic type, and for $1 + x^2 - t^2 < 0$ the equation is of elliptic type. Therefore, we have a partial differential equation of mixed type, because the discriminant depends on the independent variables, contrary to the case of the classical d'Alembert partial differential equation, which is always of hyperbolic type. Note that, as we have considered the transformation $t \to ct/R$ and $x \to x/R$, if we take the limit $R \to \infty$, we recover exactly the classical d'Alembert partial differential equation.

11.8.2 General Solution of a Partial Differential Equation

SE 11.23 Given the partial differential equation

$$2\frac{\partial^2}{\partial x^2}u(x, y) - \frac{\partial^2}{\partial x \partial y}u(x, y) - \frac{\partial^2}{\partial y^2}u(x, y) = 0,$$

(a) reduce it to the canonical form and (b) obtain its general solution.
Solution: (a) We first obtain the characteristic transformation that permits us to reduce the equation to the canonical form. This transformation, using the notation of Chap. 8, is obtained from the two first-order ordinary differential equations

$$\frac{dy}{dx} = \frac{B \pm \sqrt{\Delta}}{2A} = \frac{-1 \pm \sqrt{1 + 8}}{4} = \frac{-1 \pm 3}{4},$$

or, explicitly,

$$\frac{dy}{dx} = -1, \qquad \text{and} \qquad \frac{dy}{dx} = \frac{1}{2}.$$

Integrating these ordinary differential equations, we have

$$y + x = \xi, \quad y - \frac{x}{2} = \eta,$$

where ξ and η are the new independent variables. We must calculate the partial derivatives. The first-order derivatives are

$$\frac{\partial}{\partial x} = \frac{\partial \xi}{\partial x}\frac{\partial}{\partial \xi} + \frac{\partial \eta}{\partial x}\frac{\partial}{\partial \eta} = \frac{\partial}{\partial \xi} - \frac{1}{2}\frac{\partial}{\partial \eta}; \qquad (11.21)$$

$$\frac{\partial}{\partial y} = \frac{\partial \xi}{\partial y}\frac{\partial}{\partial \xi} + \frac{\partial \eta}{\partial y}\frac{\partial}{\partial \eta} = \frac{\partial}{\partial \xi} + \frac{\partial}{\partial \eta}. \qquad (11.22)$$

For the second-order derivatives, we get

$$\frac{\partial^2}{\partial x^2} = \frac{\partial^2}{\partial \xi^2} + \frac{1}{4}\frac{\partial^2}{\partial \eta^2} - \frac{\partial^2}{\partial \xi \partial \eta}\,; \tag{11.23}$$

$$\frac{\partial^2}{\partial y^2} = \frac{\partial^2}{\partial \xi^2} + \frac{\partial^2}{\partial \eta^2} + 2\frac{\partial^2}{\partial \xi \partial \eta}\,; \tag{11.24}$$

$$\frac{\partial^2}{\partial x \partial y} = \frac{\partial^2}{\partial \xi^2} - \frac{1}{2}\frac{\partial^2}{\partial \eta^2} + \frac{1}{2}\frac{\partial^2}{\partial \xi \partial \eta}\,. \tag{11.25}$$

Substituting Eqs. (11.21)–(11.25) into the original equation, we have

$$\left\{ 2\frac{\partial^2}{\partial \xi^2} + \frac{1}{2}\frac{\partial^2}{\partial \eta^2} - 2\frac{\partial^2}{\partial \xi \partial \eta} - \frac{\partial^2}{\partial \xi^2} + \frac{1}{2}\frac{\partial^2}{\partial \eta^2} - \right.$$
$$\left. - \frac{1}{2}\frac{\partial^2}{\partial \xi \partial \eta} - \frac{\partial^2}{\partial \xi^2} - \frac{\partial^2}{\partial \eta^2} - 2\frac{\partial^2}{\partial \xi \partial \eta} \right\} u(\xi, \eta) = 0,$$

which becomes, after simplification,

$$\frac{\partial^2}{\partial \xi \partial \eta} u(\xi, \eta) = 0$$

which is the desired canonical form.

(b) Integrating in variable η, we obtain

$$\frac{\partial}{\partial \xi} u(\xi, \eta) = F(\xi),$$

where $F(\xi)$ is an arbitrary function of ξ; integrating again, now in variable ξ, we find

$$u(\xi, \eta) = \int^{\xi} F(\xi')d\xi' + g(\eta)\,,$$

which can be written as

$$u(\xi, \eta) = f(\xi) + g(\eta)\,,$$

where $f(\xi)$ and $g(\eta)$ are arbitrary functions, twice continuously differentiable, i.e., with continuous derivatives up to second order.

Going back to the primitive variables, we finally get for the general solution:

$$u(x, y) = f(x + y) + g\left(-\frac{x}{2} + y\right),$$

which is desired result.

11.8.3   *Partial Differential Equation with Variable Coefficients*

SE 11.24 Let $x, y \in \mathbb{R}_+^*$. Consider the partial differential equation

$$xy^3 \frac{\partial^2}{\partial x^2} u(x, y) + yx^3 \frac{\partial^2}{\partial y^2} u(x, y) - y^3 \frac{\partial}{\partial x} u(x, y) - x^3 \frac{\partial}{\partial y} u(x, y) = 0.$$

(a) Classify this partial differential equation and (b) reduce it to the canonical form.
Solution: (a) The discriminant is given by $\Delta = -4x^4 y^4$. It is always negative, then this is a partial differential equation of elliptic type.

(b) To obtain the canonical form, we must first evaluate the characteristic equations, that is, the two first-order ordinary differential equations

$$\frac{dy}{dx} = \frac{0 \pm 2ix^2 y^2}{2xy^3} = \pm i \frac{x}{y},$$

whose integration furnishes

$$y^2 = ix^2 + C_1 \qquad \text{and} \qquad y^2 = -ix^2 + C_2,$$

where C_1 and C_2 are two arbitrary constants. Introducing the characteristic coordinates ξ and η, we have

$$y^2 - ix^2 = \xi \qquad \text{and} \qquad y^2 + ix^2 = \eta.$$

As we have two complex conjugate characteristic coordinates, we can use them to define two other real coordinates, α and β, given by

$$\frac{\eta + \xi}{2} = y^2 = \alpha \qquad \text{and} \qquad \frac{\eta - \xi}{2i} = x^2 = \beta.$$

Evaluating the first derivatives we get

$$\frac{\partial}{\partial x} = 2\sqrt{\beta} \frac{\partial}{\partial \beta} \qquad \text{and} \qquad \frac{\partial}{\partial y} = 2\sqrt{\alpha} \frac{\partial}{\partial \alpha}.$$

For the second derivatives, we have

$$\frac{\partial^2}{\partial x^2} = 2\frac{\partial}{\partial \beta} + 4\beta\frac{\partial^2}{\partial \beta^2} \qquad \text{and} \qquad \frac{\partial^2}{\partial y^2} = 2\frac{\partial}{\partial \alpha} + 4\alpha\frac{\partial^2}{\partial \alpha^2}\,.$$

Introducing the expressions for the first and second derivatives into the partial differential equation and simplifying, we get

$$\frac{\partial^2}{\partial \alpha^2}u(\alpha, \beta) + \frac{\partial^2}{\partial \beta^2}u(\alpha, \beta) = 0\,,$$

which is the canonical form, the desired result.

11.8.4   Proposed Exercises

PE 11.28   Write the two-dimensional Laplace equation,

$$\Delta u(x, y) = \frac{\partial^2}{\partial x^2}u(x, y) + \frac{\partial^2}{\partial y^2}u(x, y) = 0\,,$$

in polar coordinates (r, θ) and classify it according to its type.

PE 11.29   Write the three-dimensional Laplace equation,

$$\Delta u(x, y, z) = \frac{\partial^2}{\partial x^2}u(x, y, z) + \frac{\partial^2}{\partial y^2}u(x, y, z) + \frac{\partial^2}{\partial z^2}u(x, y, z) = 0\,,$$

in spherical coordinates (r, θ, ϕ) and classify it according to its type.

PE 11.30   Find the general solution of the partial differential equation

$$\frac{\partial^2}{\partial x^2}u(x, y) + 2\frac{\partial^2}{\partial x \partial y}u(x, y) + \frac{\partial^2}{\partial y^2}u(x, y) = 0\,.$$

11.9   Separation of Variables

In this section we discuss the Laplace equation in spherical coordinates, in whose solution emerge the Legendre polynomials; we solve a problem involving the so-called Poisson equation and present the Stark [1874 – Johannes Stark – 1957] effect in parabolic coordinates.

11.9.1   *Laplace Equation in Spherical Coordinates*

SE 11.25 Separate the Laplace equation written in spherical coordinates, i.e., the partial differential equation

$$\frac{\partial^2 \psi}{\partial r^2} + \frac{2}{r}\frac{\partial \psi}{\partial r} + \frac{1}{r^2}\frac{\partial^2 \psi}{\partial \theta^2} + \frac{\cot g\,\theta}{r^2}\frac{\partial \psi}{\partial \theta} + \frac{1}{r^2 \sin^2 \theta}\frac{\partial^2 \psi}{\partial \phi^2} = 0\,,$$

where r, θ, and ϕ are the usual spherical coordinates, and $\psi = \psi(r, \theta, \phi)$.
Solution: We suppose that $\psi(r, \theta, \phi) = R(r)T(\theta)S(\phi)$. Introducing these functions into the partial differential equation, we get

$$T S R'' + \frac{2T S}{r} R' + \frac{RS}{r^2} T'' + \frac{RS \cot \theta}{r^2} T' + \frac{RT}{r^2 \sin^2 \theta} S'' = 0\,,$$

where we have omitted the explicit dependence associated with each function and the prime denotes differentiation with respect to the corresponding independent variable.

Multiplying this equation by $r^2 \sin^2 \theta / RTS$ we have

$$\frac{r^2 \sin^2 \theta}{RTS}\left[T S R'' + \frac{2T S}{r} R' + \frac{RS}{r^2} T'' + \frac{RS \cot g\,\theta}{r^2} T'\right] = \frac{-S''}{S}\,.$$

As the first member does not depend on ϕ, while the second one depends only on ϕ, we must have

$$S'' + \lambda S = 0 \tag{11.26}$$

and

$$T R'' + \frac{2T}{r} R' + \frac{R \cot g\,\theta}{r^2} T' - \frac{\lambda RT}{r^2 \sin^2 \theta} = 0\,, \tag{11.27}$$

where λ is a separation constant which depends on the boundary conditions imposed on variable ϕ.

Multiplying Eq. (11.27) by r^2 / TR, we get

$$\frac{r^2}{TR}\left[T R'' + \frac{2T}{r} R'\right] = -\frac{T''}{T} - \frac{\cot g\,\theta}{T} T' + \frac{\lambda}{\sin^2 \theta}\,,$$

and, as the first member does not depend on θ and the second one depends only on θ, we have

$$r^2 R'' + 2r R' - \mu R = 0 \tag{11.28}$$

and

$$T'' + \cotg\theta\, T' - \frac{\lambda}{\sin^2\theta}T + \mu T = 0\,, \tag{11.29}$$

where μ is another separation constant.

As a particular case, we consider a situation in which $\lambda = m^2$, with $m = 0, \pm1,$ $\pm2\ldots$ and $\mu = l(l+1)$ with $l = 0, 1, 2 \ldots$ Equations (11.26), (11.28), and (11.29) are then given by

$$S'' + m^2 S = 0,$$

$$T'' + \cotg\theta\, T' + \left[l(l+1) - \frac{m^2}{\sin^2\theta}\right]T = 0\,,$$

$$r^2 R'' + 2r R' - l(l+1)R = 0\,.$$

The corresponding general solutions are

$$S(\phi) = A\cos m\phi + B\sin m\phi\,,$$

$$T(\theta) = C P_l^m(\cos\theta) + D Q_l^m(\cos\theta)\,,$$

$$R(r) = Er^l + Fr^{-l-1}\,,$$

where A, B, C, D, E, and F are constants of integration while $P_l^m(\cos\theta)$ and $Q_l^m(\cos\theta)$ are the associated Legendre functions of first and second kind, respectively.

11.9.2 *Poisson Equation in Elasticity*

SE 11.26 It is well-known in the theory of elasticity that the *stress function* $\psi(x, y) = \psi$ on a bar satisfies the Poisson differential equation, a nonhomogeneous differential equation,

$$\frac{\partial^2\psi}{\partial x^2} + \frac{\partial^2\psi}{\partial y^2} = -2\,,$$

on a region R of the xy plane, with $\psi = 0$ on the boundary of R. For the sake of simplicity, we consider a rectangular transversal section with dimensions a and b. (a) Introduce the change of independent variable

$$\psi(x, y) = u(x, y) + ax - x^2$$

into the Poisson differential equation and show that $u(x, y)$ satisfies the following boundary value problem:

$$\frac{\partial^2 u}{\partial x^2} + \frac{\partial^2 u}{\partial y^2} = 0 \quad \text{on } R,$$

$$u(0, y) = u(a, y) = 0,$$

$$u\left(x, \frac{b}{2}\right) = u\left(x, \frac{-b}{2}\right) = x^2 - ax.$$

(b) Use the method of separation of variables to show that the stress function is given by

$$\psi(x, y) = ax - x^2 - \frac{8a^2}{\pi^3} \sum_{k\ =\ \mathrm{odd}} \frac{\cosh(k\pi y/a)\sin(k\pi x/a)}{k^3 \cosh(k\pi b\, 2a)}.$$

Solution: (a) We consider $\psi(x, y) = u(x, y) + ax - x^2$. Differentiating and substituting the result into the Poisson differential equation, we verify that $u(x, y)$ satisfies the Laplace equation. From the boundary conditions, we have

$$\psi(0, y) = u(0, y) = 0 \ \Rightarrow\ u(0, y) = 0;$$

$$\psi(a, y) = u(a, y) = 0 \ \Rightarrow\ u(a, y) = 0;$$

$$\psi\left(x, \frac{b}{2}\right) = u\left(x, \frac{b}{2}\right) + ax - x^2 = 0 \ \Rightarrow\ u\left(x, \frac{b}{2}\right) = x^2 - ax;$$

$$\psi\left(x, \frac{-b}{2}\right) = u\left(x, \frac{-b}{2}\right) + ax - x^2 = 0 \ \Rightarrow\ u\left(x, \frac{-b}{2}\right) = x^2 - ax.$$

As a matter of fact, when we deal with a real problem, we write $\psi(x, y) = u(x, y) + \omega(x)$ and assume that function $u(x, y)$ satisfies the corresponding *homogeneous* differential equation. Using the fact that $\psi(x, y)$ must satisfy the nonhomogeneous equation and the boundary conditions, we arrive at an ordinary differential equation for $\omega(x)$ which, in the present case, would lead us to the solution $\omega(x) = ax - x^2$.

(b) Setting $u(x, y) = R(x)T(y)$ and introducing this function into the Laplace equation we have

$$\frac{R''}{R} = -\frac{T''}{T} = \lambda,$$

where λ is a constant.

Using the homogeneous conditions $u(0, y) = u(a, y) = 0$, we obtain the following Sturm-Liouville problem:

$$R'' - \lambda R = 0 \, ;$$
$$R(0) = R(a) = 0 \, .$$

This problem admits nontrivial solutions only for $\lambda = -\bar{k}^2$, where $\bar{k}$ is a positive real number. It is easy to see that the corresponding eigenfunctions are

$$R_k(x) = A \sin\left(\frac{k\pi}{a} x\right) ,$$

where $k = 1, 2, 3 \ldots$ Then, the equation in variable y is given by

$$T'' - \bar{k}^2 T = 0 \, ,$$

whose solution is

$$T(y) = B \cosh \bar{k} y + C \sinh \bar{k} y,$$

where B and C are constants and $\bar{k} = k\pi/a$, with $k = 1, 2, 3 \ldots$

Using the superposition principle, we find that a solution of the Laplace equation is

$$u(x, y) = \sum_{k=1}^{\infty} \left[a_k \cosh\left(\frac{k\pi}{a} y\right) + b_k \sinh\left(\frac{k\pi}{a} y\right) \right] \sin \frac{k\pi}{a} x,$$

where $a_k = AB$ and $b_k = AC$ are constants.

Now, using the nonhomogeneous conditions, we have

$$u(x, b/2) = \sum_{k=1}^{\infty} \left[a_k \cosh\left(\frac{k\pi}{2} \frac{b}{a}\right) + b_k \sinh\left(\frac{k\pi}{2} \frac{b}{a}\right) \right] \sin \frac{k\pi}{a} x = x^2 - ax \, ;$$

$$u(x, -b/2) = \sum_{k=1}^{\infty} \left[a_k \cosh\left(\frac{k\pi}{2} \frac{b}{a}\right) - b_k \sinh\left(\frac{k\pi}{2} \frac{b}{a}\right) \right] \sin \frac{k\pi}{a} x = x^2 - ax \, .$$

We then conclude that $b_k = 0$; there remains only the condition

$$\sum_{k=1}^{\infty} a_k \cosh\left(\frac{k\pi}{2} \frac{b}{a}\right) \sin \frac{k\pi}{a} x = x^2 - ax,$$

which is a sine Fourier series. Therefore, the coefficients a_k are given by

$$a_k \cosh\left(\frac{k\pi}{2} \frac{b}{a}\right) = \frac{2}{a} \int_0^a (x^2 - ax) \sin\left(\frac{k\pi}{a} x\right) dx \, ,$$

that is,

$$a_k = -\frac{8a^2}{k^3\pi^3}\left[\cosh\left(\frac{k\pi b}{2a}\right)\right]^{-1},$$

where k is an odd number.

Finally, adding the stationary solution, we get

$$\psi(x,y) = ax - x^2 - \frac{8a^2}{\pi^3}\sum_{k=\text{odd}}\frac{\cosh(k\pi y/a)\sin(k\pi x/a)}{k^3\cosh(k\pi b/2a)},$$

which is the desired solution.

11.9.3 Parabolic Coordinates and the Stark Effect

SE 11.27 Let us first introduce the parabolic coordinates, denoted by μ, ν, ψ. This is a rotational coordinate system generated by the transformation $2z = \omega^2$, where $z = x + iy$ and $\omega = \mu + i\nu$, with x and y real variables and μ and ν real functions of real variables. The relation between Cartesian coordinates and parabolic coordinates is given by Romão Martins and Capelas de Oliveira [9]

$$x = \mu\nu\cos\psi, \qquad y = \mu\nu\sin\psi, \qquad z = \frac{1}{2}(\mu^2 - \nu^2).$$

The Laplacian operator in parabolic coordinates is then given by

$$\Delta = \frac{1}{\mu^2 + \nu^2}\left(\frac{\partial^2}{\partial\mu^2} + \frac{1}{\mu}\frac{\partial}{\partial\mu} + \frac{\partial^2}{\partial\nu^2} + \frac{1}{\nu}\frac{\partial}{\partial\nu}\right) + \frac{1}{\mu^2\nu^2}\frac{\partial^2}{\partial\psi^2}.$$

Using this Laplacian operator, we will discuss the separability of the time-independent Schrödinger equation in the study of the so-called Stark effect, that is, the displacement of atomic energy levels that takes place when an atom is placed in the presence of an external electric field. We first present the general form for the potential:

$$V = \frac{v_1(\mu)}{\mu^2 + \nu^2} + \frac{v_2(\nu)}{\mu^2 + \nu^2} + \frac{v(\psi)}{\mu^2\nu^2}.$$

With this potential, the separability of Schrödinger equation in parabolic coordinates is guaranteed, so we can discuss the Stark effect. The presence of an external electric field E_0 in the direction of the positive z axis adds a term $-eE_0z$ to the potential energy in the Schrödinger equation. So, in parabolic coordinates, the Schrödinger equation becomes

$$\Delta\phi + \frac{2m}{\hbar^2}\left(\frac{e^2}{r} + eE_0 z + E\right)\phi = 0\,,$$

where the potential can be written as

$$-V = \frac{e^2}{r} + eE_0 z = \frac{1}{\mu^2 + \nu^2}\left(2e^2 + \frac{eE_0\mu^4}{2} + \frac{eE_0\nu^4}{2}\right).$$

Following the method of separation of variables, we introduce $\phi(\mu, \nu, \psi) = M(\mu)N(\nu)\Psi(\psi) = MN\Psi$ and the separation constants p^2 and q^2. For variable ψ, we can write the following ordinary differential equation:

$$\frac{d^2}{d\psi^2}\Psi(\psi) + p^2\Psi(\psi) = 0\,,$$

where p^2 is an arbitrary constant. The equation for variable μ is

$$\frac{d^2}{d\mu^2}M(\mu) + \frac{1}{\mu}\frac{d}{d\mu}M(\mu) + \left(\frac{2mE}{\hbar^2}\mu^2 + \frac{4me^2}{\hbar^2} + \frac{meE_0}{\hbar^2}\mu^4 + \frac{p^2}{\mu^2} - q^2\right)M(\mu) = 0\,,$$

with q^2 another arbitrary constant. Finally, for variable ν we have

$$\frac{d^2}{d\nu^2}N(\nu) + \frac{1}{\nu}\frac{d}{d\nu}N(\nu) + \left(\frac{2mE}{\hbar^2}\nu^2 - \frac{meE_0}{\hbar^2}\nu^4 - \frac{p^2}{\nu^2} + q^2\right)N(\nu) = 0\,.$$

The first equation, in variable ψ, is a simple equation with constant coefficients. The second and third equations can be solved by the method of power series. For $E_0 \neq 0$ we would also need to use perturbation theory, whereas in the case where $E_0 = 0$ these two ordinary differential equations have solutions in terms of Laguerre polynomials [9].

11.9.4 Proposed Exercises

PE 11.31 If a circular membrane of radius r_0 with fixed boundary is subject to a periodic force $F_0 \sin \omega t$ by unity of mass, uniformly distributed over the membrane, then the displacement function $u(r, t)$ satisfies the partial differential equation

$$\frac{\partial^2 u}{\partial t^2} = a^2\left(\frac{\partial^2 u}{\partial r^2} + \frac{1}{r}\frac{\partial u}{\partial r}\right) + F_0 \sin \omega t\,,$$

where F_0 and a^2 are constants. Substitute $u(r, t) = R(r)\sin \omega t$ in order to obtain a solution which is a periodic function of t.

PE 11.32 (a) Find the solution $u(r, \theta)$ (independent of ϕ) of the three-dimensional Laplace equation inside the sphere $r \leq a$, satisfying the following conditions:

$$u(a, \theta) = 1 \text{ if } 0 < \theta < \frac{\pi}{2} \,;$$

$$u(a, \theta) = 0 \text{ if } \frac{\pi}{2} < \theta < \pi \,.$$

(b) Show that $u(r, \frac{\pi}{2}) = \frac{1}{2}$ for $0 \leq r \leq a$.

Remember that $\int_0^1 \mathcal{P}_l(z)dz = \frac{\mathcal{P}_{l-1}(0)}{l+1}$, where $l = 1, 2, \ldots$ and $\mathcal{P}_l(z)$ is the Legendre polynomial of order l.

PE 11.33 Using **SE 11.27**, solve the ordinary differential equation in variable v, with $E_0 = 0$. To this end, first introduce the dependent variable $N(v) = x^{p/2}S(x)$, with $x = v^2$. Then, considering $E < 0$, introduce the parameter $\beta\hbar = \sqrt{-2mE}$ and another dependent variable $S(x) = e^{-\beta x/2}T(x)$. Solve the resulting ordinary differential equation.

11.10 Fractional Calculus

In this section, we discuss three exercises involving techniques of fractional calculus, namely, the Riemann-Liouville fractional integral and the Caputo fractional derivative and a particular Mittag-Leffler function.

11.10.1 Riemann-Liouville Fractional Integral

SE 11.28 (Left-/Right-Sided Riemann-Liouville Fractional Integrals) Discuss the possibilities associated with left-/right-sided Riemann-Liouville fractional integrals. Do the same for the corresponding Riemann-Liouville and Caputo fractional derivatives.

Solution: In Chap. 10, the Riemann-Liouville fractional integral (and also the corresponding fractional derivative) was introduced only by means of the left-sided Riemann-Liouville fractional integral. Nothing was said in terms of the corresponding right-sided Riemann-Liouville fractional integral. Here, we discuss briefly what we can do with left/right Riemann-Liouville fractional integral/derivative (and also for the Caputo fractional derivative). We present only the case of the Riemann-Liouville fractional integral, discussing two ways to interpret it, one of them due to Samko et al. [10] and the other due to Hilfer [5].

Fig. 11.5 ξ is on the
left/right relatively to x

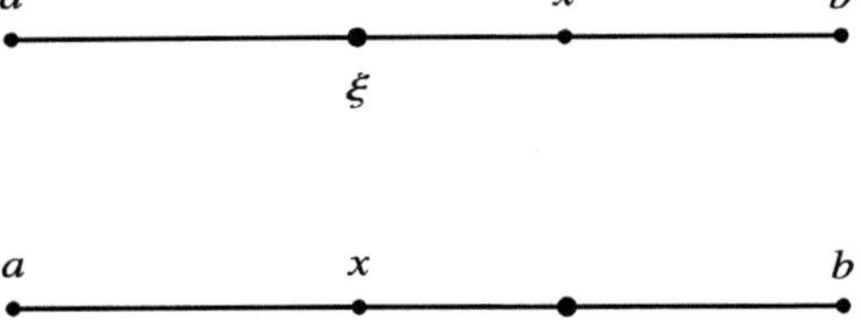

Before we introduce both definitions, we sketch a simple scheme involving an
interval. Let $a, b \in \mathbb{R}$ with $b > a$ and $x \in \mathbb{R}$ such that $a < x < b$. Considering
$\xi \in \mathbb{R}, a < \xi < b$, we have the two possibilities, shown in Fig. 11.5, for the relation
between x and ξ.

Figure 11.5 shows that we can have $\xi < x$ (ξ is on the left of x) or $\xi > x$ (ξ is
on the right of x), so that we can write

$$a < \xi < x \qquad \text{and} \qquad x < \xi < b,$$

respectively. On the other hand, using the extremes of the interval as reference, we
can also say that

$$\xi \text{ is on the right of } a \qquad \text{and} \qquad \xi \text{ is on the left of } b,$$

respectively. Given these two possibilities, we conclude that we can compare ξ
with x and we can compare ξ with one of the extremes, the starting point of
the integration. With these schemes in mind, we can introduce the following two
definitions.

Definition 11.10.1 (Left/Right Riemann-Liouville Fractional Integrals) Let
$f(x)$ be a locally integrable function on the interval (a, b) with $b > a$. The integrals

$$\left(I_{a+}^{\alpha} f\right)(x) =: \frac{1}{\Gamma(\alpha)} \int_a^x (x - \xi)^{\alpha-1} f(\xi) \, d\xi , \qquad x > a$$

and

$$\left(I_{b-}^{\alpha} f\right)(x) =: \frac{1}{\Gamma(\alpha)} \int_x^b (\xi - x)^{\alpha-1} f(\xi) \, d\xi , \qquad x < b,$$

where $\alpha > 0$ are called fractional integrals of order α. They are sometimes called
left-sided and right-sided fractional integrals, respectively. However, both integrals
can be called Riemann-Liouville fractional integrals [10].

Definition 11.10.2 (Left-Sided and Right-Sided Fractional Integrals) Let $f(x)$
be a locally integrable function on the interval (a, b) with $b > a$ and $\alpha > 0$. For
$-\infty \leq a < x < b \leq \infty$ we have

$$_aI_x^\alpha f(x) =: \frac{1}{\Gamma(\alpha)} \int_a^x (x - \xi)^{\alpha-1} f(\xi)\, d\xi, \qquad \text{(right-sided)}$$

and similarly for $-\infty < x < b \le \infty$

$$_xI_b^\alpha f(x) =: \frac{1}{\Gamma(\alpha)} \int_x^b (\xi - x)^{\alpha-1} f(\xi)\, d\xi, \qquad \text{(left-sided)}$$

where both integrals are defined for suitable f [5].

11.10.2   Memory Effect

SE 11.29 Discuss the so-called memory effect.

Solution: Let $x \in \mathbb{R}$ and $\mu \in \mathbb{R}$. The solution of a certain initial value problem, a differential equation with initial conditions, can be obtained through the Laplace transform and the convolution theorem and can be written as

$$y(x) = \frac{1}{\Gamma(\mu)} \int_0^x (x - \xi)^{\mu-1} f(\xi)\, d\xi. \tag{11.30}$$

Let us consider in Eq. (11.30) two distinct values x_1 and x_2 such that $x_1 < x_2$. We separate the integral in two intervals

$$y(x_2) = \frac{1}{\Gamma(\mu)} \int_0^{x_2} (x_2 - \xi)^{\mu-1} f(\xi)\, d\xi$$

and

$$y(x_1) = \frac{1}{\Gamma(\mu)} \int_0^{x_1} (x_1 - \xi)^{\mu-1} f(\xi)\, d\xi.$$

Take the difference $y(x_2) - y(x_1)$; we get

$$y(x_2) - y(x_1) = \frac{1}{\Gamma(\mu)} \int_0^{x_2} (x_2 - \xi)^{\mu-1} f(\xi)\, d\xi - \frac{1}{\Gamma(\mu)} \int_0^{x_1} (x_1 - \xi)^{\mu-1} f(\xi)\, d\xi,$$

which, by rearranging, can be rewritten in the form

$$y(x_2) - y(x_1) = \frac{1}{\Gamma(\mu)} \int_0^{x_1} [(x_2 - \xi)^{\mu-1} - (x_1 - \xi)^{\mu-1}] f(\xi)\, d\xi +$$

$$\frac{1}{\Gamma(\mu)} \int_{x_1}^{x_2} (x_2 - \xi)^{\mu-1} f(\xi)\, d\xi, \tag{11.31}$$

since $x_1 < x_2$. Note that these two integrals must be evaluated on two different intervals.

We see from Eq. (11.31) that the first integral involves values between 0 and x_1, while the second integral contains values between x_1 and x_2. For all values of $\mu \neq 1$, the two integrals contribute, whereas in the case $\mu = 1$ this does not happen, i.e., only the second parcel contributes.

Since the second interval does not contain values of $\xi < x_1$, we say that systems modeled by integer order equations ($\mu = 1$) do not exhibit the so-called memory effect. On the other hand, in the cases in which $0 < \mu < 1$, the first parcel also contributes, that is, $y(x)$ depends on the two parcels, so that we say that systems modeled by noninteger order equations present the memory effect, that is, they have a memory effect expressed by the integral on the interval from zero to x_1, before (in the past of) x_1.

11.10.3 Fractional Differential Equation

SE 11.30 (Fractional Differential Equation of Arbitrary Order) Let $y(x)$ be a continuous function satisfying the initial conditions $y(0) = A$ and $y'(0) = B$, where A and B are two positive constants and μ is a parameter such that $1 < \mu \leq 2$. Solve the fractional differential equation of order μ,

$$\frac{d^\mu}{dx^\mu} y(x) + k^2 \, y(x) = 0 \,,$$

where $k^2 > 0$ is a constant and the derivative is considered in the Caputo sense. *Solution*: Note that, in the case $\mu = 2$, this equation is an ordinary differential equation describing a free harmonic oscillator. Its general solution can be written in terms of trigonometric functions, and, given the initial conditions, we can obtain a particular solution satisfying the complete initial value problem.

Given the importance of this physical system, it is natural to introduce a noninteger order, μ, to obtain a fractional differential equation. It is important to observe that we exclude one of the two extremes of the range for μ, considering only $\mu = 2$. With this choice, we can recover the solution of the classical harmonic oscillator. We might also keep only the extreme $\mu = 1$, excluding the right extreme, but the particular case would not be recovered.

We use the Laplace transform to solve this fractional differential equation. Taking the Laplace transform on both sides of the fractional differential equation, we have

$$\mathscr{L}\left[\frac{d^\mu}{dx^\mu} y(x)\right] + k^2 \mathscr{L}[y(x)] = 0 \,.$$

Using the property of the Laplace transform of the derivative, we get an algebraic equation for $F(s)$,

$$\underbrace{s^\mu F(s) - y(0) \, s^{\mu-1} - y'(0) \, s^{\mu-2}}_{*} + k^2 F(s) = 0$$

where $F(s)$ is the Laplace transform of $y(x)$ with parameter s. As we have just said, in the case $\mu = 2$, the expression highlighted with a star $*$ is the expression for the Laplace transform of the second-order derivative.

We now apply the initial conditions: $y(0) = A$, the initial displacement and $y'(0) = B$, the initial speed. Solving the resulting algebraic equation, we have

$$F(s) = A \frac{s^{\mu-1}}{s^\mu + k} + B \frac{s^{\mu-2}}{s^\mu + k}.$$

We must use the complex plane to calculate the inverse Laplace transform of $F(s)$. If we do this, we arrive at the general solution of the original fractional differential equation,

$$y(x) = A \, \mathbb{E}_\mu(kx^\mu) + B \, \mathbb{E}_{\mu,2}(-kx^\mu), \tag{11.32}$$

where $\mathbb{E}_\mu(\cdot)$ is the classical Mittag-Leffler function and $\mathbb{E}_{\mu,2}(\cdot)$ is the Mittag-Leffler function with two parameters. Constants A and B are determined by the initial conditions.

It is important to note that in the case $\mu = 2$, we recover the result of the free harmonic oscillator, so that we can say that the solution obtained in Eq. (11.32) is the solution of the fractional differential equation describing the motion of a fractional harmonic oscillator. Besides, the introduction of a term involving the first-order derivative yields a differential equation associated with a damped harmonic oscillator, and we can obtain a relation between damping coefficient and the order of the fractional differential equation [3, 4]. On the other hand, if we had considered the interval $0 < \mu \leq 1$, we would have a problem associated with the so-called fractional decay, because when the parameter $\mu = 1$, the resulting equation describes a decay process.

11.10.4 *Proposed Exercises*

PE 11.34 (a) Let $f = (x - a)^\mu$ be a continuous function with $a > 0$ and $\mu \in \mathbb{R}$. Evaluate its derivative of order $\alpha \in \mathbb{R}$ in the Riemann-Liouville sense and compare it with the derivative in the Caputo sense. (b) Taking $\mu = 0$, evaluate both derivatives and compare them, to show that the derivative in the Riemann-Liouville sense of a constant is not zero.

PE 11.35 (Semigroup Property) Let $\alpha > 0$ and $\beta > 0$. Show the additive law for fractional integrals,

$$_aI_x^\alpha \, _aI_x^\beta \, f = \, _aI_x^{\alpha+\beta} \, f.$$

PE 11.36 The tautochrone problem, also known as isochrone curve problem, seems to be the first application of fractional calculus. It consists in determining a curve on which the time taken by an object to slide frictionless, in uniform gravity, to the lowest point of the curve is independent from the starting point. The problem was first solved geometrically by Huygens [1629 – Christiaan Huygens – 1695]. The analytical solution was presented by Abel [1802 – Niels Henrick Abel – 1829].

Using the principle of conservation of energy, we can obtain an integral equation, similar to the Riemann-Liouville fractional integral. This equation will be solved by means of the fractional integral, although it could be solved with the classical Laplace transform methodology.

Let m be the mass of the object, $v(t)$ its velocity at time t, y_0 the height at which it is abandoned, and $y(t)$ its height at time t. It is known that the kinetic and potential energies are given, respectively, by

$$\frac{1}{2}m\,v^2 \qquad \text{and} \qquad m\,g\,y\,.$$

Since the particle is constrained to move on the curve, its velocity is $v = \mathrm{d}s/\mathrm{d}t$, where s is the distance measured along the curve. From the principle of conservation of energy, we can write

$$\mathrm{d}t = \pm \frac{\mathrm{d}s}{\sqrt{2g(y_0 - y)}}\,.$$

As the function $s(y)$ describes the remaining distance along the curve in terms of the remaining height y and since distance and height decrease as time passes, we consider only the negative sign. Writing the expression obtained in a more adequate form, we have

$$\mathrm{d}t = -\frac{1}{\sqrt{2g}}(y - y_0)^{-1/2}\left(\frac{\mathrm{d}s}{\mathrm{d}y}\right)\mathrm{d}y\,. \tag{11.33}$$

Integrating both sides of Eq. (11.33) from y_0 to zero, we get

$$\tau = t(y_0) = \int_{y_0}^{0} \mathrm{d}t = -\frac{1}{\sqrt{2g}}\int_{y_0}^{0}(y - y_0)^{-1/2}\left(\frac{\mathrm{d}s}{\mathrm{d}y}\right)\mathrm{d}y\,,$$

which can be put in the form

$$\tau = \frac{1}{\sqrt{2g}}\int_{0}^{y_0}(y - y_0)^{-1/2}\left(\frac{\mathrm{d}s}{\mathrm{d}y}\right)\mathrm{d}y\,, \tag{11.34}$$

where τ is the descent time. This is an integrodifferential equation since the variable to be determined, $s(y)$, appears in the integrand through its derivative.

The solution proposed by Abel employs the definition of the Riemann-Liouville fractional integral. Indeed, applying the Riemann-Liouville fractional derivative of order $1/2$ to both sides of the integral equation, we have

$$\frac{\mathrm{d}^{1/2}}{\mathrm{d}y^{1/2}}\tau = \frac{\sqrt{\pi}}{\sqrt{2g}}\frac{\mathrm{d}^{1/2}}{\mathrm{d}y^{1/2}}\left[\frac{1}{\Gamma(1/2)}\int_0^{y_0}(y-y_0)^{-1/2}\left(\frac{\mathrm{d}s}{\mathrm{d}y}\right)\mathrm{d}y\right].$$

Note that the Riemann-Liouville fractional derivative operator is the left inverse operator of the fractional integral. Using the convolution theorem, it is possible to evaluate the derivative of order $1/2$ and to show that the solution of the Abel equation is given by

$$s(y) = \frac{2\tau\sqrt{2g}}{\pi}y^{1/2}.$$

Solve this exercise using the Laplace transform to obtain the same result.

11.11 Miscellaneous Problems

In this section we discuss four problems, three of them involving partial differential equations. The first problem is solved using Fourier transform techniques, while the second one employs Laplace transforms. The third problem brings up a partial differential equation solved with the method of separation of variables. The section ends with an ordinary fractional differential equation.

11.11.1 Wave Equation

SE 11.31 Let $c^2 > 0$ be a constant with dimensions of velocity. Using the Fourier transform methodology, solve the one-dimensional wave equation

$$\frac{\partial^2 u}{\partial t^2} = c^2\frac{\partial^2 u}{\partial x^2}, \quad -\infty < x < \infty,\ t > 0,$$

subject to conditions (*i*) the initial deflection is known, $u(x,0) = f(x)$; (*ii*) the initial velocity is zero, $\frac{\partial}{\partial t}u(x,t)|_{t=0} = 0$; and (*iii*) $u \to 0$ and $\frac{\partial u}{\partial x} \to 0$ as $|x| \to \infty$, for all t. Assume that $f(x)$ admits a Fourier transform.
Solution: First, we take the Fourier transform with respect to variable x, writing $\mathcal{F}[u(x,t)] = F(\omega,t)$. Using the properties of the Fourier transform, we have

$$\mathcal{F}\left(\frac{\partial^2 u}{\partial t^2}\right) = \frac{\partial^2 F}{\partial t^2} = c^2\mathcal{F}\left(\frac{\partial^2 u}{\partial x^2}\right) = -c^2\omega^2 F,$$

that is,

$$\frac{\partial^2 F}{\partial t^2} + c^2 \omega^2 F = 0,$$

whose general solution is given by

$$F(\omega, t) = A(\omega) \cos c\omega t + B(\omega) \sin c\omega t \,,$$

where $A(\omega)$ and $B(\omega)$ are independent of variable t. For $t = 0$, since $\mathcal{F}[u(x, 0)] = F(\omega, 0)$, we obtain, using the initial conditions:

$$F(\omega, 0) = A(\omega) = \mathcal{F}[f(x)];$$

$$\left. \frac{\partial}{\partial t} F(\omega, t) \right|_{t=0} = c\omega B(\omega) = 0 \,.$$

Therefore, the solution can be written as

$$F(\omega, t) = \mathcal{F}[f(x)] \cos c\omega t.$$

Expressing the cosine in terms of the exponential function and using the displacement formula, Theorem 6.9,

$$\mathcal{F}[f(x - a)] = e^{-i\omega a} \, \mathcal{F}[f(x)],$$

we have

$$\mathcal{F}[f(x)] \cos c\omega t = \frac{1}{2} \mathcal{F}[f(x)][e^{i\omega ct} + e^{-i\omega ct}] \,.$$

Evaluating the corresponding inverse Fourier transform, we finally get

$$u(x, t) = \frac{1}{2}[f(x - ct) + f(x + ct)],$$

where $f(x - ct)$ and $f(x + ct)$ are two arbitrary, twice differentiable functions. This is the well-known *d'Alembert solution* for the one-dimensional wave equation.

11.11.2   *First-Order Partial Differential Equation*

SE 11.32 Let $u = u(x, t)$. Use the Laplace transform methodology to solve the linear first-order partial differential equation

$$\frac{\partial u}{\partial t} + x \frac{\partial u}{\partial x} = 0 \,,$$

satisfying the conditions $u(0, t) = t$ and $u(x, 0) = 0$.

Solution: Taking the Laplace transform with respect to variable t, we have

$$\mathcal{L}\left[\frac{\partial u}{\partial x}\right] + x\left[s\mathcal{L}[u] - u(x, 0)\right] = 0.$$

Assuming that we can interchange the integral with a derivative and applying the initial condition $u(x, 0) = 0$, we get

$$\mathcal{L}\left[\frac{\partial u}{\partial x}\right] = \int_0^\infty e^{-st}\frac{\partial u}{\partial x}\,dt = \frac{\partial}{\partial x}\int_0^\infty e^{-st}u(x, t)\,dt = \frac{\partial}{\partial x}F(x, s),$$

where $F(x, s)$ is the Laplace transform of $u(x, t)$.

Then, the differential equation for F is

$$\frac{\partial F}{\partial x} + xsF = 0,$$

whose solution is given by

$$F(x, s) = c(s)\,e^{-sx^2/2},$$

where $c(s)$ is a function independent of x. Using the other condition and recalling that $\mathcal{L}[t] = 1/s^2$, we obtain $c(s) = 1/s^2$, thus,

$$F(x, s) = \frac{1}{s^2}\,e^{-sx^2/2}.$$

Evaluating the corresponding inverse Laplace transform, we finally obtain

$$u(x, t) = \begin{cases} 0 & \text{for } t \le x^2/2, \\ t - \dfrac{x^2}{2} & \text{for } t \ge x^2/2, \end{cases}$$

which is the desired result.

11.11.3 *Helmholtz Equation in Cylindrical-Parabolic Coordinates*

SE 11.33 The cylindrical-parabolic coordinate system is used, for example, in the analysis of the relativistic electron scattering process by a scattering center with symmetry associated with cylindrical-parabolic coordinates. Solve the so-called Helmholtz [1821 – Hermann Ludwig Ferdinand von Helmholtz – 1894] equation.

Solution: Denote by μ, v, and z the cylindrical-parabolic coordinates $0 \leq \mu < \infty$, $-\infty < v < \infty$, and $-\infty < z < \infty$; their relations with Cartesian coordinates are given by

$$x = \frac{1}{2}(\mu^2 - v^2), \qquad y = \mu v, \qquad z = z.$$

Using these relations, evaluating the first and second derivatives, and simplifying, we get for the Laplacian operator

$$\Delta = \frac{1}{\mu^2 + v^2}\left(\frac{\partial^2}{\partial \mu^2} + \frac{\partial^2}{\partial v^2}\right)\frac{\partial^2}{\partial z^2}.$$

Let k^2 be an arbitrary constant. We want to find the separation of variables of the so-called Helmholtz partial differential equation in three independent variables, that is

$$\Delta\phi(\mu, v, z) + k^2\phi(\mu, v, z) = 0,$$

with the Laplacian operator written in cylindrical-parabolic coordinates. Looking for separable solutions, we first introduce $\phi(\mu, v, z) = M(\mu)N(v)Z(z) \equiv MNZ$. We can then write, for the function in coordinate z,

$$\frac{d^2 Z}{dz^2} + (k^2 + \alpha_1)Z = 0,$$

where α_1 is an arbitrary separation constant. For variable μ we have

$$\frac{d^2 M}{d\mu^2} - (\alpha^2 + \alpha_1\,\mu^2)M = 0,$$

with α_2 another arbitrary separation constant. Finally, we find for variable v

$$\frac{d^2 N}{dv^2} + (\alpha^2 - \alpha_1\,v^2)N = 0.$$

In the particular case $\alpha_1 = q^2/4$ and $\alpha_2 = q^2(p + 1/2)$, the ordinary differential equation in variable μ can be written as

$$\frac{d^2 M}{d\mu^2} - \left[q^2(p + 1/2) + \frac{q^2\mu^2}{4}\right]M = 0,$$

which is known as cylindrical-parabolic equation and whose solution is given by the so-called cylindrical-parabolic functions [9].

Note that the solution of the ordinary differential equation in variable z is immediate, a combination of sines and cosines, depending on k^2 and α_1. Besides, in the case $\alpha_2 = 0$ and $\alpha_1 = -q^2$, the solutions of the ordinary differential equations in variables μ and ν can be expressed in terms of Bessel functions.

11.11.4 Fractional Order Viscoelastic Object

SE 11.34 Let us consider the constitutive equation with a periodic voltage

$$\sigma(t) = E^{1-\alpha}\eta^{\alpha}\frac{\mathrm{d}^{\alpha}}{\mathrm{d}t^{\alpha}}\epsilon(t),$$

where $0 < \alpha < 1$ and coefficients E and η are constants.

Using the derivative in the Caputo sense and assuming a sinusoidal voltage $\epsilon = \epsilon_0 \sin(\omega t)$ with ϵ_0 and ω positive constants, we can write for the voltage [3]

$$\sigma(t) = E^{1-\alpha}\eta^{\alpha}\left\{\frac{\epsilon_0\omega}{\Gamma(1-\alpha)}\int_0^t (t-\xi)^{-\alpha}\cos(\omega\xi)\,\mathrm{d}\xi\right\}.$$

In order to perform the integration, we will use the Laplace transform methodology with parameter s. Applying the Laplace transform, we find

$$\sigma(s) = E^{1-\alpha}\eta^{\alpha}\frac{\epsilon_0\omega}{\Gamma(1-\alpha)}\mathscr{L}\left[\frac{\epsilon_0\omega}{\Gamma(1-\alpha)}\int_0^t (t-\xi)^{-\alpha}\cos(\omega\xi)\,\mathrm{d}\xi\right],$$

with $\sigma(s) = \displaystyle\int_0^{\infty} e^{-st}\sigma(t)\,\mathrm{d}t$.

Identifying this integral with the Laplace convolution product, we have

$$\sigma(s) = E^{1-\alpha}\eta^{\alpha}\frac{\epsilon_0\omega}{\Gamma(1-\alpha)}F(s)G(s),$$

where $F(s) = \mathscr{L}[t^{-\alpha}]$ and $G(s) = \mathscr{L}[\cos\omega t]$. We know that

$$F(s) = \Gamma(1-\alpha)\,s^{\alpha} \qquad \text{and} \qquad G(s) = \frac{1}{s^2+\omega^2}.$$

It then follows for the Laplace transform of the voltage:

$$\sigma(s) = E^{1-\alpha}\eta^{\alpha}\epsilon_0\omega\frac{s^{\alpha}}{s^2+\omega^2}.$$

To recover the solution of the initial fractional differential equation, we must consider the inverse Laplace transform, that is,

$$\sigma(t) = \epsilon\omega E^{1-\alpha}\eta^{\alpha}\mathscr{L}^{-1}\left[\frac{s^{2-(2-\alpha)}}{s^2+\omega^2}\right],$$

which provides

$$\sigma(t) = \epsilon\omega E^{1-\alpha}\eta^{\alpha}\mathbb{E}_{2,2-\alpha}(-\omega^2 t^2),$$

where $\mathbb{E}_{\alpha,\beta}(\cdot)$ is a Mittag-Leffler function with two parameters.

11.11.5 Proposed Exercises

PE 11.37 Let $u = u(x, t)$. Use the Laplace transform methodology to solve the second-order linear partial differential equation

$$\frac{\partial^2 u}{\partial t^2} = c^2 \frac{\partial^2 u}{\partial x^2}, \quad x > 0, \ t > 0,$$

where $c^2 > 0$ is a constant, with the boundary conditions

$$u(0, t) = f(t), \quad \lim_{x \to \infty} u(x, t) = 0,$$

and the initial conditions

$$u(x, 0) = 0, \quad \frac{\partial}{\partial t} u(x, t) \Big|_{t=0} = 0$$

and where the function $f(t)$ is given by

$$f(t) = \begin{cases} \sin t & \text{if } 0 \le t < \infty, \\ 0 & \text{if } t \le 0. \end{cases}$$

PE 11.38

(a) Find the temperature $u(x, t)$ on a homogeneous beam laterally isolated with constant transversal section, from $x = -\infty$ to $x = \infty$, for $t > 0$, supposing that the initial temperature is given by

$$u(x, 0) = f(x), \quad -\infty < x < \infty,$$

and that, for all $t \ge 0$ the solution and its derivative at x satisfy

$$u(x, t) \to 0 \quad \text{and} \quad \frac{\partial}{\partial x} u(x, t) \to 0 \quad \text{as} \quad |x| \to \infty.$$

(b) As a particular case, find $u(x, t)$ for

$$f(x) = u_0 = \text{constant if } |x| < 1 \quad \text{and} \quad f(x) = 0 \text{ if } |x| > 1.$$

PE 11.39 Use the Laplace transform to solve the partial differential equation

$$\frac{\partial^2}{\partial t^2} u(x, t) = c^2 \frac{\partial^2}{\partial x^2} u(x, t) + f_0 \, ,$$

with c a constant, $0 < x < \infty$, $t > 0$, f_0 another constant and with $u(x, t)$ satisfying the conditions

$$u(x, 0) = 0, \qquad \frac{\partial}{\partial t} u(x, t)|_{t=0} = 0, \qquad u(0, t) = 0$$

and

$$\frac{\partial}{\partial x} u(x, t) \to 0 \quad \text{as} \quad x \to \infty.$$

PE 11.40 Consider the so-called Dirichlet problem on the upper half-plane $y > 0$, i.e., the partial differential equation

$$\frac{\partial^2}{\partial x^2} u(x, y) + \frac{\partial^2}{\partial y^2} u(x, y) = 0 \, ,$$

together with the conditions $u(x, 0) = f(x)$; $u(x, y)$ limited as $y \to \infty$; and $u(x, y)$ and $\partial u(x, y)/\partial x$ going to zero as $|x| \to \infty$. Using the Fourier transform, show that

$$u(x, t) = \frac{y}{\pi} \int_{-\infty}^{\infty} \frac{f(\xi)}{(\xi - x)^2 + y^2} \, d\xi \, .$$

PE 11.41 Let $u(\rho, z)$ be the stationary temperature on a semi-infinite cylinder $\rho \leq 1$, $z \geq 0$, whose basis $z = 0$ is isolated. If $u(1, z) = 1$ for $0 < z < 1$ and $u(1, z) = 0$ for $z > 1$, obtain, using the cosine Fourier transform, the expression

$$u(\rho, z) = \frac{2}{\pi} \int_0^{\infty} \frac{I_0(\alpha \rho)}{I_0(\alpha)} \cos \alpha z \, \sin \alpha \frac{d\alpha}{\alpha},$$

where $I_0(x)$ is the zero order modified Bessel function of the first kind.

PE 11.42 (Cylindrical-Elliptic Coordinates) The cylindrical-elliptic coordinate system is used, for example, to deal with the mathematical problem of the potential field of infinite length cylindrical-elliptic fibers [1]. Consider the cylindrical-elliptic coordinates η, ψ and z, with $0 \leq \eta < \infty$, $0 \leq \psi < 2\pi$ and $-\infty < z < \infty$, which are related to the Cartesian coordinates x, y and z by

$$x = a \cosh \eta \cos \phi, \qquad y = a \sinh \eta \sin \phi, \qquad z = z \, ,$$

where a is a positive constant. (a) Show that the Laplacian operator is given by

$$\Delta = \frac{1}{a^2(\cosh^2 \eta - \cos^2 \psi)} \left(\frac{\partial^2}{\partial \eta^2} + \frac{\partial^2}{\partial \psi^2} \right) + \frac{\partial^2}{\partial z^2} \, .$$

(b) Using separation of variables, with the notation $u(\eta, \psi, z) = H(\eta)\Psi(\psi)Z(z)$, separate the Helmholtz equation $\Delta u(\eta, \psi, z) + k^2 u(\eta, \psi, z) = 0$, where $k^2 > 0$ is a constant, to obtain three ordinary differential equations. (c) Classify the ordinary differential equations obtained.

PE 11.43 Let $u = u(x, t)$. Let β and v be two positive constants, representing the conductivity and velocity of a diffused material, respectively. Consider the heat equation with transport term

$$\frac{\partial u}{\partial t} = \beta \frac{\partial^2 u}{\partial x^2} + v \frac{\partial u}{\partial x} \, ,$$

with $t > 0$ and $-\infty < x < \infty$, whose solution satisfies the initial condition $u(x, 0) = f(x)$. Solve this problem, composed of a linear second-order partial differential equation and an initial condition, using the Fourier transform methodology.

PE 11.44 (Fractional Maxwell Model of Elasticity)   Let $\sigma(t)$ be the stress, $\epsilon(t)$ the strain, E the Young's modulus and η the shear modulus, with $\tau = \eta/E$. Suppose that the strain is described by a Heaviside (step) function whose fractional derivative is

$$\frac{d^\alpha}{dt^\alpha}\epsilon(t) = \frac{d^\alpha}{dt^\alpha} 1 = \frac{t^\alpha}{\Gamma(1 - \alpha)} \, .$$

Solve the fractional differential equation [8]

$$\sigma(t) + \tau^{\alpha-\beta} \frac{d^{\alpha-\beta}}{dt^{\alpha-\beta}}\sigma(t) = E\tau^\alpha \frac{d^\alpha}{dt^\alpha}\epsilon(t) \, ,$$

with $\alpha, \beta \in \mathbb{R}$ such that $0 < \alpha \le \beta < 1$. To do this, use the Laplace transform methodology to show that

$$\sigma(t) = E \left(\frac{t}{\tau} \right)^{-\beta} \mathbb{E}_{\alpha-\beta, 1-\beta} \left[\left(-\frac{t}{\tau} \right)^{\alpha-\beta} \right] \, ,$$

where $\mathbb{E}_{\alpha,\beta}(\cdot)$ is the Mittag-Leffler function with two parameters.

PE 11.45 Let $\alpha > 0$, $\beta > 0$ and $x, y \in \mathbb{C}$. Show the relation

$$\sum_{r=0}^{\infty} (x + y)^r \mathbb{E}_{2\alpha, \alpha r + \beta}^{r+1}(-xy) = \sum_{k=0}^{\infty} (-xy)^k \mathbb{E}_{\alpha, 2\alpha k + \beta}^{k+1}(x + y),$$

where $\mathbb{E}_{\alpha,\beta}^{\gamma}(\cdot)$ is the Mittag-Leffler function with three parameters.

PE 11.46 (Volterra Integral Equation) Solve the so-called Volterra fractional integral equation. To this end, assume that the kernel of the second-order Volterra integral equation

$$\Phi(x) = f(x) + \lambda \int_0^x \mathsf{k}(x|t)\Phi(t)\,\mathrm{d}t \qquad (11.35)$$

has the form $\mathsf{k}(x - t)$ and is a continuous function on both variables, for $t \le x$. Write $F(s)$ and $K(s)$ for the Laplace transforms of functions $f(x)$ and $\mathsf{k}(x|t)$, respectively, and then show that

$$\Phi(x) = \frac{1}{2\pi i} \int_{\gamma_0 - i\infty}^{\gamma_0 + i\infty} \frac{F(s)}{1 - \lambda K(s)}\, \mathrm{e}^{xs}\,\mathrm{d}s\,.$$

PE 11.47 Let x and y be real constants, independent of t, and $\alpha,\,\beta,\,\gamma,\,\mu$ complex parameters with $\mathrm{Re}\,(\alpha) > 0$ and $\mathrm{Re}\,(\gamma) > 0$. Consider a Volterra integral equation (11.35) whose kernel is a Mittag-Leffler function with three parameters,

$$z(t) = t^{\beta-1}\mathbb{E}_{\alpha,\beta}(xt^\alpha) + y \int_0^t (t - \tau)^{\mu-1}\mathbb{E}_{\alpha,\mu}^{\gamma}[x(t - \tau)^\alpha]z(\tau)\,\mathrm{d}\tau, \qquad (11.36)$$

in which $\mathbb{E}_{\alpha,\beta}(\cdot)$ and $\mathbb{E}_{\alpha,\beta}^{\gamma}(\cdot)$ are the Mittag-Leffler function with two and three parameters, respectively. Show that

$$z(t) = \sum_{k=0}^{\infty} (yt^\mu)^k t^{\beta-1} \mathbb{E}_{\alpha,\mu k+\beta}^{\gamma k+1}(xt^\alpha), \qquad (11.37)$$

is a solution of Eq. (11.36).

References

1. F. Bonsignori, M. Salvini, A physico-mathematical analysis of elliptical nerve muscle fibers. N. Cimento **41B**, 375–386 (1977)
2. E. Capelas de Oliveira, *Funções Especiais com Aplicações* (Special Functions with Applications), Editora (Livraria da Física, São Paulo, 2005)
3. E. Capelas de Oliveira, *Solved Exercises in Fractional Calculus* (Springer, Switzerland, 2019)
4. E. Capelas de Oliveira, W.A. Rodrigues Jr., *Funções Analíticas com Aplicações* (Analytic Functions with Applications), Editora (Livraria da Física, São Paulo, 2006)
5. R. Hilfer (ed.) *Applications of Fractional Calculus in Physics* (World Scientific, Singapore, 2000)
6. J.D. Jackson, *Classical Electrodynamics*, 3rd edn. (Wiley, New York, 2007)
7. B. Koren, M.R. Lewis, E.H. Van Burmmelen, B. van Lee, Riemann-problem and level-set approaches for homentropic two-fluids flow computation. J. Comput. Phys. **181**, 654–674 (2002)

8. M.A. Matlob, Y. Jamali, The concepts and applications of fractional order differential calculus in modelling of viscoelastic system: a primer (2017). Crit. Rev. Biomed. Eng. **47**, 249–276 (2019). DOI 10.1615/CritRevBiomedEng.2018028368
9. E. Romão Martins, E. Capelas de Oliveira, *Método de Separação de Variáveis e os Sistemas de Stäckel* (Method of Separation of Variables and the Stäckel Systems). Coleção Imecc, vol. 4 (Imecc-Unicamp, Campinas, 2006)
10. S.G. Samko, A.A. Kilbas, O.I. Marichev, *Fractional Integrals and Derivatives: Theory and Applications* (Gordon and Breach Science Publishers, Amesterdam, 1993)

Answers and Hints

Chapter 1

1.1 $y(x) = 1 + C\,\mathrm{e}^{2\cos x}$.

1.3 $y^2\left(\ln y - \frac{1}{2}\right) = \mathrm{e}^{x^2}(x^2 - 1) + c$.

1.4 $y(x) = (2\mathrm{e}^{x^3} + C)^{\frac{1}{2}}$.

1.5 $y^2(6 - 5y^2) = x^2$.

1.6 *Hint*: Write the ordinary differential equation as $\frac{\mathrm{d}x}{\mathrm{d}y} = \frac{x - \sqrt{x^2 - y^2}}{y}$.
Introduce the change $\xi^2 = x^2 - y^2$ to obtain $y(x) = \frac{\sqrt{C - x^2}}{x}$.

1.7 $\cot(x - y) + x = c,\quad$ with $\quad (x - y) \neq \pm k\pi,\quad k = 0, 1, 2, \ldots$

1.8 $y^2 x + \int^y f(\xi)\,\mathrm{d}\xi = C$.

1.9 $y^2 + 1 = xy$.

1.10 $\ln[(y - 2)^2 - (x - 2)^2] + C = \sqrt{2}\arctan\left(\frac{\sqrt{2}}{2}\frac{y-2}{x-2}\right)$.

1.11

(a) $y = c_1 \cos \omega x + c_2 \sin \omega x$;
(b) $y = (c_1 + c_2 x)\,\mathrm{e}^{-x}$;
(c) $y = c_1 \mathrm{e}^{-x} + c_2 \mathrm{e}^{-4x}$;
(d) $y = \mathrm{e}^{-x}(c_1 \cos x + c_2 \sin x)$.

1.12 $\lambda > 0$: $y = c_1 \cos \omega x + c_2 \sin \omega x$, $\lambda = \omega^2$; $\lambda = 0$: $y = c_1 + c_2 x$; $\lambda < 0$: $y = c_1 \sinh \beta x + c_2 \cosh \beta x$, $\lambda = -\beta^2$.

1.13 *Hint*: Change the independent variable to $z = 1/x$. The solution of the original equation is $y(x) = c_1 \cos(a/x) + c_2 \sin(a/x)$.

© The Author(s), under exclusive license to Springer Nature Switzerland AG 2025
E. Capelas de Oliveira, J. E. Maiorino, *Analytical Methods in Applied Mathematics*,
Problem Books in Mathematics, https://doi.org/10.1007/978-3-031-74794-6

1.14 $y = x(c_1 + c_2 x^3)$.

1.15 *Hint*: Call $y'(x) = u(x)$ in order to obtain $y(x) = 4x + x^4$.

1.17 $y = x^3(c_1 + c_2 \ln x)$.

1.18 *Hint*: Find a first-order differential equation for the Wronskian.

1.19 $y_2 = x^{-1/2} \sin x$, $W = 1/x$.

1.20 $y_2 = -1 + \frac{x}{2} \ln \left(\frac{1+x}{1-x} \right)$, $W = 1/(1 - x^2)$.

1.21 $y = c_1 + c_2 \ln x + \dfrac{x^3}{9}$.

1.22 $y = \dfrac{x^2}{4} - \dfrac{x}{2} + \dfrac{7}{8}$.

1.23 $y = -\dfrac{\pi}{2} + \left(\dfrac{\pi}{2} - x \right) \cos x + c_1 \sin x + \ln | \operatorname{tg} x + \sec x | - \sin x \ln | \sec x |$.

1.24

(a) $\omega \neq \omega_0$: $y = \dfrac{\sin \omega_0 x}{\omega_0^2 - \omega^2}$;

(b) $\omega = \omega_0$: $y = \dfrac{\sin \omega_0 x}{4\omega_0^2} - x \dfrac{\cos \omega_0 x}{2\omega_0}$.

1.25 Putting $\lambda_0^2 = \omega^2 - b^2/4$ and $b = \lambda/m$, we get:

(a) $\lambda_0^2 > 0$: $x = e^{-bt/2}(c_1 \cos \lambda_0 t + c_2 \sin \lambda_0 t)$;
(b) $\lambda_0^2 = 0$: $x = e^{-bt/2}(c_1 + c_2 t)$;
(c) $\lambda_0^2 < 0$: $x = e^{-bt/2}(c_1 \cosh \lambda_0 t + c_2 \sinh \lambda_0 t)$.

1.26 *Hint*: Use Kirchhoff's [1824 – Gustav Robert Kirchhoff – 1887] voltage law and get for the charge $Q(t)$ an equation equivalent to the one shown in **PE 1.25**. Call the initial charge in the capacitor $Q(0) = Q_0$ and the initial current $(dQ/dt)_{t=0} = I_0$. The voltage across the inductance L is given by $L\, dI/dt = L\, d^2 Q/dt^2$; it is equal to RI across the resistance and to Q/C across the capacitance. Draw an analogy between the values of L, R, and C and the data of **PE 1.25**.

1.27 $y = -\cos x \ln | \operatorname{tg} x + \sec x |$.

1.31 Show that $y_2(x) = 1 - Cx$, where C a constant, is a second linearly independent solution.

1.32 *Hint*: Use the transformation $y = xv(x)$.

1.33 Do as in the previous exercise with $y = v(x) \cdot \sin x$.

1.34 Idem to **PE 1.32**.

1.35 *Hint*: Try to eliminate the term with the first derivative by searching for a solution with the form $y(x) = x^\alpha e^{\beta x} v(x)$, with α and β to be determined.

1.36 $y = c_1 + c_2 x - \ln x$.

1.37 $y = c_1 x + c_2/x$.

1.39 $y = c_1 x + c_2 e^x + 1$.

1.40 $y = x(-x + 2\arctan x) + \dfrac{x^2 - 1}{2} \ln(1 + x^2) + c_1 x + c_2(x^2 - 1)$.

1.41 $Q(x) = 1 + \dfrac{a^2}{2} + \dfrac{1/4 - n^2}{x^2}$, $f(x) = \sqrt{x}\, e^{-ax/2}$.

1.43 Just substitute the corresponding values into the solution of the previous exercise.

1.44 $I(x) = \dfrac{1}{4x^2}\{-x^2 - (4a - 2c)x + 2c - c^2\}$.

1.45 *Hint*: Obtain the self-adjoint form $\dfrac{d}{dx}[(1 - x^2)y'] + n(n + 1)y = 0$.

1.46 $W = c_1 c_2(x^2 + 1)$.

1.47 It is a nonlinear, first-order, ordinary differential equation.

1.48 Just calculate the derivatives and substitute them into the differential equation.

1.49 $y' = \dfrac{1}{x} - \dfrac{2}{x}y + xy^2$.

1.50 $\omega'' - (x^2 - 1)\omega = 0$.

Chapter 2

2.1 Introduce the substitution $x = \frac{1}{\xi}$ and discuss the nature of singularity at the point $\xi = 0$.

2.2 The singular points are $x_0 = 0$ and $x_0 = \infty$; both are regular singular points.

2.3 Yes, two singular points, $x_0 = -1$ and $x_0 = 1$. Both are regular singular points.

2.4 (a) $R = 1$ and (b) $\frac{1}{1-x}$.

2.5 (a) $R = \infty$; (b) $\sin x$ and $\cos x$, respectively.

2.6 Singular points $x_0 = -1$, $x_0 = 0$, $x_0 = 1$, and $x_0 = \infty$, all singular regular points.

2.7 There are no singular regular points.

2.8 (a) $x_0 = -1$, $x_0 = 2$ and $x_0 = \infty$ are regular singular points. (b) $y_1(x) = x + 1$ and $y_p(x) = 2022$.

2.9 Write the equation in the form

$$\frac{d^2}{dx^2}y(x) - (x - 1)y(x) - y(x) = 0$$

and look for the solution as

$$y_1(x) = \sum_{n=0}^{\infty} a_n(x - 1)^n \, .$$

2.10 The solution is given by $y(x) = 2\cosh x + \sinh x$. We have only regular points.

2.11 $f(x) = \displaystyle\sum_{k=0}^{\infty}(-1)^k \frac{x^{2k}}{(2k)!}$; it converges for all values of x (x in radians).

2.12 (a) Use the chain rule to rewrite the equation and evaluate it at the limit $b \to 0$. (b) For $t_0 = \infty$ introduce $t = \frac{1}{\xi}$ and conclude the analysis taking the limit $\xi \to 0$.

2.15 $f(x) = \displaystyle\sum_{k=0}^{\infty}\frac{x^k}{k!}$.

2.16 $f(x) = \displaystyle\sum_{k=0}^{\infty}\frac{x^{2k+1}}{(2k+1)!}$.

2.17 $f(x) = \displaystyle\sum_{k=0}^{\infty}\frac{x^{2k}}{(2k)!}$.

2.19 Taking $x(t) = \displaystyle\sum_{k=0}^{\infty} a_k t^k$ and writing $f(t) = \displaystyle\sum_{n=0}^{\infty} b_n t^n$, we obtain the recurrence relation

$$m(n + 1)(n + 2)a_n + \lambda(n + 1)a_{n+1} + k a_n = b_n \, .$$

2.20 Taking $x(t) = \displaystyle\sum_{k=0}^{\infty} a_k t^k$ and $E(t) = \displaystyle\sum_{n=0}^{\infty} b_n t^n$, we obtain the recurrence relation

$$L(n + 1)(n + 2)a_n + R(n + 1)a_{n+1} + (1/C)a_n = b_n.$$

2.21 For $s = 1$ we have $a_{\text{odd}} = 0$; $a_k = \dfrac{k^2 - 4k + 4}{k(k+1)} a_{k-2}$, $k = 2,\ 4\ldots$, and this implies $y_1(x) = Ax$.

2.22 Recurrence relation: $a_n = -\dfrac{a_{n-2}}{n(n+2)}$; for the second solution, use the generalized series.

2.24 $y(x) = Ax + bx^4$, where A and B are constants.

2.25 $y(x) = A + B \ln x$.

2.26 $y(x) = A + B \displaystyle\sum_{k=\text{odd}} \dfrac{x^k}{k}$.

2.27 Recurrence relation: $k(k+1)a_k + a_{k-2} - a_{k-4} = 0$, $k \geq 4$; $a_2 = -\dfrac{a_0}{3!}$, $a_1 = 0$.

2.28 *Hint*: Make the change of variable $x = 1/z$ and expand around the point $z = 0$; at the end, substitute $z = 1/x$. $y(x) = A \displaystyle\sum_{k=0}^{\infty} \dfrac{(\omega x)^{-2k}}{(2k)!} + B \sum_{k=0}^{\infty} \dfrac{(\omega x)^{-2k-1}}{(2k+1)!}$, where A and B are constants.

2.29 $y(x) = \displaystyle\sum_{k=0}^{\infty} \dfrac{x^k}{k!} = e^x$.

2.30 Recurrence relation: $a_k = \dfrac{a_{k-3}}{k(k-1)}$ for $k \geq 3$ with a_1 arbitrary and $a_2 = 0$.

2.32 a_1 is arbitrary; $a_k = -\dfrac{1}{k} a_{k-2}$, $k \geq 2$.

2.33 $a_k = -\dfrac{5}{k(k-1)} a_{k-5}$, $k \geq 5$; $a_2 = a_3 = a_4 = 0$ and a_1 is arbitrary.

2.34 $a_k = \dfrac{a_{k-1}}{2k}$; $a_k = \dfrac{a_{k-1}}{2k+1}$, with $k \geq 1$.

2.35 $J_{1/3}(x)$ where $J_\mu(\cdot)$ is a Bessel function.

2.36 $a_k = -\dfrac{a_{k-3}}{k(k - 2\sqrt{2})}$, $k \geq 3$, with $a_1 = 0$ and $a_2 = 0$.

2.37 $-a_k = \dfrac{3(k-1)(k-2) + (k-1) - 1}{3k(k-1) + 2k} a_{k-1}$, $k \geq 1$.

2.38 $a_k = \dfrac{a_{k-1}}{k(2k-1)}$; $a_k = \dfrac{a_{k-1}}{k(2k+1)}$, $k \geq 1$.

2.40 $a_k = \dfrac{a_{k-1}}{8(k+s)(k+s-1) + 2(k+s) + 1}$, $k \geq 1$, with $s_1 = -1/2$ and $s_2 = -1/4$.

2.41 $y(x) = x$.

2.42 $a_k = \dfrac{a_{k-1}}{k+2}, k \geq 1.$

2.43 $a_k = \dfrac{a_{k-1}}{k^2}, k \geq 1$; the other solution is found with the generalized series.

2.44 $a_k = -\dfrac{k}{k+2}a_{k-1}, k \geq 1.$

2.45 $a_k = -\dfrac{a_{k-1}}{k+2}, k \geq 1.$

2.46 $a_k = -\dfrac{(k-2)}{3k(k-1)}a_{k-2}$, null for even k;

$a_k = -\dfrac{(k-1)}{3k(k+1)}a_{k-2}$, null for odd k.

2.47 $y_1(x) = x^2.$

2.49 $a_k = \dfrac{a_{k-2}+a_{k-3}}{k(2k-1)}, k \geq 3.$

$a_k = \dfrac{a_{k-2}+a_{k-3}}{k(2k+1)}, k \geq 3.$

2.50 $a_k = \dfrac{a_{k-2}-a_{k-1}}{k(k+2)}, k \geq 2.$

Chapter 3

3.1 $z_1 = 3+4i$ and $z_2 = 5-12i.$

3.2 *Hint*: First, write Ω as $1+2z/(1-z+z^2)$ and then show that $z+1/z = \bar{z}+1/\bar{z}.$

3.3 Rewrite the equation as $z^9 = (11-10iz)/(11z+10i)$. Prove by contradiction that $|z^9| > 1$ and $|z^9| < 1$ are not possible, and conclude the result.

3.4 $\pi/2.$

3.5 $\tan 1.$

3.6 $z = 0$ is a removable singularity; $z = \pm 2i$ are poles of order one and $z = -4$ is a branch point.

3.7 $3\pi/8.$

3.8 $\pi(4\pi^2 - 1)/4.$

3.9 $8(5+16i)/5.$

3.10 2.

3.14 (a) Essential singularity. (b) Branch point.

3.15 (a) Residue $= 1$. (b) Residue $= 0$.

3.16 $f(z) = \dfrac{1}{3!} - \dfrac{z^2}{5!} + \dfrac{z^4}{7!} - \cdots$ Removable singularity.

3.17 $f(z) = \dfrac{1}{z+2} + 1 + (z+2) + \cdots$ The point $z = -2$ is a simple pole and the series converges for $0 < |z+2| < 1$.

3.18 $f(z) = \dfrac{1}{z} + \dfrac{1}{2!z^3} + \dfrac{1}{4!z^5} + \cdots$, Residue $= 1$.

3.19 Residue $= -7/45$.

3.20 (a) $\dfrac{t-1}{2}$. (b) $\dfrac{t-1}{2} + \dfrac{e^{-t}}{2} \cos t$.

3.21 $-6i\pi^2$.

3.22 *Hint*: Use the function $f(z) = \dfrac{\ln(z+i)}{z^2+1}$ and take as contour the semicircumference around the point $z = i$.

3.23 $\pi/2$.

3.24 $\pi/\sin \pi a$.

3.25 (a) $\pi/2$; (b) $\pi/2\sqrt{2}$.

3.27 Consider the function $f(z) = \dfrac{e^{iz}}{z}$.

3.28 It is enough to show that $F(i) = 0$, where F is the numerator.

3.32

$$\frac{\partial u}{\partial r} = \frac{1}{r}\frac{\partial v}{\partial \theta}; \qquad \frac{1}{r}\frac{\partial u}{\partial \theta} = -\frac{\partial v}{\partial r}.$$

3.33 Show that $\dfrac{\partial u}{\partial x} \neq \dfrac{\partial v}{\partial y}$.

3.36 $z = 0$, pole of order 2; $z = 1$, pole of order 1; $z = \pm 4i$, poles of order 3.

3.37 (a) Pole of order 1 at $z = 1$ and at $z = \dfrac{-1\pm\sqrt{3}i}{2}$. (b) Simple poles at $z = \pi/2 + k\pi$, $k = 0, 1, 2 \ldots$ (c) Removable singularity.

3.38 Both residues are equal to 1.

3.39 *Hint*: Use the geometric series.

3.41 *Hint*: Contour the branch points and the simple pole.

3.42 *Hint*: Start with the function $f(z) = \dfrac{\ln(1 - i\, yz)}{1+z^2}$ and use as contour a semicircumference containing $z = i$ in its interior.

3.43 *Hint*: Start with the function $f(z) = \dfrac{\ln z}{(1+z^2)^4}$.

3.44 *Hint*: Start with the function $f(z) = \dfrac{e^{iz}}{z(z^2+1)^2}$.

3.45 $3\pi/8$.

3.46 *Hint*: Consider the function $f(z) = e^{-z^2} z^{2\mu-1}$.

3.47 Use the result of the previous exercise.

3.48 *Hint*: Consider $e^{i\theta} = z$.

3.49 *Hint*: Consider $x^p = t$.

Chapter 4

4.1 Use the definition of gamma function.

4.2 Evaluate the integral

$$\Lambda = \int_{-\infty}^{\infty} e^{-x^2} \, dx \cdot \int_{-\infty}^{\infty} e^{-y^2} \, dy$$

by means of polar coordinates.

4.3 π.

4.4 Use the relation $_2F_1(-a, b; b; -x) = (1+x)^{-a}$ to show that the integral is equal to $(a+1)^{-1}$.

4.5 Use the integral representation for the hypergeometric function and the definition of beta function to show that $c > a + b$.

4.6 $\frac{1}{x} \ln |1 + x|$.

4.7 $\sqrt{\pi}/2$.

4.8 (a) $\exp(x)$ and (b) $\cosh(x)$.

4.9 Use the integral representation of the confluent hypergeometric function to show that

$$_1F_1(1; 2; x) = \frac{1}{x} \left(e^x - 1 \right)$$

and then use the definition of the Mittag-Leffler function with two parameters to conclude.

4.11 Analyze the recurrence relation.

4.12 Use the integral representation and **PE 4.41**.

4.13 (a) Use the relation $\Gamma(x)\Gamma(-x) = \dfrac{-\pi}{x}\sin \pi x$, expanding it in a MacLaurin series. (b) MacLaurin series.

4.14 (a) Directly from the series. (b) Use the integral representation given.

4.15 (a) Directly from the series. (b) Use the integral representation.

4.16 $F'' + \left\{ -\dfrac{1}{4} + \dfrac{x}{z} + \dfrac{1/4 - \mu^2}{z^2} \right\} F = 0$, with $\mu = (c-1)/2$ and $x = (c-2a)/2$.

4.17 (a) In the confluent hypergeometric equation, $a = -n$ and $c = 1 + \alpha$. (b) In the confluent hypergeometric equation, with $a = -n/2$ and $c = 1/2$. (c) In the previous equation, $z^2 = 2x^2$.

4.18 (a) $\nu \to -\nu - 1$ in the ordinary differential equation. (b) Directly from the series. (c) Use the ordinary differential equation.

4.20 Let $z \to (1 - x)/2$ in the differential equation

$$(1 - x^2)u'' + [\beta - \alpha - (\alpha + \beta + 2)x]u' + n(n + \alpha + \beta + 1)u = 0.$$

(a) $(1 - x^2)u'' - 2xu' + n(n + 1)u = 0$ (Legendre).
(b) $(1 - x^2)u'' - (2\lambda + 1)xu' + n(n + 2\lambda)u = 0$ (Gegenbauer).

4.21 Expand $(x^2 - 1)^l$ using the binomial formula and differentiate it l times.

4.22 Consider the function $g(x, t) = \displaystyle\sum_{n=0}^{\infty} a_n t^n F_n(x)$. Obtain an ordinary differential equation for $g(x, t)$ and integrate it.

4.23 (a) Directly from the series. (b) Use the ordinary differential equation.

4.24 Use the result of **PE 4.20** in both itens.

4.25 (a) Manipulate the corresponding series. (b) $z^2 u'' + z u' - (z^2 + v^2)u = 0$.

4.26 Use the result of **PE 1.18** in both itens.

4.27 Directly from the series in all itens.

4.28 Show that it is equal to the equation of **PE 4.29**.

4.29 Use the result of **PE 3.49** with $p = 1$.

4.30 Use a change of variable of the type $e^{-t} = u$.

4.31 $(a)_k = \dfrac{(a + k + 1)!}{(a + 1)!}$.

4.32 Call $z - x = u$ and use **PE 4.30**.

4.33 Introduce the change of variable $x = \sin^2 \theta$.

4.34 Use the relation $[\Gamma(z)]^2 = \Gamma(2z)B(z, z)$. Introduce the change $2t = 1 - \sqrt{x}$ in the definition of beta function to obtain $[\Gamma(z)]^2 = 2^{1-2z}\Gamma(2z)B(\frac{1}{2}, z)$.

4.35 Call $\cos\theta = u$ and use **PE 4.33**.

4.36 Multiply and divide by $n!$.

4.37 Multiply and divide by $(2n)!!$ and use **PE 4.36**.

4.38 Call $x^2 = t$ and use the beta function.

4.39 Proceed as in **PE 4.38**.

4.40 Use the results of the previous exercise.

4.41 Use the identity $\displaystyle\sum_{k=0}^{\infty} \frac{\Gamma(a + k)}{k!}(xt)^k = \Gamma(a)(1 - xt)^{-a}$ and the beta function.

4.42 (a) Use the integral representation given in **PE 4.41**. (b) Use the result of the previous item and symmetry considerations.

4.43 $\dfrac{\Gamma(c)\Gamma(c - a - b)}{\Gamma(c - b)\Gamma(c - a)}$.

4.44 Use the integral representation given in **PE 4.41** with $x = -1$ and $a+c = b+1$.

4.45 Call $y = x\dfrac{r + 1}{r + x}$ and use the beta function.

4.46 In the integral representation of **PE 4.41**, let $x \to x/b$ and take the limit $b \to \infty$.

4.47 (a) Use the result of **PE 4.46**. (b) Use the integral representation given in the text.

4.48 Use the series for $_1F_1(a; c; x)$.

4.49 (a) Use the beta function, the series for $J_n(x)$ and the duplication formula given in **PE 4.34**. (b) $J_0(z) = \dfrac{2}{\pi}\displaystyle\int_0^{\pi/2} \cos(z\cos\theta)\,d\theta$.

Chapter 5

5.1 Zero.

5.2 The Kronecker symbol δ_{kn}.

5.3 Use the orthogonality of trigonometric functions.

5.4 $\dfrac{\pi - 2x}{2}$.

5.5 $f(x) = \dfrac{\pi}{4} + \displaystyle\sum_{k=1}^{\infty}\left[\dfrac{(-1)^k - 1}{k^2\pi}\cos(kx) + \dfrac{(-1)^{k+1}}{k}\sin(kx)\right].$

5.6 Use the result of **PE 5.5**.

5.7 Use the parity of the function.

5.8 $f(x) = \dfrac{4}{\pi}\displaystyle\sum_{k=1}^{\infty}\dfrac{\sin[(2k-1)x]}{2k-1}.$

5.9 $f'(x) = \displaystyle\sum_{k=1}^{\infty}\dfrac{k\pi}{\ell}\left[-a_k \sin\left(\dfrac{k\pi x}{\ell}\right) + b_k \cos\left(\dfrac{k\pi x}{\ell}\right)\right].$

5.10 $f(x) = \displaystyle\sum_{k=-\infty}^{\infty} c_k\exp\left(i\dfrac{k\pi x}{\ell}\right)$ with $c_k = \dfrac{1}{2\ell}\displaystyle\int_{-\ell}^{\ell} f(x)\exp\left(-i\dfrac{k\pi x}{\ell}\right)dx.$

5.13 $f(x) = \dfrac{\pi}{2} + \displaystyle\sum_{k=1}^{\infty}\dfrac{(-1)^k}{k}\sin kx.$

5.14 Integrate the series given from a to x; integrate it once more, from $-\pi$ to π, to find the constant.

5.16 $f(x) = i\displaystyle\sum_{k=-\infty}^{\infty}(-1)^k\dfrac{e^{ikx}}{k}.$

5.17 $f(x) = \dfrac{-\pi}{4} + \displaystyle\sum_{k=1}^{\infty}\left[\dfrac{1}{\pi k^2}(\cos k\pi - 1)\cos kx + \dfrac{1}{k}(1 - 2\cos k\pi)\sin kx\right].$

5.18 After the expansion, take $x = \pi$.

5.19 Take $x = 0$.

5.20 Add the series of exercises **PE 5.18** and **PE 5.19**.

5.21 After the expansion, take $x = \pi$.

5.22 Proceed in a way analogous to what was suggested for **PE 5.14**.

5.23 Take $x = 0$ in item (a) and $x = \pi$ in item (b).

5.24 Use the orthogonality of the trigonometric functions (sine and cosine).

5.25 $f(x) = \displaystyle\sum_{k=1}^{\infty}\dfrac{\sin kx}{k}.$

5.26 For S_2, take $x = \pi/2$ and for S_1, proceed as in **PE 5.14**.

5.27 $f(x) = \dfrac{2}{\pi} + \dfrac{4}{\pi}\displaystyle\sum_{k=1}^{\infty}\dfrac{\cos 2kx}{1 - 4k^2}.$

5.28 Differentiate the series of **PE 5.27**.

5.29 Use the identity $\displaystyle e^x = \sum_{n=0}^{\infty} \frac{x^n}{n!}$.

5.30 Substitute a_0, a_k, and b_k into the n-th partial sum, and make a convenient change of variable $\sin(A+B) - \sin(A-B) = 2\sin A \cos B$.

5.31 Use the identity given in the previous exercise.

5.32 $\displaystyle f(x) = \frac{\pi}{8} + \frac{1}{\pi}\sum_{k=1}^{\infty} \frac{1}{k^2}\left\{\left(2\cos\frac{k\pi}{2} - 1 - \cos k\pi\right)\cos kx + 2\sin\frac{k\pi}{2}\sin kx\right\}$.

5.33 Take $x = \pi/4$ to show that $S_0 = \frac{\pi^2\sqrt{2}}{16}$.

5.34 $\displaystyle u(x) = \frac{E_0}{\pi} + \frac{E_0}{2}\sin\omega x + \frac{2E_0}{\pi}\sum_{k=2,\,4,\ldots}^{\infty}\frac{\cos k\pi x}{1-k^2}$.

5.35 $\displaystyle f(x) = \frac{\pi}{2} - \frac{4}{\pi}\sum_{k=0}^{\infty}\frac{\cos(2k+1)x}{(2k+1)^2}$.

5.36 Take $x = 0$ in the result of the previous exercise.

5.37 Expand $E(t)$ in a Fourier series and suppose that the solution of the differential equation can also be expanded in another Fourier series. We get for the coefficients:
$$A_0 = 0;\ A_k = \frac{40}{\pi k^2}\frac{10-k^2}{k^4+80k^2+100}(1-\cos kt);\ B_k = \frac{400}{\pi k}\frac{1-\cos k\pi}{k^4+80k^2+100}.$$

5.38 Use term-by-term integration to show that the sum is equal to $(\pi^2 x - \pi x^2)/8$.

5.39 $\displaystyle y(x) = \sum_{k=1}^{\infty}\frac{(-1)^k/k^3}{k^2-\omega^2}\sin kx$.

5.40 Use the relation $\mathcal{J}_1(x) = -\mathcal{J}_0'(x)$.

5.41 Expand the function $f(x) = 1$ in a Fourier-Bessel series.

5.42 *Hint*: Use the integral
$$\int_0^1 x^{\mu+1}\mathcal{J}_\mu(ax)\,\mathrm{d}x = \frac{1}{a}\mathcal{J}_{\mu+1}(a),$$

with $\mathrm{Re}(\mu) > -1$, to show that $\displaystyle x^3 = \sum_{n=1}^{\infty}\frac{8}{k_{3n}}\frac{\mathcal{J}_3(k_{mn}x)}{\mathcal{J}_4(2k_{3n})}$.

5.43 $\displaystyle f(x) = \frac{2Ca}{R^2}\sum_{n=1}^{\infty}\frac{1}{k_{0n}}\frac{\mathcal{J}_1(ak_{0n})}{\mathcal{J}_1^2(Rk_{0n})}\mathcal{J}_0(k_{0n}x)$.

5.44 Consider an adequate change in the indices and use the definition of the Bessel function.

5.45 Multiply by $e^{-in\theta}$ the two members of the expression appearing in the previous exercise and integrate it.

5.46 *Hint*: Use the integral

$$\int_0^1 x^{\nu+1}(1 - x^2)^{\mu} \mathcal{J}_{\nu}(bx)\mathrm{d}x = 2^{\mu}\Gamma(\mu + 1)b^{-\mu-1}\mathcal{J}_{\mu+\nu+1}(b),$$

with $b > 0$, $\mathrm{Re}\,(\nu) > -1$ and $\mathrm{Re}\,(\mu) > -1$.

5.47 Evaluate the integrals to verify that they are equal to zero.

5.48 Use the result of the previous exercise.

Chapter 6

6.1 $\dfrac{\Gamma(\mu + 1)}{s^{n+1}}$.

6.2 Use the convolution product.

6.3 $x(t) = x_0\, e^{-at} + \displaystyle\int_0^t f(\tau)\, e^{-a(t-\tau)}\mathrm{d}\tau$.

6.4 $x(t) = t + \cos t - \sin t$.

6.5 $y(x) = \cosh x$.

6.6 $F(k) = \dfrac{1}{\sqrt{2\pi}}e^{-k^2/4\sigma}$.

6.7 Use integration by parts.

6.8 $\Lambda = \pi/4$.

6.9 Use integration by parts.

6.10 Use integration by parts.

6.11 Use adequate changes of variables and integration by parts.

6.12 Use integration by parts.

6.13 $\mathcal{L}[x(t)] = \dfrac{2}{s^2(s^2 + \omega^2)}$.

6.15 $\mathcal{L}[f(t)] = \dfrac{(s^2 - a^2)}{(s^2 + a^2)^2}$.

6.16 Use the theorem of residues.

6.17 Apply the convolution theorem.

6.18 $y(t) = C\,e^t$, where C is a constant.

6.19 Recall that $\displaystyle\sum_{k=0}^{\infty} e^{-x} = \frac{1}{1-x}$.

6.20 $\mathcal{L}[f(t)] = \dfrac{1}{s}\,\mathrm{tgh}\,s$.

6.21 $F(s) = \dfrac{\pi/2}{1+s}$.

6.22 $I(x) = \dfrac{\pi}{2}\,e^{-x}$, $I(1) = \displaystyle\int_0^\infty \frac{\cos t}{1+t^2}dt = \frac{\pi}{2\,e}$.

6.23 $\mathcal{L}[f(t)] = \mathrm{arctg}\,s$.

6.24 $I(x) = \dfrac{\pi}{2}x$.

6.25 $\mathcal{L}[J_0(t)] = \dfrac{1}{\sqrt{s^2+1}}$. To get this result, use $\mathcal{L}[t^n] = n!/s^{n+1}$.

6.26 Take the inverse transform in the previous exercise.

6.27 $f(x) = \sin x$.

6.28 $y(x) = \cosh x$.

6.29

$$\omega(x,t) = \begin{cases} 0 & \text{for } t < x^2/2\,, \\ t - x^2/2 & \text{for } t \geq x^2/2\,. \end{cases}$$

6.30 Use adequate changes of variables and integration by parts.

6.31 $F(\alpha) = \sqrt{\dfrac{2}{\pi}}\,\dfrac{1}{1+\alpha^2}$.

6.32 $F_s(\alpha) = \alpha F_c(\alpha) = \sqrt{\dfrac{2}{\pi}}\,\dfrac{\alpha}{1+\alpha^2}$.

6.33 Use integration by parts.

6.34 $\mathcal{F}_s[f'(t)] = -\alpha \mathcal{F}_c[f(t)]$; $\mathcal{F}_c[f'(t)] = -\sqrt{2/\pi}\,f(0) + \alpha \mathcal{F}_s[f(t)]$.

6.36 $f(t) = \dfrac{l^2}{3} + \dfrac{4l^2}{\pi^2}\displaystyle\sum_{k=1}^{\infty} \frac{(-1)^k}{k^2}\cos\frac{k\pi t}{l}$.

6.37 $\mathcal{F}(\alpha) = \sqrt{\dfrac{8}{\pi}}\left(\dfrac{\sin\alpha - \alpha\cos\alpha}{\alpha^3}\right)$.

6.39 Use the result of the previous exercise.

6.40 Use the result of **PE 6.38**.

6.41 With the help of the integral $\int_0^\infty e^{-px^2}\,\mathrm{d}x = \dfrac{1}{2}\sqrt{\dfrac{\pi}{p}}$ we get $\varphi(k) = \dfrac{a}{\sqrt{2\pi}}\exp(-a^2k^2/2)$.

6.42 $\Psi(x,t) = \dfrac{\exp\left\{-\left[\frac{x^2}{2}\left(a^2 + \frac{i\hbar t}{m}\right)^{-1}\right]\right\}}{\left(1 + \frac{i\hbar t}{ma^2}\right)^{1/2}}.$

6.43 $G(x|\xi) = \dfrac{1}{2k_0}\sin[k_0(x-\xi)], \quad x > \xi.$

6.44 $G(x|\xi) = \dfrac{1}{2k_0}\sin[k_0(\xi-x)], \quad x < \xi.$

6.45 $F(\omega) = -\dfrac{\Phi(\omega)}{\omega^2 + i\omega k - \omega_0^2}.$

6.46 $G(t) = \dfrac{1}{2\pi}e^{kt/2}\int_{-\infty}^{\infty} e^{itx}\,\dfrac{\mathrm{d}x}{\beta^2 - x^2}$, where $\beta^2 = \omega_0^2 - k^2/4 > 0$. The solution of this integral is analogous to **PE 6.43**.

6.47 $G(x|\xi) = -\dfrac{1}{2\pi}\int_{-\infty}^{\infty} \dfrac{e^{-i\alpha x}\,e^{i\alpha\xi}}{k^2 + \alpha^2}\,\mathrm{d}\alpha.$

6.48 $G(x|\xi) = -\dfrac{1}{2k}\exp(-k|x-\xi|).$

6.49 Use the result $\dfrac{1}{2\pi}\int_0^{2\pi} e^{ix\cos\theta}\,\mathrm{d}\theta = J_0(x).$

6.50 Use the integral $\int_0^\infty J_0(x)\,\mathrm{d}x = 1.$

Chapter 7

7.1 The Sturm-Liouville form is

$$\frac{\mathrm{d}}{\mathrm{d}x}\left[x^c(1-x)^{a+b-c+1}\frac{\mathrm{d}}{\mathrm{d}x}y(x)\right] - abx^{c-1}(1-x)^{a+b-c}y(x) = 0.$$

7.2 The Sturm-Liouville form is

$$\frac{\mathrm{d}}{\mathrm{d}x}\left[x^c e^{-x}\frac{\mathrm{d}}{\mathrm{d}x}y(x)\right] - ax^{c-1}e^{-x}y(x) = 0\,.$$

7.3 The Bessel functions of the first kind of order $\pm\mu$ are given by the series expansions

$$J_{\pm\mu}(x) = \sum_{k=0}^{\infty} \frac{(-1)^k}{k!}\,\frac{(x/2)^{2k\pm\mu}}{\Gamma(k+1\pm\mu)}\,.$$

Use the leader term of these expansions and the relation $\Gamma(x)\Gamma(1-x) = \pi/\sin(\pi x)$.

7.4 Eigenvalues $\lambda_n = [(2n+1)\frac{\pi}{2}]^2$ with $n = 0, 1, 2, \ldots$ The eigenfunctions are $y_n(x) = \sin[\frac{1}{2}(2n+1)\pi x]$.

7.5 Eigenvalues $\lambda_n = (n\pi)^2$ with $n = 0, 1, 2, \ldots$; eigenfunctions $y_n(x) = \cos(n\pi x)$.

7.6 The normalized eigenfunctions are $y_n(x) = \sqrt{2}\cos(n\pi x)$, $n = 0, 1, 2, \ldots$

7.7 The Sturm-Liouville form is

$$(x^{-1}y')' + (\lambda+1)x^{-3}y = 0, \qquad y = y(x)\,.$$

7.8 Eigenvalues $\lambda_n = (n\pi)^2$ with $n = 1, 2, \ldots$ and eigenfunctions $y_n(x) = x\sin(n\pi \ln|x|)$.

7.9 The Green's function is

$$\mathscr{G}(x|\xi) = \begin{cases} \xi & 0 \le \xi \le x, \\ x & x \le \xi \le 1. \end{cases}$$

7.10 The solution is $y(x) = \frac{x}{2}(2-x)$.

7.14 Eigenvalues $\lambda_n = n$ with $n = 1, 2, \ldots$; eigenfunctions $u_n(x) = \sin n\pi x$.

7.15 The eigenvalues are the solutions of the equation $\operatorname{tg}\lambda_n = \lambda_n$ and the corresponding eigenfunctions are $u_n(x) = \sin\lambda_n(1-x)$.

7.16 Eigenvalues $\lambda_n = n\pi$ with $n = 1, 2, \ldots$ and eigenfunctions $u_n(x) = A\sin n\pi x + B\cos n\pi x$.

7.17 Eigenvalues $\lambda_n = \frac{-3}{4} + n^2\pi^2$ for $n = 1, 2, \ldots$ and eigenfunctions $u_n(x) = e^{-x/2}\sin n\pi x$.

7.18 Eigenvalues $\lambda_n = \frac{1}{12}(4n^2 - 3)$ for $n = 1, 2, \ldots$ and eigenfunctions $u_n(x) = e^{3x/2}\sin nx$.

7.19 Eigenvalues $\lambda_n = \dfrac{1}{4} + \left(\dfrac{n\pi}{\ln 3}\right)^2$ for $n = 1, 2, \ldots$ and eigenfunctions

$$u_n(x) = (2+x)^{-1/2} \sin\left[n\pi \frac{\ln(2+x)}{\ln 3}\right].$$

7.20 The eigenvalues are all positive real numbers $\lambda > 0$, and the corresponding eigenfunctions are given by $u_\lambda(x) = \sin(\lambda \ln x)$.

7.21 The eigenvalues are all positive real numbers $\lambda > 0$, and the corresponding eigenfunctions are given by $u_\lambda(x) = \sin \lambda x$.

7.22 $f(x) = \dfrac{8}{\pi} \displaystyle\sum_{n=1}^{\infty} \dfrac{n}{4n^2 - 1} \sin 2nx.$

7.23 $f(x) = \dfrac{4}{\pi} \displaystyle\sum_{n=1}^{\infty} \dfrac{n}{4n^2 - 1} \sin 2nx.$

7.24 $\mathscr{G}(x|\xi) = \begin{cases} x(1 - \xi) & 0 < x < \xi, \\ \xi(1 - x) & \xi < x < 1. \end{cases}$

7.25 $\mathscr{G}(x|\xi) = \begin{cases} x & 0 < x < \xi, \\ \xi & \xi < x < 1. \end{cases}$

7.26 $\mathscr{G}(x|\xi) = \dfrac{1}{\omega \sin \omega} \begin{cases} \sin \omega x \sin \omega(1 - \xi) & 0 < x < \xi, \\ \sin \omega \xi \sin \omega(1 - x) & \xi < x < 1. \end{cases}$

7.27 Imposing $y(0) = y'(1) = 0$, we have

$$\mathscr{G}(x|\xi) = \begin{cases} x + x^3/3 & 0 < x < \xi, \\ \xi + \xi^3/3 & \xi < x < 1. \end{cases}$$

7.28

$$\mathscr{G}(x|\xi) = \begin{cases} \ln \xi & 0 < x < \xi, \\ \ln x & \xi < x < 1. \end{cases}$$

7.29

$$\mathscr{G}(x|\xi) = \begin{cases} \ln \xi & 0 < x < \xi, \\ \ln x & \xi < x < 1. \end{cases}$$

7.30 Use the definition of Green's function.

7.32 $\mathscr{G}(x|\xi) = \begin{cases} x(1 - \xi) & 0 < x < \xi, \\ \xi(1 - x) & \xi < x < 1. \end{cases}$

7.33 $y(x) = -\displaystyle\int_0^{x_0} \mathscr{G}(x|\xi) g(\xi) y(\xi) \, d\xi.$

7.34 $\mathscr{G}(x|\xi) = \dfrac{1}{\eta} \begin{cases} \sinh \eta x (\sinh \eta\xi - \cosh \eta\xi) & 0 < x < \xi, \\ \sinh \eta\xi (\sinh \eta x - \cosh \eta x) & \xi < x < x_0. \end{cases}$

7.36 $\mathscr{G}(x|\xi) = \begin{cases} \sinh \eta x (\sinh \eta\xi - \cosh \eta\xi) & 0 < x < \xi, \\ \sinh \eta\xi (\sinh \eta x - \cosh \eta x) & \xi < x < x_0. \end{cases}$

7.37 $y(x) = \dfrac{V_0}{\eta} \displaystyle\int_0^\infty \mathscr{G}(x|\xi) \dfrac{e^{-\xi}}{\xi} y(\xi)\mathrm{d}\xi$, where

$$\mathscr{G}(x|\xi) = \begin{cases} e^{-\eta\xi} \sinh \eta x & 0 < x < \xi, \\ e^{-\eta x} \sinh \eta\xi & \xi < x < \infty. \end{cases}$$

7.39 $y(x) = \lambda \displaystyle\int_0^1 \mathscr{G}(x|\xi) y(\xi)\,\mathrm{d}\xi$, where

$$\mathscr{G}(x|\xi) = \begin{cases} \ln \xi & 0 < x < \xi, \\ \ln x & \xi < x < 1. \end{cases}$$

7.40 $y(x) = -\lambda \displaystyle\int_0^1 \mathscr{G}(x|\xi)\xi y(\xi)\,\mathrm{d}\xi$, where

$$\mathscr{G}(x|\xi) = 2 \begin{cases} \sqrt{x}(1 - \sqrt{\xi}) & 0 < x < \xi, \\ \sqrt{\xi}(1 - \sqrt{x}) & \xi < x < 1. \end{cases}$$

7.41

$$\mathscr{G}(x|\xi) = \dfrac{1}{n} \begin{cases} \sin nx \cos n\xi & 0 < x < \xi; \\ \sin n\xi \cos nx & \xi < x < 2\pi; \end{cases} \quad n = 1, 2, \ldots$$

7.42 $u(x) = -\lambda^2 \displaystyle\int_0^1 \mathscr{G}(x|\xi) u(\xi)\,\mathrm{d}\xi$, where

$$\mathscr{G}(x|\xi) = \begin{cases} x(1 - \xi) & 0 < x < \xi, \\ \xi(1 - x) & \xi < x < 1. \end{cases}$$

7.43 $u(x) = \lambda^2 \displaystyle\int_0^1 \mathscr{G}(x|\xi)\xi u(\xi)\,\mathrm{d}\xi$, where

$$\mathscr{G}(x|\xi) = -\dfrac{1}{2} \begin{cases} x(\xi - 1/\xi) & 0 < x < \xi, \\ \xi(x - 1/x) & \xi < x < 1. \end{cases}$$

7.44 $u(x) = \displaystyle\int_0^1 f(\xi)\mathscr{G}(x|\xi)\,\mathrm{d}\xi$, where

$$\mathscr{G}(x|\xi) = \begin{cases} x(1-\xi) & 0 < x < \xi, \\ \xi(1-x) & \xi < x < 1. \end{cases}$$

7.45 $u(x) = \dfrac{1}{\omega_0^2 \sin \omega_0 \pi} \{\sin \omega_0 \pi (1 - \cos \omega_0 x) + \sin \omega_0 x (\cos \omega_0 \pi - 1)\}.$

7.46 $u(x) = (\cotg 1 - \cosec 1) \sin x - \cos x + 1$

7.47 $u(x) = -\dfrac{x}{6}\left(x^2 - \dfrac{7}{3}\right).$

7.48

$$\mathscr{G}(x|\xi) = \begin{cases} x^3\xi/2 + x\xi^3/2 - \frac{9}{5}x\xi + x & 0 < x < \xi, \\ x^3\xi/2 + x\xi^3/2 - \frac{9}{5}x\xi + \xi & \xi < x < 1. \end{cases}$$

7.50 $u(x) = \displaystyle\int_{-1}^{1} f(\xi)\mathscr{G}(x|\xi)\,\mathrm{d}\xi$, where $\mathscr{G}(x|\xi)$ is given in **PE 7.49**.

Chapter 8

8.1 Nonlinear first-order partial differential equation.

8.2 $u(x, y) = \ln(xy).$

8.3 $yx = C$, where C is a constant.

8.4 $u(x, y) = -\ln y + f(xy)$, where $f(xy)$ is an arbitrary function.

8.5 $u(x, y) = 1 + \ln x.$

8.6 $u(x, y) = -\dfrac{1}{y} + f\left(\dfrac{xy}{y-x}\right)$, where $f\left(\dfrac{xy}{y-x}\right)$ is an arbitrary function.

8.7 $u(x, y) = -\dfrac{1}{y} + 1 + \dfrac{y-x}{xy}.$

8.8 Quasilinear first-order partial differential equation.

8.9 $u(x, y) = -\dfrac{1}{y + f(y-x)}$, where $f(y-x)$ is an arbitrary function.

8.10 Nonhomogeneous partial differential equation of the parabolic type.

8.14 Make the change of variables given in **PE 8.23** and calculate the adequate derivatives.

8.15 Proceed as in the previous exercise.

8.16 Proceed as in the previous exercise.

8.17 Just write the equations for the characteristics and solve them.

8.18 Equation of mixed type: elliptic for $x > 0$, hyperbolic for $x < 0$ and parabolic if $x = 0$.

8.19 Elliptic.

8.20 Parabolic.

8.23 Elliptic for $y > 0$, parabolic for $y = 0$ and hyperbolic for $y < 0$.

8.24 Hyperbolic on the entire plane.

8.25 Parabolic on $x^2 - y^2 = 0$, hyperbolic on $x^2 - y^2 > 0$ and elliptic on $x^2 - y^2 < 0$.

8.26 Characteristic equations: $\dfrac{dy}{dx} = \pm i$; curves: $y - ix = c_1$ and $y + ix = c_2$; coordinates: $\alpha = \dfrac{1}{2}(\xi + \eta) = y$ and $\beta = \dfrac{1}{2i}(\xi - \eta) = -x$.

8.27 Characteristic equations: $\dfrac{dy}{dx} = \pm iy$; curves: $ix + \ln y = c_1 \text{ e} -ix + \ln y = c_2$; coordinates: $\alpha = \ln y$ and $\beta = x$.

8.28 $\dfrac{\partial^2 u}{\partial x^2} - \dfrac{\partial^2 u}{\partial y^2} + \dfrac{7}{2}\dfrac{\partial u}{\partial x} - \dfrac{3}{2}\dfrac{\partial u}{\partial y} - 2u = 0.$

8.29 $\dfrac{\partial^2 u}{\partial \eta^2} - 2\dfrac{\partial u}{\partial \xi} + 3\dfrac{\partial u}{\partial \eta} + u = 0.$

8.30 $\dfrac{\partial^2 u}{\partial \alpha^2} + \dfrac{\partial^2 u}{\partial \beta^2} + 4\dfrac{\partial u}{\partial \alpha} - \dfrac{4}{\sqrt{3}}\dfrac{\partial u}{\partial \beta} - \dfrac{u}{3} = 0.$

8.31 Just substitute the function and its derivatives into the equation.

8.32 Do as in the previous exercise.

8.33 Do as in **PE 8.31**.

8.34 Identify $g = f \exp(-ax - by)$, $h = c_3 + a^2 + b^2$, $a = c_1/2$ and $b = c_2/2$.

8.35 $g_1 = f \exp(-ax - by)$, $h_1 = a_3 + a_1 a_2$.

8.38 $\dfrac{\partial^2 v}{\partial x^2} + \dfrac{\partial^2 v}{\partial y^2} = \dfrac{9}{2}v.$

8.39 $\dfrac{\partial^2 v}{\partial \eta^2} = 2\dfrac{\partial v}{\partial \xi}.$

8.40 $\dfrac{\partial^2 v}{\partial \alpha^2} + \dfrac{\partial^2 v}{\partial \beta^2} = \dfrac{17}{3}v.$

8.41 $\dfrac{\partial^2 u}{\partial \xi \partial \eta} + \dfrac{1/2}{\xi - \eta}\left(\dfrac{\partial u}{\partial \xi} - \dfrac{\partial u}{\partial \eta}\right) = 0.$

8.42 Use the result of the previous exercise.

8.45 $u(x, t) = f(x + ct) + g(x - ct)$.

8.46 $u(x, y) = yf(y/x) + g(y/x)$.

8.47 $u(x, y) = f(y - 3x) + g(y - x/3)$.

8.48 Consider $v(r, t) = ru(r, t)$.

8.49 $u(\xi, \eta) = \xi f(\eta) + g(\eta)$.

8.50 $u(\xi, \eta) = f(\xi) + g(\eta)$.

Chapter 9

9.1 Evaluate the derivatives and substitute into the two-dimensional Laplace equation.

9.2 Yes.

9.3 Evaluate the derivatives and substitute into the one-dimensional wave equation.

9.4 Evaluate the derivatives and substitute into the one-dimensional heat equation.

9.5 Yes.

9.6 $T''(\theta) + \lambda^2 T(\theta) = 0$, whose solution is $T(\theta) = A \cos \lambda\theta + B \sin \lambda\theta$, with A and B arbitrary constants.

9.7 $u(r) = A + \frac{B}{r}$, with A and B arbitrary constants.

9.8 Nonhomogeneous partial differential equation of the hyperbolic type.

9.9 With positive sign, always parabolic; with negative sign, always hyperbolic.

9.10 $T(\theta) = \alpha\theta + \beta$, with α and β arbitrary constants.

9.11 (a) Linear; (b) linear; (c) nonlinear; (d) linear.

9.12 (a) Homogeneous; (b) nonhomogeneous; (c) nonhomogeneous.

9.13 (1a) Second; (1b) second; (1c) second; (1d) third; (2a) second; (2b) second; (2c) second.

9.14 The three functions are solutions.

9.15 Both equations are linear, homogeneous, and of first order. The general solutions are: (a) $u(x, y) = C \exp[\lambda(x^2/2 - y)]$; (b) $u(x, y) = C \exp[\lambda(x - y)]$.

9.17 (a) $u(x, y) = F(x)$; (b) $u(x, y) = F(y)$; (c) $u(x, y) = e^y F_1(x) + F_2(y)$. (d) $u(x, y) = f_1(x) + f_2(y)$; (e) $u(x, y) = xf_1(y) + f_2(y)$; (f) $u(x, y) = yf(x) + g(x)$.

9.18 $u(x, y) = f(x + y) + g(3x + y)$.

9.19

$$T'' + \lambda^2 T = 0;$$

$$R'' + \frac{1}{r} R' - \frac{\lambda^2}{r^2} R = 0.$$

9.20

$$\frac{\partial^2 u}{\partial r^2} + \frac{2}{r} \frac{\partial u}{\partial r} + \frac{1}{r^2} \left(\frac{\partial^2 u}{\partial \theta^2} + \cotg\,\theta \frac{\partial u}{\partial \theta} + \frac{1}{\sin^2 \theta} \frac{\partial^2 u}{\partial \phi^2} \right) = 0;$$

$$S'' + m^2 S = 0;$$

$$T'' + \cotg\,\theta\, T' - \frac{m^2}{\sin^2 \theta} T + l(l + 1)T = 0;$$

$$R'' + \frac{2}{r} R' - l(l + 1)R = 0.$$

9.21

$$T'' + k^2 T = 0;$$

$$\nabla^2 R + k^2 R = 0.$$

9.22

$$T'' + k^2 T = 0;$$

$$\frac{\partial^2 R}{\partial r^2} + \frac{1}{r} \frac{\partial R}{\partial r} + \frac{1}{r^2} \frac{\partial^2 R}{\partial \theta^2} - k^2 R = 0.$$

9.23

$$T'' + k^2 T = 0;$$

$$\nabla^2 R + k^2 R = 0.$$

9.24

$$\frac{1}{r^3} \frac{\partial}{\partial r} \left(r^3 \frac{\partial u}{\partial r} \right) + \frac{1}{r^2} \frac{1}{\sin^2 \theta} \frac{\partial}{\partial \theta} \left(\sin^2 \theta \frac{\partial u}{\partial \theta} \right) +$$

$$\frac{1/r^2}{\sin^2\theta\,\sin\phi}\frac{\partial}{\partial\phi}\left(\sin\phi\frac{\partial u}{\partial\phi}\right)+\frac{1/r^2}{\sin^2\theta\,\sin^2\phi}\frac{\partial^2 u}{\partial\psi^2}=0;$$

$$S''+m^2S=0;$$

$$\frac{\mathrm{d}}{\mathrm{d}\phi}\left(\sin\phi\frac{\mathrm{d}T}{\mathrm{d}\phi}\right)+\left(\lambda-\frac{m^2}{\sin^2\phi}\right)T=0;$$

$$\frac{1}{\sin^2\theta}\frac{\mathrm{d}}{\mathrm{d}\theta}\left(\sin^2\theta\frac{\mathrm{d}T}{\mathrm{d}\theta}\right)+\left(\Omega-\frac{\lambda}{\sin^2\theta}\right)T=0;$$

$$\frac{1}{r^3}\frac{\mathrm{d}}{\mathrm{d}r}\left(r^3\frac{\mathrm{d}R}{\mathrm{d}r}\right)-\frac{\Omega}{r^2}R=0.$$

9.26 $u(x,t)=\dfrac{4}{\pi}\displaystyle\sum_{n=1}^{\infty}\dfrac{1-\cos n\pi}{n}\cos\dfrac{n\pi c}{l}t\,\sin\dfrac{n\pi}{l}x.$

9.27 *Hint*: Suppose that $u(x,t)=v(x,t)+w(x)$ where $w(x)$ is the *stationary state solution* and $v(x,t)$ satisfies the homogeneous equation and boundary conditions which are also homogeneous. Then,

$$u(x,t)=\frac{u_0}{l}x-\sum_{n=1}^{\infty}A_n\exp\left(-K\frac{n^2\pi^2}{l^2}t\right)\sin\frac{n\pi}{l}x,$$

where

$$A_n=\frac{2}{l}\int_0^l\left[f(x)-\frac{u_0}{l}x\right]\sin\frac{n\pi}{l}x\,dx.$$

9.28 Put $u(x,t)=v(x,t)+w(x)$ where $v(x,t)$ satisfies homogeneous conditions and $w(x)$ is determined in such a way as to satisfy the boundary conditions.

9.29 Proceed in a way analogous to the previous exercise.

9.30

$$u(x,t)=t+\frac{x}{l}(1-t)+\sum_{n=1}^{\infty}\left[\frac{2l/c}{(n\pi)^2}\sin\frac{n\pi c}{l}t+\frac{2}{n\pi}(-1)^n\cos\frac{n\pi c}{l}t\right]\sin\frac{n\pi}{l}x.$$

9.31 $u(x,t)=\dfrac{4}{\pi}\displaystyle\sum_{n=1}^{\infty}\dfrac{\sin\left[\left(\frac{\pi}{2}+2nx\right)\frac{x}{l}\right]}{1+4n}e^{-t}.$

9.33

$$u(x,t) = \frac{x^3}{6kl} - \frac{lx}{6k} + \frac{xt}{l} - \frac{2l}{k}\sum_{n=1}^{\infty}\frac{(-1)^n}{(n\pi)^2}\exp\left(-k\frac{n^2\pi^2}{l^2}t\right)\sin\frac{n\pi}{l}x.$$

9.34 It is enough to impose the initial condition $u(x,0) = f(x)$.

9.35

$$u(x,t) = \frac{x}{l} + \frac{2}{\pi}\sum_{n=1}^{\infty}\frac{(-1)^n}{n}\exp\left(-k\frac{n^2\pi^2}{l^2}t\right)\sin\frac{n\pi}{l}x.$$

9.36

$$u(x,t) = \frac{2l}{\pi}\sum_{n=1}^{\infty}\frac{(-1)^n}{n}\left\{-1+\exp\left[-k\left(\frac{n\pi}{l}\right)^2 t\right]\right\}\sin\frac{n\pi}{l}x.$$

9.37 It is enough to show that $x = \dfrac{2}{\pi}l^2\sum_{n=1}^{\infty}\dfrac{(-1)^n}{n}\sin\dfrac{n\pi}{l}x.$

9.38 $u(x,y) = -\dfrac{4}{ab}\sum_{m=1}^{\infty}\sum_{n=1}^{\infty}\dfrac{h(m,n)}{\left(\frac{m\pi}{b}\right)^2 + \left(\frac{n\pi}{a}\right)^2}$, where the function $h(m,n)$ is a Fourier transform.

9.39 The answer is analogous to the answer of the previous exercise, plus a constant.

9.40 $u(x,t) = \dfrac{1}{2\pi i}\displaystyle\int_{\gamma-i\infty}^{\gamma+i\infty} F(0,s)\exp\left[s\left(t-\dfrac{x}{c}\right)\right]\mathrm{d}s.$

9.43 $u(r,t) = \displaystyle\sum_{n=1}^{\infty}\mathcal{J}_0(\lambda r)(A_n\sin\lambda ct + B_n\cos\lambda ct)$, where

$$\lambda_n c A_n = \frac{2/R^2}{\mathcal{J}_1^2(\lambda_n R)}\int_0^R rg(r)\mathcal{J}_0(\lambda_n r)\mathrm{d}r$$

and

$$B_n = \frac{2/R^2}{\mathcal{J}_1^2(\lambda_n R)}\int_0^R rf(r)\mathcal{J}_0(\lambda_n r)\mathrm{d}r.$$

9.44

$$u(x,t) = \frac{xt^2}{2} + \frac{tx^2}{2} - \frac{xt}{2} - \frac{1}{\pi^2} \sum_{n=1}^{\infty} \frac{(-1)^n}{n^2} \sin n\pi t \sin n\pi x.$$

9.45 $\displaystyle u(r,z) = u_0 \sum_{n=1}^{\infty} \frac{1}{\lambda_n} \frac{\sinh[\lambda_n(4-z)]}{\sinh 4\lambda_n} \frac{\mathcal{J}_0(\lambda_n r)}{\mathcal{J}_1(2\lambda_n)}.$

9.47 $\displaystyle u(x,t) \equiv \psi(x) = u_0 + \frac{u_1 - u_0}{1+\pi}x.$

9.48 $\displaystyle u(x,y) = \frac{1}{\pi} \int_{-\infty}^{\infty} \frac{y}{y^2 + (x-\xi)^2} f(\xi)\,\mathrm{d}\xi.$

9.49

$$u(\rho,\phi,t) = u(\rho) + \sum_{n=1}^{\infty} A_n \mathcal{J}_0\left(\frac{\rho}{R}\lambda_n\right) \exp\left(-\frac{\lambda_n^2}{R^2}kt\right),$$

where

$$A_n = \frac{2}{\mathcal{J}_1(\lambda_n)} \left(\frac{T_b - T_a}{\lambda_n} - \frac{2\Omega R^2}{k\lambda_n^3}\right)$$

and

$$u(\rho)T_a + \frac{\Omega}{4k}(R^2 - \rho^2).$$

9.50 $\displaystyle u(r) = \frac{Bb}{r}\frac{r-a}{b-a} + \frac{Aa}{r}\frac{b-r}{b-a}.$

Chapter 10

10.1 See reference [2] of chap. 5.

10.2 $\displaystyle \mathcal{J}^\mu(x^n) = \frac{n!}{\Gamma(\mu+n+1)}x^{\mu+n}.$

10.3 $\displaystyle \mathcal{J}^{\frac{1}{2}}(\sqrt{x}) = \frac{\sqrt{\pi x}}{2}.$

10.4 $\displaystyle \mathcal{D}^\mu(x^\nu) = \frac{\Gamma(\nu+1)}{\Gamma(\nu-\mu+1)}x^{\nu-\mu}.$

10.5 $\mathcal{D}^{\frac{1}{2}}(x) = \dfrac{2\sqrt{x}}{\sqrt{\pi}}$.

10.6 Using the definition of fractional derivative, we obtain in the numerator $\Gamma(0)$, which is not defined. Otherwise, all integer derivatives are defined for $x \neq 0$.

10.7 Use the definition of error function to obtain

$$\mathcal{D}^{\frac{1}{2}}(e^x) = \frac{1}{\sqrt{\pi x}} + e^x \, \mathrm{erf}(\sqrt{x}).$$

10.8 $^{C}\mathcal{D}^{\mu}(x^{\nu}) = \dfrac{\Gamma(\nu + 1)}{\Gamma(\nu - \mu + 1)} x^{\nu - \mu + 1}$.

10.9 Same as in **PE 10.6**.

10.10 $^{C}D^{\frac{1}{2}}(e^x) = e^x \, \mathrm{erf}(\sqrt{x})$.

10.11 $^{C}\mathcal{D}^{\frac{1}{2}}(\sqrt{x}) = \dfrac{\sqrt{\pi}}{2}$.

10.12 Using integration by parts and the relation

$$-\frac{\nu}{\Gamma(1 - \nu)} = \frac{1}{\Gamma(-\nu)} \, ,$$

the result follows.

10.13 Using **PE 10.12**, we conclude that for $f(0^+) = 0$, both derivatives are equivalent.

10.14 Use the definition of Laplace transform.

10.15 Using the Riemann-Liouville fractional integral, the definition of beta function and the relation between beta and gamma functions, we obtain

$$\mathcal{J}^{\mu}\mathbb{E}_{\alpha}(-x^{\alpha}) = x^{\mu}\mathbb{E}_{\alpha,\mu+1}(-x^{\alpha}) = \mathcal{E}_{\alpha,\mu+1}(-x^{\alpha}) \, ,$$

where we have introduced, in the second equality, the Prabhakar function.

10.16 Evaluate the Caputo fractional derivative of the Mittag-Leffler function with two parameters and show that it is equal to the second member.

10.17 $^{C}\mathcal{D}^{\nu}(x + 1)^{m} = \dfrac{\Gamma(m + 1)}{\Gamma(m + 1 - \nu)} x^{m-\nu}$.

10.18 Introducing Eq.(10.3) into Eq.(10.5) and changing the order of integrations, the result follows.

10.19 $^{C}\mathcal{D}^{\nu}(C) = 0$ and $\mathcal{D}^{\nu}(C) = \dfrac{C}{\Gamma(1 - \nu)}(x - a)^{-\nu}$.

10.20 Substitute the series expansion of the Mittag-Leffler function, change the order of integration with the sum and use the following relation:

$$\mathbb{E}_{\alpha,\beta}(z) = \frac{1}{\Gamma(\beta)} + z\mathbb{E}_{\alpha,\alpha+\beta}(z).$$

10.21 Use the Laplace transform methodology to obtain

$$f(x) = \frac{\sin(\pi\alpha)}{\pi}x^{\alpha-1}.$$

10.22 Evaluate the derivative; integrate using the definition of beta function and the relation between beta and gamma functions.

10.23 Use the definitions of the Laplace transform and the gamma function.

10.24 For $0 \leq \mu < 1$, we have

$$\mathscr{L}\left[{}_{0}^{\mathsf{RL}}\mathbb{D}_{x}^{\mu}f(x); p\right] = p^{\mu}F(p) - \left[{}_{0}^{\mathsf{RL}}\mathbb{D}_{x}^{\mu-1}f(x)\right]_{x=0}$$

and, in the case $1 \leq \mu < 2$, we have

$$\mathscr{L}\left[{}_{0}^{\mathsf{RL}}\mathbb{D}_{x}^{\mu}f(x); p\right] = p^{\mu}F(p) - p\left[{}_{0}^{\mathsf{RL}}\mathbb{D}_{x}^{\mu-2}f(x)\right]_{x=0} - \left[{}_{0}^{\mathsf{RL}}\mathbb{D}_{x}^{\mu-1}f(x)\right]_{x=0}.$$

10.25 For $0 \leq \mu < 1$, we have

$$\mathscr{L}\left[{}_{0}^{\mathsf{C}}\mathbb{D}_{x}^{\mu}f(x); p\right] = p^{\mu}F(p) - p^{\mu-1}f(0)$$

and, in the case $1 \leq \mu < 2$, we have

$$\mathscr{L}\left[{}_{0}^{\mathsf{C}}\mathbb{D}_{x}^{\mu}f(x); p\right] = p^{\mu}F(p) - p^{\mu-1}f(0) - p^{\mu-2}f'(0).$$

10.26 Using the series expansion for the Mittag-Leffler with two parameters, the definition of gamma function and the geometric series, the result follows. This is a special property of the Prabhakar function with two parameters.

10.27 Using the series expansion for the Mittag-Leffler function with two parameters and changing the order of the integration with the derivative, the result follows.

10.28 Taking $n = 1$ in the result obtained in **PE 10.27**, we have the following expression:

$$\frac{\mathrm{d}}{\mathrm{d}z}\left[z^{\beta-1}\mathbb{E}_{\alpha,\beta}(z^{\alpha})\right] = z^{\beta-2}\mathbb{E}_{\alpha,\beta-1}(z^{\alpha}).$$

10.29 Proceed as in **PE 10.28** and use the definition of the Pochhammer symbol.

10.30 Proceed as in **SE 10.9**. Note that this case is a generalization whose particular case, discussed in the text, is recovered by taking $\gamma = 1$.

10.31 Introducing the series expansion for the classical Mittag-Leffler function, changing the order of integration with the sum and using the definition of beta function, we obtain for the integral

$$\Gamma(\beta)x^{\beta}\mathbb{E}_{\alpha,\beta+1}(x^{\alpha}).$$

10.32 Using the definition of the Laplace transform and the definition of gamma function, the first result follows. For the corresponding inverse, we take the inverse Laplace transform on both sides of the result obtained previously.

10.33 Use the series expansion for the classical Mittag-Leffler function, and change the order of the derivative with the sum.

10.34 For $n = 1$, we obtain

$$x(t) = \mathbb{E}_{\alpha}(-t^{\alpha})\, x(0) - q(t) \star \mathbb{E}'_{\alpha}(-t^{\alpha}),$$

and in the case $n = 2$, we have

$$x(t) = \mathbb{E}_{\alpha}(-t^{\alpha})\, x(0) + t\, \mathbb{E}_{\alpha,2}(-t^{\alpha})\, x'(0) - q(t) \star \mathbb{E}'_{\alpha}(-t^{\alpha}).$$

10.35 The result is in the text.

10.36 Similar to **PE 10.32**.

10.37 Multiply by x the Mittag-Leffler function $\mathbb{E}_{\alpha,\alpha+1}(x)$ and show that $x\,\mathbb{E}_{\alpha,\alpha+1}(x)$ is equal to $-1 + \mathbb{E}_{\alpha}(x)$.

10.38 Putting $\alpha = 1$ in Newton's law of cooling, discussed in **SE 10.11**, and using the result obtained in **PE 10.37**, we obtain

$$T(t) = T + (T_0 - T)\, e^{-kt}.$$

10.39 Use the series expansion for the classical Mittag-Leffler function, and change the order of the derivative with the sum.

10.40 Separate the classical Mittag-Leffler function in two other series expansions, an odd and an even series, and rearrange.

10.41 Evaluate the first derivative of the series expansion of the Mittag-Leffler function with two parameters and rearrange the index to obtain the expression in the second member.

10.42 Consider the Mittag-Leffler function with two parameters $\mathbb{E}_{1,3}(x)$, and manipulate the index to write it as a sum.

10.43 Consider the series expansion of the confluent hypergeometric function, rearrange the index, and use the definition of gamma function.

10.44 Proceed as in **PE 10.43**.

10.45 Introduce the series expansion for the Mittag-Leffler function with two parameters, change the order of the integration with the sum, and use de definition of beta function.

10.46 Use the series expansion for the Mittag-Leffler function with two parameters, evaluate the difference, and rearrange to obtain $\dfrac{1}{\Gamma(\beta)}$.

10.47 Consider the series expansion for the Mittag-Leffler function with three parameters, and proceed as in **PE 10.27**.

10.48 Take the Laplace transform on both sides to obtain an identity associated with the geometric series.

10.49 Use the series expansion for the Mittag-Leffler function with two parameters and the definitions of gamma and beta functions.

10.50 Proceed as in the text and use the relation

$$\frac{\mathrm{d}}{\mathrm{d}x}\mathbb{E}_{\alpha,\beta}(x) = x\frac{\mathrm{d}}{\mathrm{d}x}\mathbb{E}_{\alpha,\alpha+\beta}(x) + \mathbb{E}_{\alpha,\beta}(x)$$

to obtain the relation

$$\int_0^1 t^{\gamma-1}\mathbb{E}_{\alpha,\gamma}(xt^\alpha)(1-t)^{\beta-1}\mathbb{E}_{\alpha,\beta}(x(1-t)^\alpha)\,\mathrm{d}t = \mathbb{E}'_{\alpha,\beta+\gamma-\alpha}(x).$$

Chapter 11

11.2 (b) Use the method of undetermined coefficients. (d) Use the result obtained in item (a) to get

$$Q(t) = \exp\left(-\frac{R}{2L}t\right)\left\{A\exp\left(\frac{\mu t}{2L}\right) + B\exp\left(-\frac{\mu t}{2L}\right)\right\}$$

$$Q(t) = (A + Bt)\exp\left(-\frac{R}{2L}t\right)$$

$$Q(t) = \exp\left(-\frac{R}{2L}t\right)\left\{\tilde{A}\cos\left(\frac{\mu t}{2L}\right) + \tilde{B}\sin\left(\frac{\mu t}{2L}\right)\right\}$$

where $\mu^2 = R^2 - 4L/C$.

11.3 (a) $U_1(r) = a\,r^\alpha$ with $\alpha = (1 - 2C)/C$. (b) $U_1(r) = \frac{a}{r}$, limited motion for $a < 0$ and $U_1(r) = a\,r^2$, limited motion for $a > 0$. In both cases, the other linearly independent solution is a constant. (c) In this case the constant C must be equal to $1/2$. (d) For $\alpha \neq 0$ we have $U(r) = D + a\,r^\alpha$ where D is a real constant; for $\alpha = 0$ we have $U(r) = E + b\,\ln r$ where E is a real constant.

11.4 (a) $y(x) = \displaystyle\sum_{k=0}^{\infty} (k + 1)x^k$. (b) Converges for $|x| < 1$. $y(x) = (1 - x)^{-2}$.

11.5 The roots of the indicial equation are $s = 0$ and $s = 1$, which provide respectively the following recurrence relations: (a) $\forall a_1,\ 2a_2 + \beta a_0 = 0;\ k(k-1)a_k + \beta a_{k-2} - a_{k-4} = 0$. (b) For $a_1 = 0,\ 6a_2 + \beta a_0 = 0;\ k(k+1)a_k + \beta a_{k-2} - a_{k-4} = 0$.

11.6 $y_P(x) = 2 + 2\sin x + \cos x$.

11.11 Obtain a linear, first-order partial differential equation for $G(x, t)$ and integrate it.

11.12

$$u(x, t) = \frac{f_0 v^2}{c^2 - v^2}\left\{\left(t - \frac{x}{v}\right)\theta\left(t - \frac{x}{v}\right) - \left(t - \frac{x}{c}\right)\theta\left(t - \frac{x}{c}\right)\right\}$$

for $v \neq c$ and

$$u(x, t) = \frac{-f_0}{2c}x\theta\left(t - \frac{x}{c}\right)$$

for $v = c$, where

$$\theta(a - b) = \begin{cases} 0 \text{ for } a < b, \\ 1 \text{ for } a \geq b, \end{cases}$$

is the Heaviside function (or step function).

11.13 $\gamma(\mu, x) = \frac{1}{\mu}x^\mu \, {}_1F_1(\mu; \mu + 1; -x)$.

11.15 Use the Rodrigues formula and the relation $\dfrac{d^{2n}}{dx^{2n}}(x^2 - 1)^n = (2n)!$

11.16 (a) $2A_\ell = \dfrac{2\ell + 1}{\ell(\ell + 1)}[(1 - a^2)\mathcal{P}'_\ell(a) - (1 - b^2)\mathcal{P}'_\ell(b)],\ \ell \geq 1$, with $A_0 = \frac{1}{2}(b - a)$ because $\mathcal{P}_0(z) = 1$. (b) $1 = \displaystyle\sum_{n=0}^{\infty} \frac{4n + 3}{(2n + 1)(2n + 2)}\mathcal{P}'_{2n+1}(0)\mathcal{P}_{2n+1}(z) = \frac{3}{2}\mathcal{P}_1(z) - \frac{7}{8}\mathcal{P}_3(z) + \frac{11}{16}\mathcal{P}_5(z) + \dots$

11.19 Use the Laplace transform.

11.20 $\mathcal{L}[f(t)] = \dfrac{k}{s}\,\text{tgh}\left(\dfrac{as}{2}\right).$

11.21 $\Lambda(t) = \dfrac{\pi}{2}e^{-t}.$

11.22 $f(x) = \dfrac{2}{\pi}\dfrac{1}{1+x^2}.$

11.23 Eigenvalues $\lambda_n = n^2$ for $n = 1, 2, 3 \ldots$; eigenfunctions $y_n(x) = \sin(n \ln x)$.

11.24 $u(x) = \displaystyle\int_0^1 G(x|x')f(x')\mathrm{d}x'$, where

$$
G(x|x') = \begin{cases} \dfrac{x^3}{2}x' + \dfrac{x'^3}{2}x - \dfrac{9}{5}xx' + x, & 0 \le x < x'; \\[2mm] \dfrac{x^3}{2}x' + \dfrac{x'^3}{2}x - \dfrac{9}{5}xx' + x', & x' < x \le 1. \end{cases}
$$

11.25 $u_0(x) = a_0$ and $u_k(x) = a_k \cos k\pi x$ with $k = 1, 2, 3, \ldots$

11.26 $u_0(x) = 1$ and $u_k(x) = \sqrt{2}\cos k\pi x$ with $k = 1, 2, 3, \ldots$

11.27 (a) The Green's function is given by

$$
G(x|\xi) = \begin{cases} x(1-\xi) & 0 < x < \xi, \\ \xi(1-x) & \xi < x < 1. \end{cases}
$$

(b) $u(x) = x(x^3 - 1).$

11.28 $\dfrac{\partial^2 u}{\partial r^2} + \dfrac{1}{r}\dfrac{\partial u}{\partial r} + \dfrac{1}{r^2}\dfrac{\partial^2 u}{\partial \theta^2} = 0$, with $u = u(r, \theta)$. This is an equation of elliptic type.

11.29 $\dfrac{\partial^2 u}{\partial r^2} + \dfrac{2}{r}\dfrac{\partial u}{\partial r} + \dfrac{1}{r^2}\dfrac{\partial^2}{\partial \theta^2} + \dfrac{\cot g\,\theta}{r^2}\dfrac{\partial u}{\partial \theta} + \dfrac{1}{r^2 \sin^2 \theta}\dfrac{\partial^2 u}{\partial \phi^2} = 0$ with $u = u(r, \theta, \phi)$. This is an equation of elliptic type.

11.30 $u(x, y) = xf(y + x) + g(y - x)$, where f and g are twice continuously differentiable.

11.31 $u(r) = \dfrac{F_0}{\omega^2} + AI_0\left(\dfrac{\omega}{a}r\right)$, where A is constant and $I_0(\mu)$ is a modified Bessel function of order zero.

11.32

(a) $u(r, \theta) = \dfrac{1}{2} + \displaystyle\sum_{l=1}^{\infty} \left(\dfrac{r}{a}\right)^l \dfrac{l + 1/2}{l + 1}\mathcal{P}_{l-1}(0)\mathcal{P}_l(\cos \theta).$

(b) Use $\mathcal{P}_{l-1}(0)\mathcal{P}_l(0) = 0.$

11.33 $N(v) = A\,v^p \exp(-\beta v^2/2)L_n^p(\beta v^2)$ where A is an arbitrary constant and $L_n^p(\cdot)$ are the generalized Laguerre polynomials.

11.37

$$u(x, t) = \begin{cases} \sin(t - x/c) & \text{if } t \geq \frac{x}{c}; \\ 0 & \text{if } t \leq \frac{x}{c}. \end{cases}$$

11.38 *Hint*: Solve the heat equation, shown in Chap. 8, using the Fourier transform.

(a) $u(x, t) = \dfrac{1}{2\sqrt{xt}} \displaystyle\int_{-\infty}^{\infty} f(\xi) \exp\left[-\dfrac{1}{4}(x - \xi)^2\right] d\xi$

(b) $u(x, t) = \dfrac{u_0}{2} \left\{ \mathrm{erf}\left(\dfrac{x + 1}{2}\sqrt{t}\right) - \mathrm{erf}\left(\dfrac{x - 1}{2}\sqrt{t}\right) \right\}$, where

$\mathrm{erf}(x) = \dfrac{2}{\sqrt{\pi}} \displaystyle\int_{0}^{x} e^{-u^2}\, du$ is the error function.

11.39

$$u(x, t) = \frac{f_0}{2} \begin{cases} t^2 - \left(t - \frac{x}{c}\right)^2 & t \geq x/c; \\ t^2 & t \leq x/c. \end{cases}$$

11.42 (b) $\dfrac{d^2}{d\psi^2}\Psi(\psi) + (\alpha + \beta a^2 \cos^2 \psi)\Psi(\psi) = 0$; $\dfrac{d^2}{d\eta^2}H(\eta) - (\alpha + \beta a^2 \cosh^2 \eta)H(\eta) = 0$; $\dfrac{d^2}{dz^2}Z(z) + (k^2 + \beta)Z(z) = 0$, where α and β are two separation constants. (c) All equations are second-order ordinary differential equations; two are nonlinear and one is linear.

11.43 $u(x, t) = \dfrac{1}{\sqrt{2\pi}} \displaystyle\int_{-\infty}^{\infty} e^{-ikx} e^{-k^2 \beta t} e^{-ivkt} F(k)\, dk$ where $F(k) = \dfrac{1}{\sqrt{2\pi}} \displaystyle\int_{-\infty}^{\infty} f(\xi)\, e^{ik\xi}\, d\xi$.

Index

E. Capelas de Oliveira, J. E. Maiorino, *Analytical Methods in Applied Mathematics*,
Problem Books in Mathematics, https://doi.org/10.1007/978-3-031-74794-6

If you have any concerns about our products,
you can contact us on
ProductSafety@springernature.com

In case Publisher is established outside the EU,
the EU authorized representative is:
Springer Nature Customer Service Center GmbH
Europaplatz 3, 69115 Heidelberg, Germany

Printed by Libri Plureos GmbH
in Hamburg, Germany